Managing Internet-Driven Change in International Telecommunications

For a listing of recent titles in the *Artech House Telecommunications Library*, turn to the back of this book.

Managing Internet-Driven Change in International Telecommunications

Rob Frieden

Artech House
Boston • London
www.artechhouse.com

Library of Congress Cataloging-in-Publication Data
Frieden, Rob.
Managing Internet-driven change in international telecommunications / Rob Frieden.
p. cm. – (Artech House telecommunications library)
Includes bibliographical references and index.
ISBN 1-58053-019-2 (alk. paper)
1. Telecommunications policy—International cooperation. 2. Multimedia systems.
3. Internet. I. Title. II. Series.
HE7645.F744 2001
384.3—dc21 00-068931
CIP

British Library Cataloguing in Publication Data
Frieden, Rob
Managing Internet-driven change in international telecommunications. — (Artech House telecommunications library)
1. Telecommunications policy—International cooperation 2. Multimedia systems
3. Internet
I. Title
384.3
ISBN 1-58053-019-2

Cover design by Wayne McCaul

International Standard Book Number: 1-58053-019-2
Library of Congress Catalog Card Number: 00-068931

10 9 8 7 6 5 4 3 2 1

For Katie, Alex, and Elizabeth

Contents

Introduction

When I wrote *International Telecommunications Handbook* in 1996, it was much easier to map the telecommunications landscape. The Internet had only begun to gain momentum, and much of what now triggers substantial change stood on the periphery. Entrepreneurs had targeted telecommunications and information processing, but analysts had primarily forecasted changes rather than consider their consequences. In only a few years, billions of dollars have flowed into a variety of market segments, spawning both megamergers and new, well-capitalized ventures keen on uprooting incumbents. Indeed, incumbents as a whole now have more to worry about, in terms of revenues and market share lost to competition, but they also have more opportunities to thrive.

The pace of change in the telecommunications industry has so accelerated that more change and tumult have occurred in the last decade than in the 150 years before, beginning from the invention of the telegraph. In *International Telecommunications Handbook*, I was able to erect a somewhat sustainable dichotomy between the past up to the mid-1980s and something akin to a "new world order." Present trend lines surely lack order, but they have so deviated from the status quo as to qualify for a monumental title. If the "old world order" evidenced stability and centralized management by the state, then the new world order manifests extreme volatility, change, uncertainty, and opportunity. Technological innovation and new attitudes toward the role of government in the sector have combined to make telecommunications as dynamic and robust as the information-processing industries.

Telecommunications and information-processing markets and technologies have merged, adding complexity and the juxtaposition of cultures. The old world telecommunications order had plenty of rules and a steep learning curve, but (generally speaking) one could crack a single code and extrapolate from it. The post telephone and telegraph (PTT) carrier model predominated, because governments had yet to relinquish belief in the essentialness of centralized state management. Even where private industry thrived, a "Phonehead" culture established certain uniform expectations and assumptions, (e.g., that government regulation created certain rights, responsibilities, opportunities, and rules within which everyone had to play).

The new world order reflects the increasingly important role of the new set of players who have little tolerance for government involvement and little patience for rules in general. The libertarian, "Nethead" culture is substantially juxtaposed with the Phonehead culture, even though individuals and companies reflective of one or the other have to work with counterparts in a convergent world.

This book offers a road map for tracking developments and trends in both international telecommunications and Internet-mediated communications. I have attempted to write in a manner accessible to Phoneheads and Netheads alike, as well as to offer both a primer for novices and some higher-level analysis.

1

Telecommunications in an Internet-Driven Information Economy

1.1 Overview

After years of stable obscurity and predictability, international telecommunications are growing increasingly complex, volatile, and unpredictable. Just 20 years ago, a rather thin book could set out uniform rules of the road and be embraced and enforced by a small, clubby group of government-owned or franchised monopolies. More has changed in the last 20 years than in the preceding century. Indeed, much has changed since the publication of this book's predecessor, *International Telecommunications Handbook.*

This book offers a substantially different view of international telecommunications: one driven more by the Internet, technological innovation, and entrepreneurialism, and less by regulation, managed change, and incrementalism occurring in the transition from a centrally managed environment to a competitive marketplace. Despite substantial upheaval and flux, one can get a handle on changes and accrue the benefits of such knowledge. This book provides a road map for tracking those changes with an eye toward offering insight that can provide advantages in managing careers, employees, investment portfolios, and the like. However, the volatility of the marketplace forecloses the use of single descriptive models or even a single set of descriptive works and concepts. Nevertheless, this book can provide both a frame of reference and a set of baseline concepts with which to work.

Today, we see the real social, political, financial, and cultural effects of words like *convergence*, *globalization*, *deregulation*, and *liberalization*. Some "blue-sky" projections have come to fruition, and newcomers can hardly believe that just a few years ago government employees ran the telephone company in many nations. Entrepreneurs offer satellite parking places for sale, nations run clearance sales to "privatize" their telephone company services, and hundreds of companies search the world for new profit centers. Just a few years ago, a small group of carriers held what seemed to be permanent and exclusive franchises to what many thought were "natural" impenetrable monopolies.

Until just recently, the study of international telecommunications involved a relatively straightforward process of learning the "system is the solution" world of government-owned or highly regulated public utilities. Whether on national security or economic rationales, most governments could not envision a competitive marketplace that would satisfy consumer requirements while also meeting public policy goals such as universal access to affordable, possibly subsidized services. The government role in telecommunications paralleled its sole provision of postal services. The reasoning was that governments could balance, even subordinate, profit-maximizing decisions with countervailing public policy missions such as providing subsidized services to rural residents and disadvantaged members of society. Few thought that a long-distance telephone company could and would provide flat-rate, distance-insensitive services, the same kind of "postalized" rates we expect from the post office. Fewer still thought competition would prove both sustainable and essential, even in market segments and localities considered unprofitable.

In the old world order, which existed until the mid-1980s, telecommunications carriers operated in a more comfortable, less competitive environment. Whether owned and operated by government or pervasively regulated by government, telecommunications carriers willingly traded many of the operational freedoms available to commercial enterprises in exchange for insulation from competition and most financial risks. Customers were neither right nor sovereign. They took what carriers offered with few options on price, service, and choice of carrier. Carriers could not charge what the market would bear. Instead, rates were subject to government scrutiny to ensure that they adequately compensated the carrier.

In a relatively short time span, governments and carriers have since accepted the need for a change in assumptions and the laws, policies, and regulations derived from those assumptions. Technological innovations disputed the natural monopoly assumption, as entrepreneurs lined up for the opportunity to invest in the telecommunications sector and vie for previously captive customers. Incumbent carriers no longer can impose restrictive policies, rules, and prices when customers can easily choose the services of competitors. Fast-changing technology, new market access initiatives, and liberalizing policies have combined to alert incumbent operators and regulators alike to a growing need for change.

Such a shock to the system provides interesting case studies in comparison to the long-established old world order, which lacked drama, dynamism, customer service, and a diverse cast of characters. Fewer and fewer nations continue to support exclusive local and international telecommunications franchises, and just about every nation now allows some market entry to leased lines, including Internet access, financial services, and other information-processing systems. This is particularly true in noncore, niche markets, for example, mobile services like cellular radio and information processing.

Compelling case studies can examine the tensions caused by technological innovation and the promises made by prospective entrants versus the fears raised by incumbents. Incumbents warn of reduced employment, threats to national security, and the inability or unwillingness of newcomers to achieve public policy objectives. Newcomers promise faster, better, smarter, cheaper, and more convenient service. This book examines why and how the forces for change achieved critical mass and succeeded in dislodging incumbents and the natural monopoly rationale. In the vernacular of the economist, most telecommunications and information-processing markets are either competitive by nature or contestable, that is, ripe for market entry in the near term. The sleepy, insular world of public utility telecommunications has become but a memory in most nations.

1.2 An Interdisciplinary Undertaking

The convergent, Internet-driven world requires nimble, interdisciplinary thinking. To be successful, practitioners, students, and educators have to be equally proficient in both telecommunications and information processing. We must keep track of numerous, seemingly countervailing factors that quickly change and involve a wide range of disciplines. These characteristics can be complicated, and few people have adequate training in several of the fields involved; most educational programs have yet to offer an interdisciplinary approach. One may face new situations and work assignments without a solid foundation and, worse yet, lack even threshold credentials for employment. Corporate downsizing and pressures to economize place a premium on the ability of employees to perform several functions.

Universities have just begun to address whether they should provide an introductory international telecommunications course or revamp the curriculum to reflect the Internet's ascendancy. Few undergraduate programs skillfully blend previously discrete course tracks into an eclectic approach for developing competency and skills in a number of career areas. Questions arise, such as whether a course belongs in the political science department, because it involves government ownership and policy making, or whether it belongs in the economics department, because the driving force for deregulation and market entry involves application of economic principles. If such a course finds its way into communications, how do issues involving access to satellite parking places, infrastructure development, spectrum management, and service tariffing mesh with the more traditional courses in broadcasting and journalism? Some universities have a school or college of information science, perhaps an ideal home for such interdisciplinary pursuits. The new curriculum should include softer science issues, including law, regulation, business, and social impact of information technologies.

This book has two primary goals: (1) to identify subjects that affect broad international telecommunications and information-processing careers and (2) to examine some of the more focused substantive issues being debated in policy-making forums, courts, and the marketplace. The book also provides some insight into the kinds of skills required to excel in telecommunications and information-processing careers. Telecommunications studies often involve an integrated set of subjects, not always viewed as complementary or within the same ambit of investigation. For example, one could assess a nation's decision to embrace (or not to embrace) procompetitive deregulation in terms of the following factors:

- Political policies that include a national commitment to make service available throughout a nation, regardless of traffic density and the potential to recoup investments;
- Economic policies that view telecommunications as a natural monopoly and that favor the incumbent retaining exclusive control of all or most service sectors;
- Industrial policies that insulate domestic manufacturers and service providers from foreign competition;
- Technology policies that endorse the view that no single enterprise can provide all available innovations at an optimal scale;

- Social policies that mandate government intervention to support universal service objectives or reliance on the free interplay of market forces even if some prices rise;
- Laws that permit market entry or reserve markets to a regulated monopoly;
- Treaties or other multilateral commitments that preclude procompetitive initiatives;
- Foreign relations concerns that foreclose efforts to foster competition if it might force lesser developed nations to incur higher costs;
- National security interests that preclude market access liberalization;
- Labor relations issues that jeopardize any initiative that could result in reduced employment in the telecommunications sector;
- Consumer protection needs to guard against the potential for more fraud and deception, given the expanded number of involved ventures;
- Competition policy assessments of whether business transactions like mergers and acquisitions will adversely affect the continuing viability of competition.

If we do not consider telecommunications issues in terms of their numerous component parts, we may fail to address all of the elements that stimulate formation of a particular policy and why some particular technologies and companies succeeded where others failed. Strategies for successful national or corporate policies need balanced assessments, which are possible only when the participants are well versed in most, if not all, of the components involved.

A partial list of the disciplines typically involved in international telecommunications would include economics, law, engineering, international relations, business, and technology management.

1.2.1 Economics

Policy makers increasingly rely on economic analysis to evaluate arguments concerning efficiency and equity. Telecommunications policy often constitutes one of the many aspects in a nation's overall campaign to spur industrial development and employment. Initiatives governing a decision to sell some or all of the government monopoly telephone company may parallel similar decisions in other industries, for example, civil aviation, petroleum, and public utilities such as electric companies.

Economic analysis provides the calculus for measuring social welfare enhancements and efficiency gains. The forum for such decision-making often is the legislature or ministry of posts and telecommunications, in which economic arguments such as the virtues of private enterprise and competition are considered against the political clout of incumbent institutions like the post, telegraph, and telephone company (PTT), the incumbent carrier, and labor unions.

Economic analysis constitutes a key method for assessing whether and how to reform the telecommunications sector. As such, anyone seeking full understanding of the field should comprehend certain fundamental economic concepts. Economists face some compelling questions in this field, including the following:

- Can competition thrive in telecommunications and information-processing sectors, or would a natural monopoly accrue economies of scale and scope that put it in the best position to achieve such social goals as ubiquitous and affordable telephone service?
- Would a monopoly exploit its market power to preempt competition made possible by technological innovations?
- Can an unregulated marketplace achieve social goals relating to the price and availability of service, or must government mandate subsidies and prescribe rates of return?
- Is telecommunications a public good that unregulated private enterprises will not produce in socially desirable quantities at affordable prices?
- Can the marketplace achieve a balance that matches supply with demand?
- Can policy makers rely on competition to diversify services, reduce rates, and stimulate demand?
- Will short-term price competition result in destructive competition, in which the retained earnings of the incumbent enable it to underprice services and survive an unprofitable period, after which it can recoup the losses when competitors exit the market?

- Will market entrants serve only the most profitable routes, leaving the incumbent carrier as the only one available for service to low-income users or ones in remote locations?
- Would using the Internet as the major medium for telecommunications services change fundamental assumptions about the marketplace?

1.2.2 Law

Lawyers provide an essential function by drafting telecommunications and Internet contracts, policies, laws, and regulations. Although they often lack training in economics, engineering, or business, lawyers provide essential advice on making policy and marketing decisions. Because they are involved primarily in legal analysis and document drafting, lawyers have become adept at incorporating (but not succumbing to) analytical models of other disciplines. Many lawyers have developed a working knowledge of the telecommunications infrastructure and how its component parts operate. Similarly, many attorneys can incorporate economic terms to support a client's legislative or regulatory agenda.

The application of other disciplines and the work of nonlawyers may become subordinate to a legislative, administrative, and policy-making process managed by lawyers. That process may emphasize procedural fairness and opportunities for interested parties to participate, but in the final analysis, lawyers may dominate the process. That means much time, money, and effort may apply to questions on organization, jurisdiction, the intent of a law or treaty, and what procedures will ensure fairness and opportunities for all parties to participate.

1.2.3 Engineering

Technology has a substantial impact on policy making and marketplace prospects for both telecommunications and Internet products and services. The development of technology forces policy makers to consider whether and how much the status quo should change to accommodate new service options and prospective market entrants. For example, more sensitive and selective satellite Earth station equipment makes it possible to point orbiting satellites closer to each other. Policy makers must balance the ability to accommodate the demand for more satellite capacity with the extra expense that most users would incur in upgrading Earth station receiving equipment to distinguish between the target satellite and the now more closely spaced satellites capable of causing interference.

The Internet's transmission and addressing protocols may provide a more efficient medium for transmission and delivery of both voice and data services. However, the overlay of regulatory and other policies may enhance or blunt a technology-driven competitive advantage. For example, Internet-mediated telephone services, a subject examined in greater detail in another chapter, offer the promise of international calls charged at a few cents per minute, substantially less than what conventional services now cost. Part of the financial savings results from Internet telephony operators qualifying for a regulatory qualification that exempts them from having to pay interconnection charges to other carriers and to contribute to universal service funding. The blending of technology-delivered enhancements with regulator-conferred financial opportunities requires keen and clear analysis.

Legislators and regulators often try to classify technologies with an eye toward developing markets and regulatory programs according to semantic line drawing. For example, government officials must decide whether a new service is basic telecommunications transport subject to conventional regulation or enhanced, value-added services, which often are subject to less regulatory oversight. Telecommunications regulators and providers think of technology in terms of whether it supports journalism, entertainment, broadcasting, closed-circuit transmissions like cable television, information processing, telephony, and other market segments. Even those classifications and more legalistic ones, like *common carrier* and *private carrier*, increasingly converge. For example, telephone companies, which typically are regulated as common carriers and public utilities, vie to become cable television operators. Cable television operators, which typically are subject to less regulation and which are not classified as common carriers or public utilities, have targeted new telephone and Internet access markets. Interactive television and multimedia transmission of voice, data, audio, and video over a single medium frustrate a desire to think in terms of discrete and mutually exclusive industries and service classifications.

The Internet presents even more daunting technology classification challenges, given the breadth of services available to users. Does the fact that the Internet can provide a medium for real-time "streaming" of audio and video content make Internet service providers (ISPs) the functional equivalent to radio and television stations? The technologies that make it possible for Internet-mediated commerce also jeopardize legal and regulatory classifications that make distinctions among, for example, banks, stock brokers, and insurance companies. Technology makes it possible for one Internet site to offer one-stop shopping for a variety of heretofore discrete financial services offered by different companies.

1.2.4 International Relations

International telecommunications and Internet communication blend public diplomacy and private commerce. Delegations to bilateral or multilateral telecommunications conferences have addressed such esoteric but critical issues as parking places for domestic and international satellites, the amount of radio spectrum allocated for particular services, and fundamental rules of the road like the number and sequence of digits one must dial to reach a particular telephone. On the other hand, carriers negotiate business alliances and agree to match circuits and deliver the traffic generated by "correspondents."

In the Internet environment, the percentage of commercial versus government-orchestrated policy making is nearly opposite the percentage for telecommunications. While governments have actively incubated Internet development, most do not currently operate networks and manage the offering of services. Many of the Internet's rules of the road resulted from open and voluntary user forums, without formal recognition or participation by governments. For many Internet stakeholders, the thought of government involvement is anathema, an intrusion likely to interfere with the creativity and vast benefits accruing from unfettered entrepreneurialism. For those players, Internet governance constitutes an oxymoron. Yet a maturing, commercialized Internet requires the very kinds of standards and uniformity that have made international telecommunications user friendly and seamless even as it traverses borders and transits different vintages and types of equipment. Internet ventures will find that they cannot provide the next best thing that offers a faster, better, smarter, cheaper, and more convenient enhancement without some degree of certainty over baseline technical and operational issues. For example, the Internet's domain numbering system provides a user-friendly way to convert easily remembered names into an Internet protocol numbering sequence (e.g., www.psu.edu specifies the Penn State University server).

Currently, many nations are rethinking the balance among government regulation, management, and ownership in telecommunications and the permissible scope of marketplace resource allocation. Some nations rely on marketplace forces to assign access to spectrum through auctions and to establish de facto operating standards. The Internet has an even greater reliance on marketplace forces, but there are plenty of instances in which courts provide a necessary forum to resolve disputes (e.g., multiple claims to the same domain name or the need to judge whether two similar addresses would result in consumer confusion and infringement of previously registered copyrights and trade names).

No national telecommunications administration has embraced a totally laissez faire view, and, despite the libertarian wishes of some, no national government has abdicated entirely a regulatory role over the Internet. Many governments have rejected aspects of marketplace resource allocation and decision-making as inconsistent with the pursuit of broader national interests. Even as more nations opt to replace government ownership with private enterprise in telecommunications, they do not cede to individual companies' decision-making on issues that affect nationhood, sovereignty, and key social issues. As such, one should understand the role of intergovernmental agreements and the forums where nations meet to establish global rules and policies.

1.2.5 Business

International telecommunications and information processing may "contribute to world peace and understanding" [1], but they still constitute a business, with enterprises vying for billions of dollars in annual revenues. The fundamental elements of business administration apply to telecommunications, particularly now that many nations have "corporatized" or privatized the incumbent carrier to make it more businesslike and less bureaucratic. Governments incubated the Internet, but most ventures operate in a commercial, largely unsubsidized and unregulated environment. It has taken more time for governments to opt out of the telecommunications sector.

In telecommunications, social and public policy objectives persist in preventing companies from maximizing profits. Regulators may require carriers to subsidize some services (e.g., domestic local telephone service) from revenues generated by other, more profitable services (e.g., international long-distance telephone service). Carriers may have universal service obligations that require them to extend service to remote areas at below-cost rates, regardless of the potential for recouping such capital-intensive investments. Telecommunications regulation can establish governments as gatekeepers, deciding who can enter what markets and how much they can charge. Service providers and government officials apply accounting and financial skills to determine rates rather than allow the marketplace to establish an equilibrium between supply and demand.

An increasingly liberalized and competitive telecommunications marketplace requires contract negotiation skills. As the sector becomes more business-oriented, users need not tolerate a take-it-or-leave-it attitude on the part of equipment suppliers or service operators. The greatest degree of negotiating clout will lie with users who generate large traffic volumes and who can migrate to other suppliers or install their own equipment. Carriers and manufacturers will need to customize deals; so too will regulators need to permit nontariffed service arrangements and adjustments in the process by which they evaluate the fairness of negotiated terms and conditions.

1.2.6 Technology Management

Consumers, rather than technology or governments, have an increasingly dominant role in shaping the telecommunications industry. Technological innovations may trigger changes in regulation and particular policies. The pace and direction of innovations force improvements in the ability of public and private organizations to manage technology and change.

Technological innovations necessitate modifications in the way managers and regulators consider the telecommunications infrastructure. Users no longer have to segregate voice and data networking when a single digital network can serve most telecommunications requirements. Similarly, it is increasingly difficult to separate the transmission facilities used to disseminate news and entertainment, like broadcasting and newspapers, from the conduits used to transmit and process data, for example, fiber optic cables, satellite links, and intelligent terminals.

Telecommunications professionals have begun to employ technological developments in ways that previously would have violated regulations designed to segregate markets and, as well, to insulate them from some types of competition. For example, managers and users of cable television once viewed their distribution plant as a one-way vehicle for delivering video channels to homes. With minor modification and investment, the plant can provide two-way distribution of voice, data, facsimile, and video traffic. Analysts now consider coaxial cable more advanced and flexible than the telephone company's pair of twisted wires that provide "first and last mile" access to telecommunications users. Many cable television operators plan on providing telecommunications services, including local and long-distance telephone services and high-speed Internet access. Although technologically feasible, interconnection of cable television and incumbent carrier facilities may require legislative and regulatory changes. Such promising market opportunities challenge the status quo in terms of both conventional thinking in each industry and the regulatory structure erected to oversee that industry.

1.3 Appreciating Different Cultures

Technological and marketplace convergence expands the set of stakeholders in telecommunications. Such expansion introduces a variety of new philosophies, market orientations, and perspectives. Collectively, the new cultures have a profound influence on how carriers provide telecommunications and information services, an influence significantly different from those that previously drove industry. Players in the old world order understood regulation and government involvement as a given, and operated successfully in that environment. Critics considered the regulator as captive to incumbent stakeholders, but with regulation deemed a necessary evil, market countervailing policies, like universal service subsidies, met with acceptance, no matter how grudgingly given.

Many new world order players have no perspective on how to thrive in a pervasively regulated environment. They bridle at the thought of having to work within such a system, notwithstanding clear evidence that incumbents had gamed the system to their advantage. The juxtaposition of cultures has resulted in conflict, just as the convergence of technologies and markets has produced volatility. The successful practitioner, student, and investor understand the strengths and weaknesses of the several cultures now in play.

1.4 Clash of the Phoneheads and the Netheads

Perhaps the greatest clash of culture lies in the vastly divergent mind-sets of telecommunications incumbents and their Internet counterparts [2]. The Phonehead culture starts with a regulated telecommunications industry perspective modified by recent deregulation and liberalization. That old world order orientation emphasizes telecommunications networks as the focal point for meeting consumers' requirements. The Nethead culture starts with a libertarian predisposition against any government involvement in information processing. Phoneheads typically work in large hierarchical organizations that tend to respond slowly and incrementally to changed circumstances and opportunities. Netheads typically work in smaller, more nimble organizations that embrace change and thrive on chaos.

Nethead information technologists and Phonehead telecommunications professionals have plenty of shared interests and increasingly operate in the same, converging industry. They have to learn to work together and to appreciate, if not accept, the other's perspective. Telecommunications and information-processing industries integrate what had constituted largely discrete and mutually exclusive market segments, including voice, data, broadcasting (radio and television), and closed-circuit media like cable television.

In a converging marketplace, preexisting cultures, assumptions, and government policies may quickly become inappropriate, perhaps even harmful. The Netheads' culture, which supports individualism and entrepreneurship substantially free of government meddling, juxtaposes with the Phoneheads' culture, which recognizes government involvement as a necessary evil and perhaps a vehicle to secure a competitive advantage. Without significant accommodation, neither orientation can apply in an Internet-mediated telecommunications and information-processing environment. Telecommunications offers more than computer bit delivery; on the other hand, the Netheads correctly emphasize that consumers expect a convergence of basic telecommunications transport and information processing.

1.5 Challenges to Incumbents

Incumbents have to accept the new world order, and—predictably—many do not have the inclination or wherewithal to change. Many incumbents and plenty of their customers believe the adage that "if it ain't broke, don't fix it," that is, the current system works just fine. Incumbents have mastered the task of balancing exploitation of monopolies, or at least limited market access, with partial acquiescence to the edicts of regulators. In fact, some critics believe that regulators became captive to their regulated clients in view of how the political process allocates funds and rewards regulators.

The new world order requires different skills and talents, with less emphasis on exploiting the regulatory process and more on thriving in a competitive world where government presents fewer opportunities and impediments to success. In many ways, the world appears to have turned upside down:

- Instead of exploiting inelastic demand with high margin service rates, incumbents typically have to reduce their rates, making up in volume what they lose in margin.
- Rather than "working" the legislative, regulatory, and judicial systems, incumbents must improve marketing, customer service, back-office functions, and a host of market-driven activities that they may have underemphasized.

- Having had a nearly captive customer base, incumbents must work overtime to retain and win back customers who bolt simply when afforded the opportunity to try another company.
- Often content to serve domestic markets, incumbents must pursue international opportunities to compensate for lost market share at home when competition is introduced.
- Having worked in a predictable environment where change occurs incrementally, incumbents must adapt to volatility, risk, uncertainty, and flux.
- Incumbents must respond to technological and marketplace changes that challenge the baseline assumptions they used to accept as unimpeachable.

1.6 Challenges to Market Entrants

Ventures that exploit market access opportunities available in the new world order have their own challenges to master. They start with zero market share and have to build or lease from incumbents the facilities and links to users needed to provide services. While some customers will readily embrace an alternative, many others will resist change, particularly if the newcomer has few incentives to offer. Newcomers must show customers that they will deliver a faster, better, smarter, cheaper, and more convenient service.

Newcomers typically enter markets without having constructed a network completely duplicative to what the incumbent offers. As such, newcomers must lease some elements of the service they offer from an incumbent with every incentive to drive a hard bargain, delay the conclusion of negotiations, raise competitors' costs, discriminate in favor of affiliates that compete with the newcomer, and otherwise exploit their incumbency.

In many ways, the new world order appears to offer less than what the newcomer had expected:

- Regulators may eliminate safeguards and scrutiny well before full marketplace competition evolves.
- Newcomers' zealously libertarian and procompetitive culture does not jibe with a regulated environment only beginning the transition to competition.
- Incumbents have plenty of ways to stall, thwart, and otherwise delay the inevitable transition to marketplace competition.

- Bottlenecks may persist despite the rhetoric and projections of robust, facilities-based competition.
- Limited access to capital will force newcomers to target selectively the markets they enter.
- Significantly wide gaps may develop between the promise of change, technological innovations, and market opportunities and current reality.

1.7 Managing Change

Telecommunications and information processing have merged, as evidenced by the growing importance of the Internet and data communications. That environment precludes application of preexisting telecommunications rules of the road, because the services that consumers require have widely diversified beyond simple voice communications. Many services now combine a basic telecommunications transport function with computerized enhancements that add value, new features, and greater complexity. Data services now require more telecommunications bandwidth than voice services.

On the other hand, because convergent services require telecommunications transport functions, they do not operate in a closed network environment. That means some of the rules, procedures, conventions, protocols, standards, and policies addressing telecommunications services will extend to telecommunications transmission of information services. Such rules will require modification, but they cannot be ignored or summarily dismissed as inapplicable.

Internet-driven change requires accommodation and coordination between the rules, the culture, and the mind-set of data processing and telecommunications. Predictably, individuals will balk at having to make such an undertaking. Some will simply conclude that doing so contaminates and degrades networks and makes it more difficult to serve consumers. They may be correct in that conclusion. But they will be wrong in thinking they will not have to make such accommodations.

References

[1] *Agreement Relating to the International Telecommunications Satellite Organization "Intelsat" (Intelsat Agreement)*, Preamble, Aug. 20, 1971, entered into force Feb. 12, 1973, 23 U.S.T. 3813, T.I.A.S. No. 7532.

[2] T. M. Denton Consultants, *Netheads Versus Bellheads: Research into Emerging Issues in the Development and Deployment of Internet Protocols*, Final Report, 1999. Available online at http://www.tmdenton.com/netheads.htm.

2

What Is Driving Change?

2.1 The Digital Imperative: Moore's Law and Metcalfe's Law

Technological and marketplace convergence, the growing importance of the Internet, and conversion from analog to digital network technologies are the primary drivers for change in telecommunications. We should no longer think of telecommunications separately from information processing and vice versa. Telecommunications provides the basic infrastructure for the delivery of digital bitstreams, but the information-processing functions needed to develop intelligent network applications can be found anywhere. Intelligent network applications, like most Internet and electronic commerce services, cannot be divided simply between the telecommunications transport function and computerized enhancements.

Likewise, many of the technological and financial achievements in information processing apply equally to telecommunications. Digital technologies accrue increasing returns and lower prices as they expand in scope and reach. Gordon Moore, one of Intel's founders, quantified that outcome by noting that information-processing technologies (like computer microprocessors) double in capability with a halving of cost in the span of 18 months [1]. Robert Metcalfe, one of the inventors of the Ethernet data delivery system, claims that the value of a network increases with the square of the number of people connected to it, or what economists would call increasing positive network externalities [2].

Information-processing technologies offer a double consumer benefit of increasing value and lower costs over time. When information-processing technologies are an integral part of intelligent networks, the technologies that apply to computing environments work just as well in telecommunications environments.

Figures 2.1 and 2.2 show the impact of Moore's law on computing power and telecommunications capacity.

2.2 Internet-Driven Telecommunications Development

During the 1990s, the Internet became the "killer application" for the information age. The Internet is the focal point for telecommunications development because consumer demand for data services now drives infrastructure investments. Put another way, telecommunications networks increasingly serve data as opposed to voice applications. Growing at triple-digit rates throughout the 1990s, the Internet and other data services are now the primary drivers for facilities investments. Demand for broadband services, such as real-time delivery (streaming) of video- and graphics-intensive World Wide Web (WWW) pages, has stimulated incredible increases in available

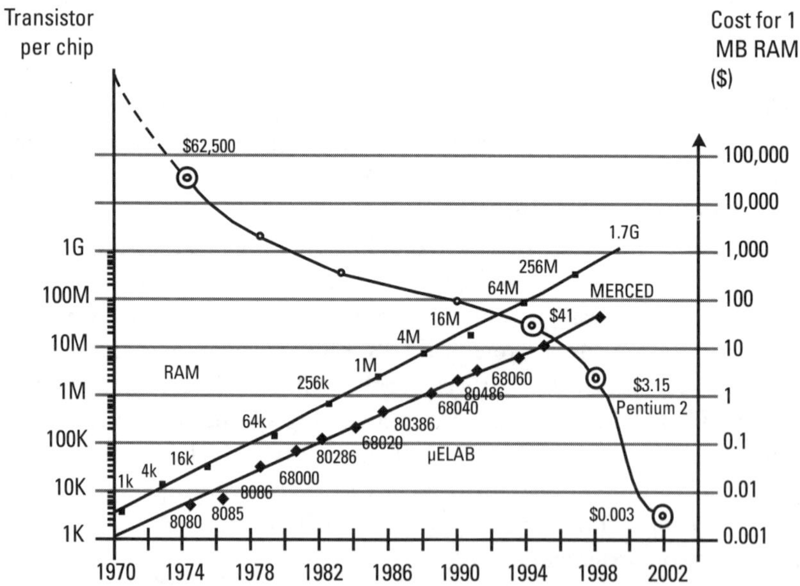

Figure 2.1 The impact of Moore's law on computing power.

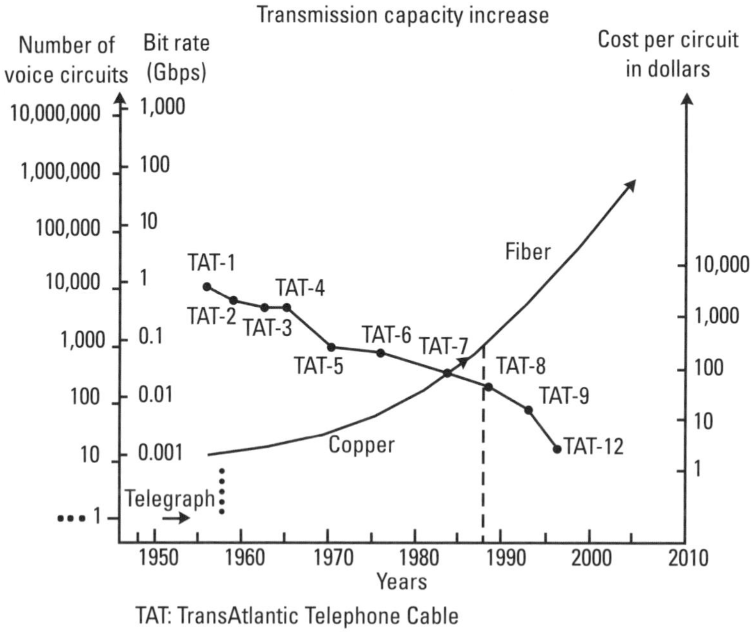

Figure 2.2 Moore's law applied to telecommunications.

transmission capacity and in the deployment of data-oriented transmission capacity.

While blue-sky prognosticators continue to extrapolate and project even greater accomplishments, the current Internet offers consumers faster, better, cheaper, smarter, and more convenient services. It has reached what Andrew Grove, the Chairman of the Board at Intel, terms the "critical inflection point" (or critical mass) for a number of reasons. In a nutshell, the Internet developed into a major communications and commercial medium because many long-touted concepts became a reality and provided the foundation on which to build a thriving network of networks.

The Internet could not become a vibrant and substantial medium without these factors:

- Global proliferation of high-bandwidth telecommunications capacity to deliver and route desirable content;
- Technological innovations that promote the convergence of previously discrete telecommunications and information services;

- Wisdom in government decisions to incubate new technologies but also to refrain from regulating and managing Internet applications;
- Creativity and drive of entrepreneurs, activists, and citizens of cyberspace to develop innovations, persuading consumers to buy the technology and to ascend information technology learning curves.

This section examines the Internet and its role in our private and public lives. It also considers how the Internet provides faster, better, cheaper, smarter, and more convenient services and runs the engine for the digital economy. It also explores how convergence became real and how the Internet became a legitimate and growing medium for commerce.

2.2.1 The Four Stages of Internet Development

The Internet provides a fundamentally new and possibly better way for conducting many types of commercial transactions. That outcome results from developments in global networking technologies, digitization, widespread diffusion of information technologies, and low barriers to market entry:

> Today, for a few thousand dollars, anyone can become a merchant and reach consumers throughout the world. As a result, electronic commerce has expanded from business-to-business transactions between known parties to a complex web of commercial activities that can involve vast numbers of individuals, many of whom may never meet. [3]

The Internet passed through four integrated stages to become the engine for an emerging digital economy.

2.2.1.1 Stage 1: Privatization and Infrastructure Buildout

Stage 1 was the privatization of the Internet and its physical networks. From its debut in 1969 to its privatization in the early 1990s, the Internet served mostly government and educational users. Governments incubated Internet development directly and indirectly. The U.S. government underwrote the original infrastructure buildout directly, for example, the Defense Department's ARPANET, an electronic mail system that linked military installations and university contractors. The U.S. government indirectly funded Internet projects through the National Science Foundation. By the 1990s, most governments had eliminated or largely reduced financial support. Sensing a vast and untapped market, commercial ventures readily replaced governments. The companies, some of which previously had leased the

Internet's telecommunications transmission capacity to governments, became the private Internet's backbone network providers.

The stage 1 buildout of the Internet stimulated demand for personal computers, electronic components (such as central processor units), remote access devices (such as modems), telecommunications transmission capacity, software and the routers, servers, and other devices that seamlessly integrate networks into a global network of networks. Companies such as Microsoft, Intel, Cisco Systems, Compaq, Dell Computer, America Online, and Netscape experienced triple-digit growth in revenues and stock prices based on the Internet's buildout. A growing and proliferating Internet, with vastly more data sources (servers) and the physical links between servers and users, accrued term-positive networking externalities by increasing in value and utility even as the cost of use and the cost per user stayed the same or declined.

2.2.1.2 Stage 2: Business-to-Business Commerce and Freebies

Even now, the vast majority of Internet-mediated electronic commerce links businesses to businesses (B2B) rather than businesses to consumers (B2C) or consumers to consumers (C2C), sometimes referred to as peer-to-peer (P2P). Most businesses have the requirements, incentives, and financial wherewithal to exploit the efficiency gains, productivity enhancements, and cost savings accruing from improved inventory control, accelerated cycle times for product development, and streamlined procurement. In stage 2, businesses began experimenting with Internet applications designed to reach consumers. The lack of simple ways to secure payment for low-ticket goods and services, coupled with consumers' concerns about sending credit card information over the Internet, stimulated the creation of two kinds of Web pages:

1. High-value pages with plenty of free content for which consumers otherwise might pay, thereby creating a windfall for early adopters;
2. Low-value pages offering little more than infomercial-type marketing and promotion.

2.2.1.3 Stage 3: On-Line Sale of Consumer Goods and Services

Internet mediation often can serve end users better than traditional channels of distribution. The World Wide Web provides a channel for the immediate electronic delivery of goods or services in digital form, as well as direct collaboration between two or more individual users. That serves consumers' desire for the earliest possible receipt of goods and businesses' interest in streamlining operations, saving money, developing new distribution channels, and stimulating greater demand. First-to-market businesses in stage 3 included providers of financial services (e.g., banking, stock market trading, newspaper and magazine publishers), ticket vendors for entertainment and sporting events, airlines and travel agents, purveyors of adult entertainment (e.g., gambling and pornography), and software distributors.

In stage 3, the Internet became more robust and rich with content and options. Stage 3 development also involved the aggregation of content and the simplification of the maneuvering process through the Internet. Companies aggregated content and access to other Web sites in an attractive and user-friendly manner. That portal function proved quite lucrative financially. A gatekeeper to the Internet can aggregate subscriber eyes and ears for advertisers and Web page operators. Ironically, as the Internet proliferates, a significant portion of the user population seeks simplicity and ease of access provided by one-stop shop service, such as those of America Online, and by portal operators such as Yahoo!.

Further stage 3 development involved the proliferation of "push" and "pull" features in the Internet. Push features include the bulk delivery of content to end users in a mass-market, noncustomized, and scattershot manner. At the other end of the continuum, consumers specify their interests and receive a personalized, daily delivery of preferred content. Pull technologies provide consumer-specified information on a customized basis. For example, advertiser-supported search engines provide consumers with a computerized vehicle to cull through countless information sources as they search for specific types of information.

2.2.1.4 Stage 4: Widespread Acceptance of E-Commerce

In stage 4, Internet mediation provided consumers with an option for acquiring tangibles and intangibles. A variety of ventures already have entered the marketplace for books, compact discs and tapes, greeting cards, flowers, apparel, toys, alcoholic beverages, automobiles, and even lower-ticket items such as drugs, vitamins, and groceries. The Internet provides a medium for bricks-and-mortar companies to establish a virtual marketing presence 24 hours a day, 7 days a week (24/7). Providers of goods and services offer consumers point-and-click convenience, time savings, and possibly even cost savings in place of preexisting options. In stage 4, the Internet became a vehicle for consumers to exploit the same sort of logistical and delivery factors that businesses addressed in stage 2. Figure 2.3 outlines the growth in e-commerce as estimated by three major market forecasting firms.

2.2.2 The Internet Mantra

If the Internet changes everything, as its advocates claim, killer applications must find a home in the hearts and computers of many. To reach such a mass market, the Internet must provide services for which consumers will voluntarily pay and give their attention. Consumers flock to the Internet because it provides desirable services in innovative, entertaining, and user-friendly ways. Internet-mediated services can generate repeat patronage, usually after an initial period of curiosity and intrigue, but that happens only if such services improve on previously available services.

The mantra "faster, better, cheaper, smarter, and more convenient" identifies some of the key features Internet-mediated services must have to find and sustain market share. While the service need not offer improvements in each category, on balance, users should perceive a qualitatively better outcome. The Internet provides consumers opportunities to change the balance of power in their favor with the ability to acquire and process information (e.g., automobile costs) and to use that information to prompt a potentially larger collection of sellers to compete more vigorously (e.g., securing bids for the sale of an automobile from a number of car dealers). The Internet expands the scope of a buyer's search and the geographical reach of a seller. It provides a medium that threatens the ability of intermediaries to mark up prices but also supports the development of intelligent intermediaries who can parse through, process, and prioritize the boundless information available.

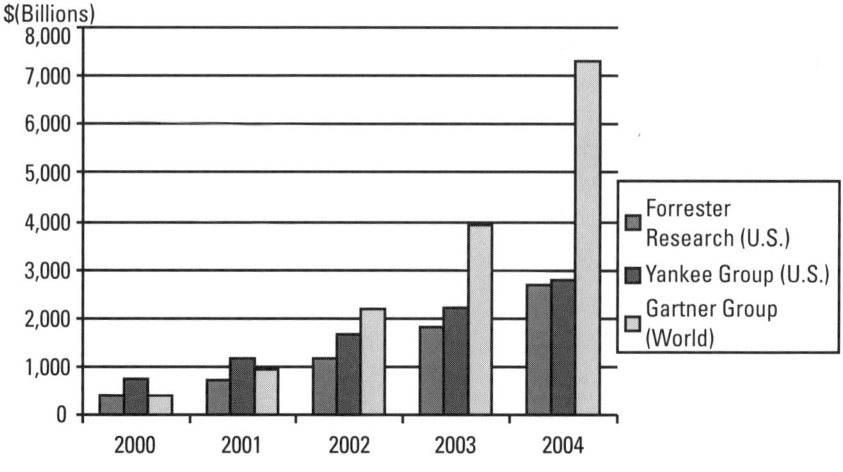

Figure 2.3 Market opportunity in e-commerce growth.

2.2.2.1 Faster

Internet time operates 24/7 with the capacity to serve whenever the consumer chooses to log on. With no closing time and the capacity to operate at the speed of electrons, Internet-mediated commerce can accelerate most transactions and make them more convenient. Enhanced speed can trigger efficiency gains as the need for fewer people, chains of marketing, management, and customer care is less. Consider, for example, the difference between on-line air travel reservation systems and human travel agents. Once a Web surfer ascends the learning curve of Internet-mediated reservation systems like Expedia, Travelocity, Cheap Tickets, Inc., and Priceline.com, access to much of the search and ticketing power previously available only to travel agents flows to consumers. Such access makes it possible for the educated and conscientious traveler to secure air travel quickly, cheaply, and with greater control. In short order, consumers can book travel even to the extent of examining the available seating options on selected flights. Travelers can avoid telephone tag with live travel agents and secure electronic tickets, thereby avoiding the need to arrange for timely delivery of paper tickets.

Obviously, the Internet-mediated option can excel only if the consumer knows how to maneuver through various reservation systems and has enough confidence in a system's security, authentication procedure, and privacy protection to file credit card information on-line. Similarly, no guarantee exists that an on-line reservation system provides the absolutely best fares and discloses all available options in an unbiased manner. Congestion on the Internet and limitations in the throughput available through telephone lines can slow the transaction. On balance, if the traveler does not need the advice and perspective offered by a seasoned travel consultant, Internet mediation provides a faster and more efficient link between consumer and airline. That is accomplished by reducing the number of distribution chains between the airline and the air traveler. The elimination of some distribution channels, such as those that travel agents typically perform, occurs when consumers in a broad geographical area can tap into the speed, data warehousing, and processing capabilities of the Web pages maintained by the airlines and other travel ventures.

2.2.2.2 Better

People differ in their perception of what makes something qualitatively better. For example, few newspaper readers consider Internet mediation, by itself, an improvement, because of the awkwardness and time it takes to download content and to maneuver from page to page. Other enhancements to the Internet version, however, provide almost real-time updates, including the ability to search a daily paper and archives using a keyword, and the use of customizing features to tailor the newspaper product to one's own tastes and interests.

Internet mediation requires the conversion of content into a somewhat attenuated visual and auditory context. Because of current limitations on the bandwidth available to most users, bottlenecks and data speed limitations currently preclude full-motion video displays. Similarly, the display medium requires the use of a video monitor and speakers that have processing limits in their ability to reproduce images and sounds. Notwithstanding those limitations, Internet access does provide consumers with substantially more options and the opportunity to pursue virtually any interest. If more means better, then access to thousands of newspapers and news sites provides greater opportunities to stay informed. If knowledge means power, then the Internet empowers almost anyone to acquire data and information. Knowledge still requires individual, noncomputerized, processing skill.

2.2.2.3 Cheaper

The Internet provides manifold opportunities to streamline, economize, and render almost any transaction more efficient. As a global medium, the World Wide Web increases in value to consumers, advertisers, and sellers as additional points of communication and sources of information become available. Consumers may achieve lower costs in Internet-mediated transactions by eliminating intermediaries, a process known as disintermediation. Intermediaries, who typically mark up prices, risk losing their niche in the distribution chain if they fail to provide value to justify the markups. In turn, suppliers, whether the actual producers or the aggregators of purchase orders, can exploit the Internet to reduce inventory to just-in-time levels to closely align production to what consumers want and to deal directly with end users. In theory, a computer manufacturer (e.g., Dell, which heavily relies on Internet-mediated transactions) can quickly produce computers tailor-made to just-received purchase orders. For their part, advertisers may accrue savings or have greater opportunities to target and reach selective audiences.

2.2.2.4 Smarter

The power of information technology includes the ability of computers to store, process, and search through massive amounts of data. Artificial intelligence, data mining, intelligent agents, and a host of information management techniques provide Internet users with new tools with which to work, play, and learn. Artificial intelligence provides ways for computers and the

Internet to respond and adapt to individual interests and requirements. Coupled with intelligent agents, Internet services can canvass and cull through massive amounts of information to collect precisely what an individual wants to see. Data mining and other search tools provide ways to understand both individual and group consumer behavior by processing massive amounts of collected information.

Consumers can enhance the Internet's value by linking its vast array of information, commercial, and entertainment sources with computerized information-processing functions. For example, on some on-line auction sites a bidder can activate a proxy bidding option to raise offers as needed up to a bidder-imposed cap. That feature obviates the need for frequent monitoring of the bidding process by enabling automatic revisions when necessary.

2.2.2.5 More Convenient

The Internet enhances corporate and consumer welfare only when it offers a medium that enhances a business or personal transaction. Greater convenience refers to the ability of the Internet to provide services and features whenever and wherever a requirement exists. The term also refers to "user friendly" access, that is, access in a manner no less convenient and achievable than in a non-Internet–mediated environment. For example, many people rejected the first attempts by retailers to convert paper catalogs in an Internet Web site. Lengthy downloading times via telephone lines, coupled with difficulty in maneuvering between pages, made the Internet option less convenient and pleasurable than portability, tactile contact, and ease in thumbing through the hard copy option. On the other hand, many book buyers appreciate the ability of Web sites like Amazon.com to suggest additional reading options based on data mining, comprehensive analysis of book buying choices of other people displaying similar tastes.

2.3 Uncovering Market Niches

In the old world order in telecommunications, entrepreneurs could not exploit market access opportunities without first securing regulatory approval. Now a new breed of entrepreneur, perhaps less willing to play by the old rules, aggressively enters markets using technologies that make it possible to violate regulations and market-entry barriers with little risk of detection. In many nations, both scenarios have occurred simultaneously. Regulators typically work incrementally and promote competition initially in noncore market niches like wireless and information services that add computerized enhancements to leased lines (e.g., credit card verification services). Meanwhile, entrepreneurs look for high-margin opportunities that, while possibly not long term, provide immediate profits and the opportunity to work toward greater legitimacy and market penetration.

Legitimate market niches created in the 1980s include the opportunity for market entrants to enhance leased basic transmission capacity with information processing that adds value to the traffic transport function (e.g., credit card verification services). Regulators authorize international value-added networks (IVANs) to operate on the assumption that they stimulate greater demand for basic transmission capacity leased from incumbent carriers, rather than migrate portions of existing traffic and revenues to a newcomer. In the early stages of promoting competition, regulators typically reject international simple resale (ISR), the repackaging of basic leased lines without enhancements. ISR promotes competition by allowing entrants to exploit the arbitrage opportunity in buying bulk transmission capacity reserved for single, large-volume users, and subdividing it among numerous users who individually would not usually qualify for the bulk discount.

2.3.1 Illegal Market Niches

The gray-market and illegal market niches that were created in the 1980s and later came to include ISR masquerading as some form of IVAN service were designed to provide a cheaper voice service alternative. Because international calling rates substantially exceeded cost in many nations, the incentive to exploit an arbitrage opportunity was substantial. That was particularly true for newcomers lacking in patience or appreciation for broad public policy objectives or rationales for incremental introduction of competition. In the 1990s, entrepreneurs found ways to "import" dial tone from nations like the United States, where international calling rates were substantially lower. These callback and call-reorigination services proliferated in nations that prohibited ISR and had comparatively higher international telephone toll charges.

Creative packaging of services and illegal or gray-market services created pressure on regulators to pick up the deregulatory pace. Technological innovations provided consumers with self-help opportunities and new options to reduce telecommunications costs. Once they had access to those options, consumers had little tolerance for persistently high rates or for the policies that made it possible for incumbents to charge such rates.

2.3.2 Fewer Barriers to Market Entry

Whether through incremental and legitimate means or through black- or gray-market opportunities, newcomers have managed to worm their way into previously captive markets. Technological innovations made it possible for organizations without deep pockets to provide international services, particularly when coupled with resale market access opportunities. Declining costs in satellite Earth stations, small-scale telecommunications switches, and computers make it possible for individuals to establish mom-and-pop international operations. Market entrants did not have to invest in their own international transmission facilities, but could start off small and increase their inventory when demand for their services grew. The opportunity to resell the capacity of other carriers made it possible for competition to arise before a business case supported the construction of additional, parallel capacity. By leasing small portions of submarine cable and satellite transponder capacity, resellers could narrow the margin between the rate paid by large consumers and the much higher rate individuals previously were charged. Figure 2.4 tracks the transition from a monopoly-based industry structure in telecommunications to one driven by privatization and competition.

2.4 Liberalization Bias

Consumer expectations and technological innovations combined during the 1980s to make networks and markets more porous. Market-access trade commitments made in the 1990s contributed to market accessibility so that most nations became predisposed toward favoring both resale and facilities-based (multiple network) competition. At some point during that time span, nations began to recognize that the telecommunications infrastructure could accrue a comparative advantage or disadvantage in economic development and trade. A liberalization bias developed as nations saw convincing empirical evidence that competition would enhance consumer welfare, thereby encouraging nations to support telecommunications-intensive industries.

Although economists and entrepreneurs may have incubated the drive for telecommunications liberalization and deregulation, consumers' expectations and political necessity combined to generate the pressure necessary for change. Having experienced or heard of telecommunications innovations and cheap rates elsewhere, consumers expected such developments at home. They rejected rationales for maintaining the status quo, particularly in nations where incumbents perennially failed to deliver dependable service.

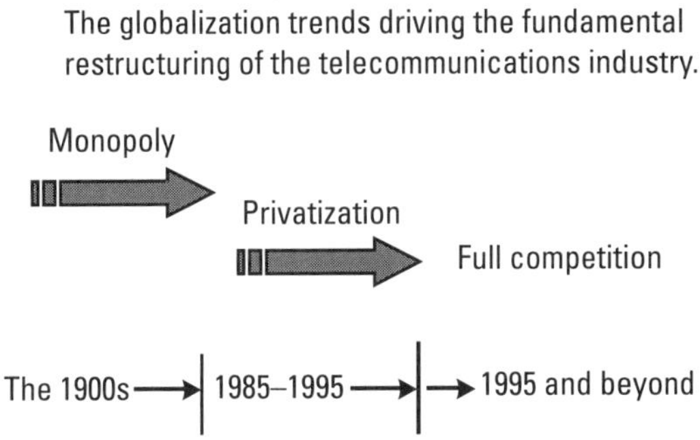

Figure 2.4 Key drivers to collapsing monopolies.

Government-owned and government-sanctioned monopolies typically failed to meet consumer requirements, either because they could not generate or borrow the funds needed to install sufficient infrastructure, or because earnings had to flow to the general treasury or the pockets of elites.

2.5 Migration to an Information Economy

Telecommunications provides the underlying infrastructure for transmitting and processing information, a commodity that is becoming as important as physical resources. Pundits may have overstated the importance of the Internet and electronic commerce, but no one can deny the importance of information as an aspect of a nation's economic well-being.

Computer mouse clicks and electronic impulses will not completely replace bricks, mortar, and physical matter, but processing information has become a high-priority, low-pollution source of employment and wealth. It primes the economic pump often by facilitating commerce rather than by substituting a virtual transaction for a previously physical one. For example, the ability to process a credit card transaction and verify a signature electronically promotes a commercial transaction that might not otherwise take place but for the convenience in having credit quickly and unobtrusively extended to a creditworthy individual keen on executing a transaction without cash.

2.6 Wireless Ascendancy

The convenience-enhancing function of telecommunications and the Internet also appears in the growing importance of wireless applications, such as cellular telephones and next-generation services, which include Internet access. A mobile society doing business 24/7 requires cheap and user-friendly access to the rest of the world, even in mobile environments.

Wireless services have become very accessible and affordable. In some developed nations, wireless services have become a competitive alternative to preexisting wireline services as carriers offer large minutes of use "baskets" that cost a few cents per minute and are available for local and long-distance calling. In developing nations, the wireless alternative provides a comparable or even lower cost per subscriber, particularly in remote locations.

The third generation of wireless services presents the potential for broadband high-speed access from mobile and remote locations. Previous mobile services provided only narrowband voice and data service. Broadband services, via terrestrial cells or satellites, should stimulate even more demand for wireless services. Mobile telephone companies are so convinced of that demand that they are risking billions of dollars in national spectrum auctions for the privilege of providing next-generation services. Figure 2.5 provides an estimate of instances where wireless options offer lower-cost options than traditional wireline services. Figure 2.6 tracks the growth of mobile telecommunications relative to fixed services. Figure 2.7 shows that competition has begun to occur in the mobile telecommunications sector.

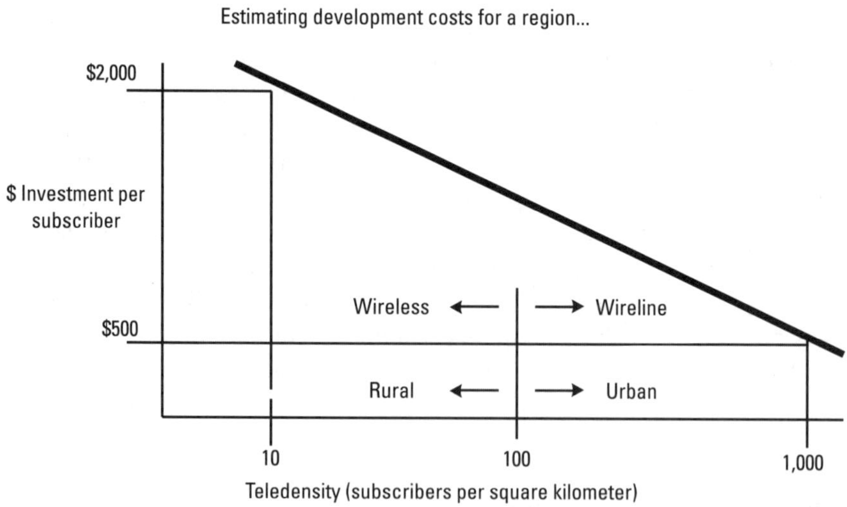

Figure 2.5 Comparison of wireline versus wireless costs.

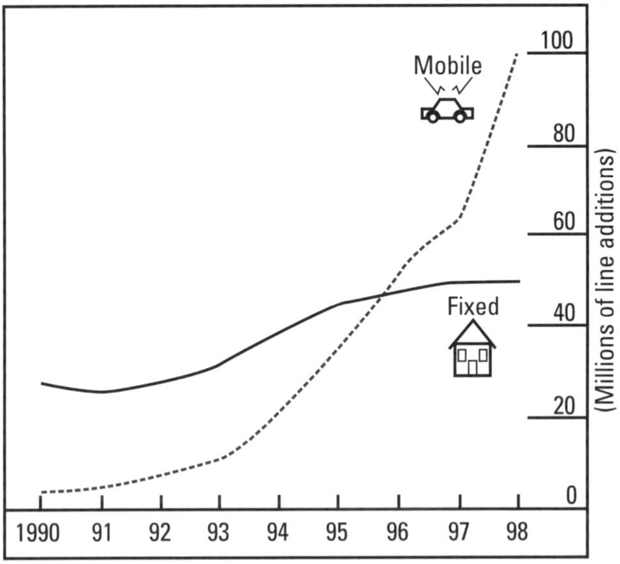

Figure 2.6 Mobile versus fixed telephone line additions. (*Source:* ITU.)

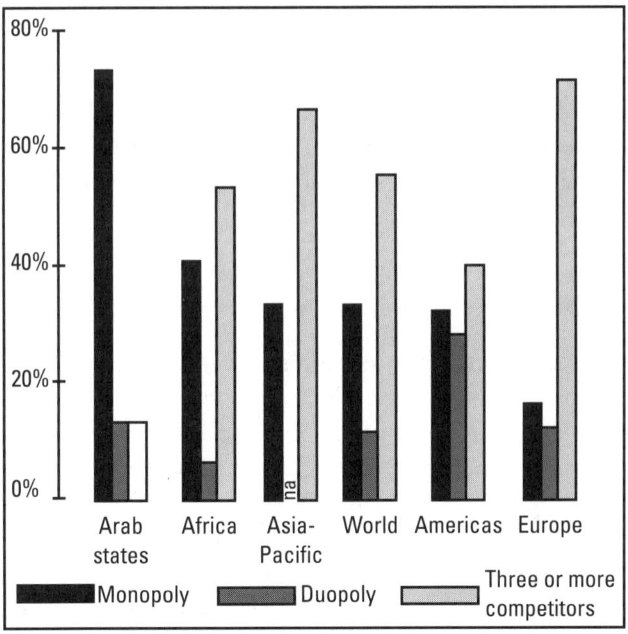

Figure 2.7 Scope of mobile radio telephone service competition. (*Source:* ITU.)

2.7 Electronic Commerce

In the near future, we might have difficulty identifying any kind of personal and interpersonal activity that the Internet cannot mediate. Already the Internet provides a medium for many commercial and private transactions that previously occurred only face to face or via less robust electronic means. Throughout the World Wide Web, the Internet operates in a vast array of functions, including those of bank, casino, pharmacy, stock market, shopping mall, pornography shop, radio station, newspaper, magazine, television station, jukebox, travel agent, inventory manager, car dealer, bookstore, grocery, pet supply store, jeweler, record store, sporting goods supplier, and auction house.

Some of the most desired functions performed by the Internet are those that provide added convenience, the opportunities to evade laws, and an environment free of public or personal impediments. For example, a person can easily obtain medication lawfully prescribed by his or her personal physician or, almost as easily, secure delivery of powerful medications prescribed by an on-line doctor after a brief, albeit costly, "consultation." The barriers to lawful gambling in the physical world involve the expense of travel, but such impediments cease to exist when a virtual casino awaits at any time and almost immediately via the Internet. In a nutshell, the Internet offers unprecedented opportunities to enhance as well as hinder one's emotional, spiritual, physical, and financial welfare. Predators, crooks, and liars compete with honest people for our attention, affection, and money. The Internet serves as a virtual microcosm of the real, material world with all the attendant risks and rewards.

2.7.1 The Consequences of E-Commerce

The impact of e-commerce on consumers blends significant positives with the risk of substantial negatives. Consumers stand to share in the benefit when companies accrue operational efficiencies from Internet mediation. However, consumers are more likely to face situations in which they risk greater privacy invasions, potential for fraud and deception, and diminished opportunities for redress. Even market-oriented advocates for e-commerce acknowledge that "[m]any businesses and consumers are still wary of conducting extensive business over the Internet because of the lack of a predictable legal environment governing transactions" [4].

While e-commerce has begun to make inroads on traditional retail activities,[1] many consumers are still waiting for a more user-friendly, reliable, trustworthy, and safe environment. A study prepared for the Organization for Economic Cooperation and Development (OECD) reported that the estimated 1995–1997 revenues generated by e-commerce comprised the equivalent of 37% of U.S. mail order catalog sales, 3% of U.S. purchases using credit or debit cards, and only 0.5% of total retail sales in seven major OECD nations.[2] Like most studies, the one prepared for OECD projects an incredible increase in e-commerce to levels exceeding $1 trillion in value by 2003–2005. That sum represents 15% of the total retail sales in seven major OECD nations. Such a massive increase in use presumes that Internet mediation provides friction-free commerce, that is, significant improvements in the efficiency and operation of markets that both businesses and consumers eagerly seek to exploit. E-commerce has the potential to accrue such performance enhancements, while actual marketplace practices may fail to achieve maximum gains.

2.7.2 The Promise of E-Commerce

Like the Internet in general, electronic commerce may create impressions of either too much change or not enough. Because well over 50% of the consumer base even in developed nations like the United States does not have access to the Internet, at least for the time being we cannot say that electronic commerce will change how we all do business. On the other hand, we can reasonably conclude that many businesses and consumers will seek to tap the benefits of cheap, fast, simple, and inexpensive access to a commercial network that defies time zones and operates continuously. "The growing popularity of this new media is driving development of new kinds of information-based products and services, while attracting business and

---

1. "Starting from basically zero in 1995, total electronic commerce is estimated at some $26 billion for 1997; it is predicted to reach $330 billion in 2001–2002 . . . and $1 trillion in 2003–2005," according to the Organization for Economic Cooperation and Development in *The Economic and Social Impacts of Electronic Commerce: Preliminary Findings and Research Agenda*, Chap. 1, p. 27; available at http://www.oecd.org//subject/e_commerce/summary.htm (viewed July 22, 1999).

2. *The Economist* ("Business and the Internet: The Net Imperative," June 26, 1999, p. 6) reported that total e-commerce amounted to $43 billion in 1998, with a projected total value of $1.3 trillion in 2003. B2C e-commerce constituted only $8 billion in 1998, with a projected value of $108 billion in 2003.

consumers alike to the electronic marketplace" [5]. The investment banking firm of Morgan Stanley Dean Witter projects that by 2004 e-commerce will constitute 20% ($2 trillion) of the U.S. economy [6].[3]

E-commerce holds the promise for reducing cost, inconvenience, and other impediments to commercial transactions. Already, plenty of case studies provide empirical evidence corroborating that point. By extrapolation, blue-sky visionaries predict Internet mediation as promoting an economy with few, if any, barriers and burdens to conducting business, presumably on a global basis. While the total, friction-free elimination of transaction costs and economic drag may overstate the case, we can expect e-commerce to stimulate efficiency gains in terms of consumer access to information and markets, business access to consumers, and in general opportunities for interaction and open, unfettered commerce.

The positive consequences of e-commence provide consumers and businesses with more and better resources. These tools empower decision makers by providing an enhanced ability to acquire and analyze data that, in turn, should support one's decision-making process. The Internet provides a multimedia forum for individual consumers to access resources that either previously did not exist or commanded prices only large businesses could afford. The Internet provides a global set of prospective consumers, thereby aggregating the demand necessary to support a service at the scale and scope necessary to maximize efficiency. That one-to-many multicasting function means a service developer can make available an Internet-mediated feature at low cost in terms of both the cost to provide the service to a user and the user's cost of access. Consumers often have free access to many high-powered Internet functions that required millions of dollars to develop. The one-to-many function also makes it possible to send the same content to a large number of simultaneous users.

2.7.3 Efficiency Gains Through E-Commerce

Internet mediation provides a vehicle for businesses to lower costs, reduce capital-intensive inventories, lower cycle times for processing and completing transactions, improve customer service, and generate new sales

3. Prudential Securities projects a slightly less steep global ramp-up: $55 billion in 1998, $330 billion in 2001–2002, and $1 trillion in 2003–2005. Charles A. Gabriel, Jr., and James P. Lucier, Jr., "Net Assessment," Prudential Securities, April 26, 1999.

opportunities.[4] Both businesses and consumers can buy products and services through a method that reduces complexity and intervening steps. For example, Dell Computer reported that in the three months ending April 30, 1999, its on-line sales of personal computers and other products averaged $18 million per day and accounted for 30% of the company's $5.5-billion first-quarter revenues [7]. On-line sales lower the cost of selling by reducing processing errors, providing better coordination between suppliers and purchasers, allowing suppliers to reduce the time it takes to bring a product to market, and tailoring products to meet consumer requirements. For example, rather than drawing from an inventory of several product lines, Dell can respond to an Internet-mediated order with a customized computer.

2.7.4 Individual Empowerment Through Disintermediation

Disintermediation is a complex-sounding term for a simple concept: the ability of the Internet to serve as the vehicle to forge closer, if not direct, links between commercial stakeholders (e.g., buyers and sellers). If an intermediary or middleman does not add value to a transaction, then the consumer can bypass that option via the Internet. Going back to the airline reservation example, several Internet-mediated travel reservation search engines offer the capacity to find the lowest fare, sometimes ones available only through the Internet. Air travelers might pursue the Internet option just because it offers greater control over the transaction and all preliminary steps leading to the purchase of a ticket. Travelers also might seek the Internet option to bypass the travel agency commercial arrangement where the amount of financial compensation flowing to the agent depends on the ticket cost, as well as the possibility of additional inducements for booking a particular carrier, hotel chain, and car rental company. After recouping the cost of establishing a Web presence and developing an on-line reservation system, airlines stand to save millions in reduced travel agent commissions with the ability to process a reservation and ticket at lower cost by rendering tickets and receiving payments independent of travel agents.

4. For a more complete discussion of these efficiency gains, see U.S. Department of Commerce, *The Emerging Digital Economy*, April 15, 1998 (available on-line at http://www.ecommerce.gov/emerging.htm); *The Emerging Digital Economy II*, June 22, 1999 (available on-line at http://www.ecommerce.gov/ede/).

2.7.5 Empowerment Through More Intermediation

Even as some middlemen lose business and market share to Internet options, other intermediary functions will develop. Because the Internet offers so much data and so many Web sites, consumers may find value in relying on an intelligent, software-driven broker or agent. Two new intermediary functions on which many Internet users increasingly rely involve search engines and portals. Search engines (e.g., Google, AltaVista, Lycos, WebCrawler, Excite, Snap, Ask Jeeves) offer the ability to canvass a portion of the millions of Web pages for sites likely to contain information of interest to the consumer. User-friendly search engines allow users to type in descriptive words or commands that the software then uses to poll cyberspace. Portals, such as Yahoo!, and the sites bundled with Web browsers, such as Netscape and Explorer, provide users with simple and easy access to a package of desirable Web sites in lieu of more time-consuming and complex searches. In the future, artificial intelligence software will enhance the ability of users to configure intelligent agents, with robot-like endurance, to cull from the onslaught of information, entertainment, commercial, and news options a customized selection tailored to the user's particular interests. Fuzzy logic and other innovations will make it possible for those agents to fine-tune and calibrate selections not just by user specifications but other criteria, such as the amount of time spent at Web sites, whether or not they are offered by the agent software.

2.7.6 Free-Rider Opportunities

Many Web sites offer attractive content free of charge and typically without conditioning access or limiting it to consumers manifesting the intention and wherewithal to buy the goods and services offered. That strategy parallels the advertiser-supported broadcast radio and television model: Web brokers prospective consumers' eyes, ears, and apparent attention for compensation. Banners and other advertisements bracket desirable content and operate as pointers to advertisers' Web sites, which contain more information about the products and services pitched. Advertisers hope that consumers will click onto the banner or at least that exposure to the banner will promote favorable perception of the advertiser and possibly greater predisposition to consume the product or service.

The Internet lends itself to both mass-market and narrowly targeted advertising, because it approaches a mass-market consumer electronic service and has intelligent applications tracking consumers' choices and tastes. The diverse and narrow nature of the content available via the Web favors narrow market advertising, for example, classical music, travel packages to the Antarctic, financial services to individuals with net worth in excess of $1 million. Regardless of their scope, advertiser-supported Web sites make it possible for free riders to consume the valuable content without also consuming the product or services advertised. All stakeholders win under this scenario: Advertisers find a new and attractive medium to get their messages across, consumers receive more information and may find Internet mediation a faster, better, cheaper, smarter, and more convenient medium for conducting business, and free riders have access to more sources of news, information, and entertainment.

2.7.7 Pitfalls and Risks of E-Commerce

With all the enthusiasm about a digital, friction-free economy, we can easily forget or ignore very real factors that impede progress. We can blithely assume that "a rising tide will raise all ships," that is, everyone will have an opportunity to exploit the many benefits of Internet-mediated commerce. Such universal opportunity presumes ubiquitous access to the Internet, an excessively optimistic presumption. Despite triple-digit increases in usage, number of server sites, and many other measures of growth, the fact remains that a relatively small percentage of the world's citizens have access to the Internet. Nua, an Internet strategy firm, estimates that 171 million people had such access as of May 1999, with more than half residing in the United States and Canada [7]. Even in developed nations, the percentage of the population with Internet access at home or the office does not exceed 37% [7]. Universal access to Internet-mediated services remains an unachieved goal, much like that of universal access to basic telecommunications services.

The term *digital divide* refers to a widening gap in access between a number of social, economic, geographic, and demographic groups [8]. In the United States, the gap in Internet usage has increased over the last few years between people in upper income brackets and people in lower income brackets, between whites and persons of color, and between residents in urban areas and those in rural locales. The access gap is even more pronounced between developed and developing nations. For example, as of June 1999, only 1.1 million African residents had Internet access versus 97 million in the United States and Canada. In Mexico, a nation with a population of about 100 million, 0.1% of the population had access to the Internet [7].

Clearly, individuals, businesses, and countries with comparatively less Internet access run the risk of losing out on the opportunities afforded by Internet-mediated commerce. Internet-only specials increasingly highlight the opportunity for individual consumers to acquire products and services at lower costs and with greater convenience. For example, the combination of better information technologies to predict seat sales coupled with the cost savings in electronic ticketing has created incentives for airlines to offer last-minute, drastically reduced fares on selected routes. The airline wins by filling seats that probably would go unfilled otherwise, while the consumer with Internet access wins by having access to travel opportunities at a fraction of the retail cost. One airline imposes a $20 additional fee for the use of its telephone reservation system in lieu of an on-line system complete with immediate debiting of credit cards. It is quite possible that in the future airlines will create a fare system with lower charges for on-line consumers and higher charges for consumers opting to use the airlines' telephone reservation system or a travel agent. As unfair as that may seem to the disenfranchised consumer lacking the resources or inclination to become computer savvy, such pricing would not constitute illegal discrimination. The airlines could justify a "digital discount" on economic grounds[5]:

> It is cheaper for an airline to process a ticket sale online than to use a travel agent or a reservations center. Not only are transaction fees reduced, but savings are also realized when cheaper electronic tickets can be substituted for more expensive paper tickets. [7]

The "productivity potential of the digital economy offers both hope and worry: with sufficient investment, it can provide the means to accelerated development, but without the needed investment, developing nations [in particular] may find themselves falling even further behind in an increasingly wired world" [7].

2.8 Unfinished Business

Few people will dispute the potential for telecommunications and information processing to enhance consumer welfare by means of tipping the bargaining scale somewhat closer to parity between buyer and seller and to "shrink the economic distance between producers and consumers" [3]. However such a favorable outcome depends on plenty of unfinished work and unresolved issues. Consumers require access to a cheap, reliable, and user-friendly physical and information infrastructure. Additionally, governments must establish a balance between erecting safeguards against fraud, invasions of privacy, and other predatory practices and in promoting a largely unregulated environment where marketplace forces can operate unfettered. Consumers also expect a commercial environment that offers simple billing and collection procedures while reducing risk and uncertainty. Technological innovations, entrepreneurialism, and marketplace outcomes may go far in making electronic commerce a reality, but limited government stewardship may foster better outcomes and perhaps reach those outcomes without the delay, litigation, and dislocation resulting from a completely unregulated, libertarian environment.

On-line electronic commerce, like many Internet-mediated activities, requires a robust, widely accessible, integrated, digital, and broadband link. Only recently have consumers begun to have even one technological option properly sized to meet future requirements. In many areas, primarily urban locales in developed nations, telephone companies, cable television operators, eclectic utilities, ISPs, and other ventures have begun the necessary retrofit or new construction.

It appears that the pace has quickened for the deployment of digital subscriber links, cable modems, and other high-speed, broadband links to the Internet. However, the diffusion of such cutting-edge technologies substantially lags behind that for basic "plain old telephone service" (POTS). Few localities currently offer the financial promise needed for ventures to invest in "pretty advanced (or amazing) new stuff" (PANS). The evolution of POTS to PANS is both essential and expensive, and for the time being most residential consumers still access the Internet via a modem and the POTS twisted-wire pair supporting no more than 56,000 bps in throughput.

2.9 Change

The Internet has evolved quickly to a point where it affects how we communicate and engage in commerce. Throughout this book, we will see how the convergence of telecommunications and information processing changes how both markets function. The locus of control in telecommunications has moved significantly from incumbent carriers to consumers. The market operates more efficiently, more competitively, and with greater diversification than it did only a few years ago. If the Internet becomes the focal point for most types of telecommunications, then even greater change and flux will ensue. From the vantage point of extrapolating from what has recently occurred, we can expect the future to blend the rules, conventions, culture, and business climate of both telecommunications and information processing. Accordingly, we should be able to understand and work within both worlds.

References

[1] http://www.intel.com/intel/museum/25anniv/hof/moore.htm.

[2] Gilder, G., "Metcalfe's Law and Legacy," *Forbes ASAP*, Sept. 13, 1993. Available on-line at http://www.forbes.com/asap/gilder/telecosm4a.htm.

[3] Organization for Economic Cooperation and Development, "Dismantling the Barriers to Global Electronic Commerce," discussion paper presented at conference held in Turku, Finland, Nov. 1997, p. 1. Available on-line at http://www.oecd.org//dsti/sti/it/ec/prod/DISMANTL.HTM.

[4] President William J. Clinton and Vice President Albert Gore, Jr., *A Framework for Global Electronic Commerce*, available on-line at http://www.iitf.nist.gov (viewed July 22, 1999).

[5] Organization for Economic Cooperation and Development, Directorate for Science, Technology and Industry, Committee on Consumer Policy, *Consumer Protection in the Electronic Marketplace*, DSTI/CP(98)13/Rev2 (Sept. 21, 1998).

[6] Morgan Stanley Dean Witter, *Internet Data Services Report*, Aug. 11, 1999, p. 2.

[7] U.S. Department of Commerce, *The Emerging Digital Economy II*, June 22, 1999; available on-line at http://www.ecommerce.gov/ede/.

[8] National Telecommunications and Information Administration, *Falling Through the Net: Defining the Digital Divide*, available on-line at http://www.ntia.doc.gov/ntiahome/fttn99/contents.html.

3

The Old World Order: Past and Present Models in International Telecommunications

Chapters 3 and 4 establish a foundation for understanding the often complex, fast-changing, and interdisciplinary world of international telecommunications. The chapters examine the dichotomy between the marketplace and the regulatory conditions that existed up to the 1980s and the substantially different environment that developed thereafter. A number of descriptive words expose the breadth of the subject and provide an easy frame of reference for differentiating between the old and the new environment. Additionally, we consider several explanatory models with an eye toward understanding how nations organized and then substantially changed the telecommunications sector.

3.1 The Rule of Multiple C's

Until recently, the rules of the road in international telecommunications were absolute, straightforward, and unavoidable. To understand how so much has changed in the last few years, we should consider descriptive key terms and underlying operational models. Only after such a foundation can we begin to see how conditions have changed, making it possible for new models and options to evolve after so many years of successful resistance.

A number of words, which begin with the letter *c*, describe many of the still applicable but no longer absolute key terms in international telecommunications.

3.1.1 Consensus and Compromise

The terms and conditions under which carriers operate international telecommunications networks require consensus decision making to establish common rules of the road. Even in a competitive marketplace with multiple carriers and equipment manufacturers, nations and private ventures still often recognize the benefit of having uniform technical and operational standards and agreements on what services operators will provide over which portions of the radio spectrum. The incentive to establish consensus remains largely undiminished even when nations adopt cutting-edge deregulatory policies. However, the ability to reach closure becomes more difficult as nations and private companies differ on which standards to adopt and what frequencies to use for a particular service.

No single nation or company can expect to have its policies and philosophies universally adopted, particularly when stakeholders perceive a financial advantage in pressing for a particular outcome. Telecommunications equipment manufacturers may not agree to a single technical standard if one or more manufacturers think they have a marketplace advantage and can establish a de facto standard before the conclusion of consideration in a multilateral forum. Individual nations may have industrial policies that attempt to favor national manufacturers in the process of setting standards and allocating spectrum.

Additionally, companies and nations may not embrace change if they do not perceive direct benefits. For example, in the late 1970s and early 1980s, the U.S. Federal Communications Commission (FCC) allowed entrepreneurs to lease domestic bulk long-distance calling capacity and to resell the capacity in smaller, discounted increments to users who individually would not generate the necessary demand for wholesale services. The FCC authorized shared use and resale of leased private lines and wide-area telephone service (WATS). Those initiatives met with universal opposition from incumbent carriers abroad, who objected to their potential loss of control over long-distance rates and the ability to charge on a usage and mileage-sensitive basis (i.e., toll charges by the minute or 10-second pulse and priced as a function of the distance between caller and call recipient) and to who could qualify for bulk capacity discounts.

Nations typically do not respond favorably to the unilateral initiatives of single nations, particularly ones that are not introduced and widely ventilated through multinational forums like the International Telecommunication Union (ITU).[1] Similarly, nations do not welcome campaigns originating in other countries that may adversely affect revenue streams.

The ITU, a specialized agency of the United Nations, provides the primary forum for consensus building among nations. First organized at the onset of telegraphy in 1865, the ITU establishes technical and operating standards, defines services, allocates spectrum, and promulgates international regulations and recommendations that make it easier to achieve connectivity and compatibility via telecommunications and information networks. Other bilateral and multilateral forums, often convened on a regional basis, augment the ITU process.

On occasion, nations deviate from the ITU-promulgated consensus. The governing documents of the ITU acknowledge the sovereign right of nations to disassociate from any decision and to enter into bilateral special arrangements that deviate from ITU rules or address issues not yet considered by the ITU. Nations also can take a "reservation" to a consensus-reached decision, in effect opting out of the general agreement to implement the ITU solution. Alternatively, nations can seek insertion of a footnote to a spectrum allocation or other regulation stating how they will deviate from the consensus.

Generally, nations recognize that universally accepted terms and conditions serve enlightened self-interest by reducing costs and expanding market access for equipment manufacturers and service providers. However, that recognition does not eliminate entirely the incentive to favor one nation's carriers and manufacturers. The consensus-building process often requires nations to modify their initial positions and to accept a compromise that constitutes a least common denominator as opposed to a technologically superior solution. The ITU provides a structure and formality to telecommunications standard setting as opposed to the less structured and more fragmented process for Internet-related matters. When applied, standards of any formulation achieve the same outcome for both telecommunications and information processing. Converging technologies may spur efforts to merge

1. See General Secretariat of the International Telecommunication Union, *Constitution and Convention of the International Telecommunication Union, Decisions Resolutions and Recommendations, Adopted by the Plentipotentiary Conference* (Kyoto, Japan, 1994) (Geneva: ITU, 1995).

the formal and informal standard-setting processes. However, many Internet carriers and service providers typically object to centralized decision-making.

3.1.2 Collaboration

International telecommunications is a bilateral, often multilateral, exercise engineered by two or more foreign correspondents. Carriers establish a correspondent relationship that relies on their joint capabilities and resources to engineer complete routes between caller and call recipient. International carriers execute operating agreements that specify the terms and conditions by which they will accept traffic and deliver it to the intended recipient or to another carrier.[2] Correspondents match transmission circuitry and coordinate the provision of service between nations.[3] Telecommunications networking requires end-to-end, complete engineering by carriers. Internet networking typically does not require service providers to anticipate and secure routing arrangements for every possible link. The Internet uses best-effort routing and greater reliance on the transport services of other ISPs, which may not always have a direct, contractual relationship with all other participating ISPs.

International carriers pool financial resources in multimillion-dollar transmission facility investments.[4] They allocate investment and operational responsibility on a half-circuit basis, with correspondents theoretically matching half-circuits into whole circuits at an international boundary or at the midpoint of a route. One half-circuit of a carrier's inventory represents

2. "The term 'operating agreement' traditionally has been used to refer to the contract or other arrangement in which an overseas administration agrees to operate with another administration or U.S. carrier and which sets out all the terms and conditions under which service is offered," according to the International Communications Policies Governing Designation of Recognized Private Operating Agencies, Grants of IRUs in International Facilities and Assignment of DNICs, CC Docket No. 83-1230, Notice of Proposed Rulemaking, 95 FCC 2d 627, 630 (1983), Report and Order, 104 FCC 2d 208 (1986), *recon. den.*, 2 FCC Rcd. 7375 (1987).

3. See also R. Frieden, "International Telecommunications and the Federal Communications Commission," *Columbia Journal of Transnational Law*, Vol. 21, 1983, p. 423. "Various foreign PTTs are equal partners with U.S. carriers and decide with which U.S. carrier(s) to connect 'half-circuits.' The financial stakes are extremely high in engineering a 'whole circuit' link between nations," *id.*, p. 425, n. 8.

4. The cost to construct, launch, and operate a geostationary telecommunications satellite ranges from $150 million to $200 million. See, for example, PanAmSat Corp., Applica-

the smallest common unit of capacity corresponding to one analog voice channel. For digital fiber optic submarine cables, carriers originally designated capacity in terms of minimum assignable units of ownership (MAUOs),[5] a pathway comprising 64,000 bps of throughput and corresponding to approximately four voice-grade channels with the use of conventional circuit multiplication technology. In view of the demand for ever larger amounts of bandwidth, many carriers now assign ownership interests in increments of synchronous transport module-ones (STM-1), equivalent to 155 Mbps.[6] Carriers must coordinate the planning, design, investment, construction, and maintenance of international facilities. Such pooling of resources can lead to a shared view that the status quo should persist. In application, policy inertia tends to foreclose competition by retaining monopolies or barriers to market entry. Figure 3.1 depicts an international half-circuit representing the telecommunications infrastructure needed to originate or terminate an international call. Carrier correspondents "match" half-circuits to establish a complete end-to-end link.

3.1.3 Consultation

Consultation regularly occurs between and among international carriers and their regulators. In many instances the line blurs between carrier-to-carrier business negotiations versus intergovernmental policy coordination. Many nations only recently have separated the regulatory and operational functions, having performed both under the auspices of a single ministry of posts and telecommunications. Even after nations spin off the operating function

---

tion for Authority to Launch and Operate a Replacement C/Ku Hybrid Fixed Satellite Service Space Station at 99 degrees W.L., Order and Authorization, DA-00-867, 2000 WL 381632 (F.C.C.) (rel. April 14, 2000)(PanAmSat's estimated cost of $168 million to construct, launch, and operate Galaxy IV-R for one year). Submarine cable costs vary as a function of the distance and number of fiber pairs installed. Current-generation transoceanic cables typically cost several hundred million dollars.

5. "A MAUO is the Minimum Assignable Unit of Ownership and provides an equivalent digital channel operating at 64,000 bits per second," American Telephone and Telegraph Co., 5 FCC Rcd. 7331, 7338, n. 5 (1990) (authorizing construction of the PacRim East fiber optic submarine cable).

6. See, for example, Atlantica USA, LLC, Application for a License to Land and Operate in the United States a Private Fiber Optic Submarine Cable System Extending Between The United States, Venezuela, Brazil and Bermuda (The Atlantica-1 Network), 14 FCC Rcd. 20,787 (1999).

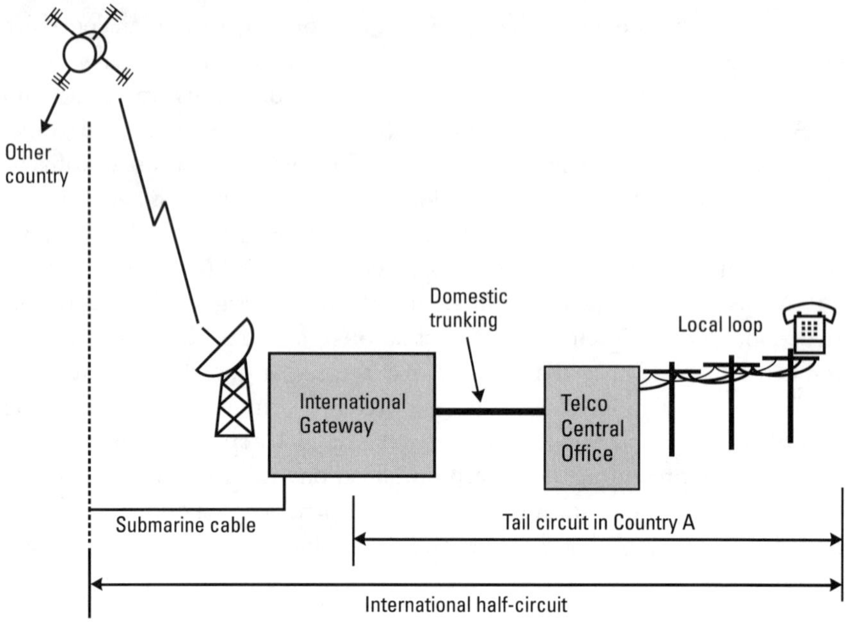

Figure 3.1 The international half-circuit.

into a separate government-operated entity or a privatized commercial one, government oversight remains extensive. Governments typically retain a single "golden share" in privatized carriers that they can cast to veto any action deemed inconsistent with the national interest. Most delegations to the ITU and other forums include both carrier representatives and regulators. Likewise, when carriers assemble to address business or policy matters, the group of participants typically includes both operational and regulatory constituencies.

The combination of regulatory and business delegates in a variety of forums has the consequence of merging points of view and institutionalizing a role for government even in matters affecting the strategic business planning of individual carriers. For example, in some nations, carriers and government regulators continue to meet in what is called the consultative process to collaborate on demand forecasting and planning the future deployment of international transmission facilities.

FCC delegates in the 1970s actively brokered deals on when and where carriers should deploy new international submarine cable transmission facilities. The FCC justified such intervention on the basis of its public interest mandate to ensure that new transmission facilities meet real, near-term demand. The commission secured a delay in construction of the TAT-7 transatlantic submarine cable until such time as it believed that demand justified deployment [1]. Even though other delegations included government representatives, most participants viewed the FCC's activism as intrusive and heavy-handed. An appellate court held that the FCC violated U.S. administrative law that requires "government in the sunshine," because its actions in consultative process were not subject to public observation or participation [2]. The FCC no longer interferes with carriers' traffic forecasting, nor does the commission closely scrutinize applications to construct and operate new international transmission facilities.

Collaboration promotes the formation of facility cooperatives, like the International Telecommunications Satellite Organization (INTELSAT) and consortia that pool investments in submarine cables. Whether for efficiency or convenience grounds, it also means that a strong incentive exists for most governments to restrict the number of operators to a manageable number. For example, many nations take incremental steps in fostering a liberalized, deregulated environment. Formerly monopolized markets may first transition to competition with the authorization of a single, new competitor. That duopoly or shared monopoly may deprive consumers of the complete benefits accruing from more open competition, but many nations proceed with caution when embracing a deregulatory philosophy.

The FCC and most regulators no longer actively participate in official consultations with carriers. However, some regulators continue to perform extensive, independent analysis to support its approval of carrier applications to construct or participate in the construction of new facilities. To satisfy its public interest mandate, the FCC used to require international carriers under its jurisdiction to submit traffic statistics[7] and to provide evidence of increasing demand for service and new transmission facilities. In conjunction with those filings, what it derives from its observer role in carrier consultations and its own analysis, the FCC prepared to consider facility construction applications filed by carriers as required by Section 214 of the Communications Act of 1934.

7. See Amendment of Sec. 43.61 of the Commission's Rules, CC Docket No. 91-22, Report and Order, 7 FCC Rcd. 1379 (1992) (simplifying and streamlining reporting requirements but adding reporting requirements for traffic to and from Canada, Mexico, and St. Pierre Miquelon).

Additionally, the FCC used to consider the Section 214 review process as an opportunity to scrutinize carrier submissions of demand projections and evidence supporting the need for additional transmission capacity. The commission now refrains from undertaking a thorough demand assessment, on the grounds that it should not second-guess carriers on capacity requirements, particularly because it believes that no carrier can afford to tie up capital in unnecessary projects. In the late 1990s, the FCC streamlined its review process to create a refutable presumption that construction applications should be granted. The commission now believes that carriers propose construction of new facilities in response to marketplace requirements, rather than anticompetitive reasons like warehousing capacity to preempt market entry.[8]

Prior to its decision to monitor the consultative process rather than actively participate in it, the FCC predicted when carriers subject to its jurisdiction would need to activate additional transmission facilities and by what ratio they must activate circuits from their inventory of cable and satellite capacity. The FCC made such projections and loading decisions for a specified time period, by ocean region. The commission no longer interferes, because it now believes that carriers should freely activate transmission media on the basis of their assessment of consumer requirements and on comparative technological features.

3.1.4 Culture

The culture of international telecommunications and information processing has become more complex and more diverse. The old world order supported greater uniformity, because a small number of operators and their government regulators had established a mutually beneficial compact whereby each party avoided inconveniencing or embarrassing the other. That environment favored a patient, long-term orientation that does not work well in Internet time. Operators in the old world order had the luxury of working at a slower, more measured pace and being able to accommodate legislative and regulatory objectives, like universal subsidies, even though those objectives may have adversely affected short-term financial results.

Despite significant differences in politics and overall culture, nations and their telecommunications service providers usually agreed on shared

8. See Streamlining the International Section 214 Authorization Process and Tariff Requirements, IB Docket No. 95-118, 11 FCC Rcd 12884 (1996).

goals articulated in forums like the ITU and facilities cooperatives and consortia. Until nations like the United States and the United Kingdom undertook an aggressively deregulatory approach, governments and carriers throughout the world agreed on common objectives like universal service, interconnected networks, price averaging between dense and sparse routes, and single composite rates to users that average possibly different cable and satellite circuit costs.

While many nations continue to support long-standing goals, deregulation and privatization have spawned a variety of new players who lack the view that their corporate objectives should parallel governmental or social goals. Because market entrants typically have no exclusive service franchise and lack government-mandated insulation from competition, they seek to maximize profits and object to any requirement to provide some services at below-cost rates. New regulatory classifications, like private, noncommon carriage, exempt operators from most public service duties in recognition of the fact that they lack the power to control price or availability of service.

Successful market entry by new international telecommunications players challenges the viability of the view that a single PTT monopoly must operate with an exclusive franchise. Incumbents may have well served both national and international goals with a public utility culture and, as a benign or enlightened monopolist, serving shared corporate and public objectives. However, recent legislative, regulatory, technological, and marketplace changes have fostered the onset of a corporate, profit-maximizing culture. Corporate players will have less to gain in serving social goals when they no longer enjoy insulation from competition. Likewise, technological innovation makes it easier for outsiders to infiltrate previously closed markets. Lacking a public service culture, private ventures may object to the imposition of mandatory service obligations and pricing limitations designed to achieve social policy objectives.

3.1.5 Clubbiness

The relative paucity in the old world order of telecommunications manufacturers and carriers may have dampened the incentive and enthusiasm for competition and "sleepless afternoons." Not until market entry initiatives reached critical mass did the marketplace show the characteristics typical of a competitive marketplace. Instead, carriers largely respected national borders and refrained from poaching the customers of other carriers. Equipment manufacturers showed less reticence to compete, but incompatible standards forced them to assess whether a business case supported creating a new production line.

Until facilities-based competition occurred in international telecommunications and information services grew in importance, the environment looked more like a club than a robustly competitive marketplace. The new world order looks far less clubby as the numbers and types of operators grow. Different facilities ownership models have evolved, as have business and strategic plans.

3.1.6 Cooperatives, Consortia, Cartels, and Collusion

The extent of investment involved in international satellite and submarine cable facilities requires spreading risks and costs over a broad set of carriers. International carriers pool satellite investments and aggregate consumer circuitry requirements through global and regional cooperatives, like INTELSAT and the European Telecommunications Satellite Organization (EUTELSAT). A parallel regional investment structure for submarine cables involves the formation of consortia. Investment shares in such ventures track anticipated usage.

At some point, however, investment pooling in transmission facilities led to cartel formation and club-like thinking. New international satellite and submarine cable systems, separate from the conventional consortium investment group, had difficulty securing operating agreements with incumbent carriers, because market entry threatened to migrate traffic and revenue away from the incumbent cooperatives and consortia. While economists suggested that technological innovation could support market entry in international telecommunications, incumbents worked mightily to retain uniform and restrictive rules of the road.

The closed group of incumbent carriers had every incentive for maintaining the comfortable status quo, but technology, consumer demand, and some activist governments, responsive to consumer requirements, made it unsustainable. Over time, cartel behavior became unenforceable as individual carriers began to perceive greater financial rewards in serving new markets than in implicitly agreeing to refrain from competing. Even as competition provides new market access opportunities, incumbent carriers still dominate. Table 3.1 shows the annual revenues of the key multinational players.

3.1.7 Constraints and Comity

Participants in forming the rules of the road in international telecommunications recognize the need to temper any one nation's policies and

Table 3.1
Who Are the Global Telecom Giants?

Annual Telecom Revenues in US$	
NTT (Japan)	$47 billion
AT&T (U.S.)	$60 billion (Includes other revenues)
Deutsche Telekom (Germany)	$33 billion
WorldCom (U.S.)	$33 billion
Bell Atlantic (U.S.)	$31 billion
British Telecom	$28 billion
France Telecom	$26 billion
SBC (U.S.)	$26 billion
BellSouth (U.S.)	$25 billion
Telefonica (Spain)	$24 billion
Total	$310 billion—50% of world's market

philosophies. The broad goal of international comity means that nations may agree to a least common denominator solution to achieve consensus even at the expense of consumer welfare. International telecommunications users accustomed to vigorous competition in domestic markets find frustrating the high cost and limitations on network control and flexibility occurring in many domestic markets. Often such constraints represent unnecessary expense and inconvenience built in by the decision of nations and carriers to mandate compliance with a single protocol or standard. That default solution frustrated telecommunications managers, who have mastered the art of negotiating domestic service arrangements customized to meet particular user requirements.

The quest to be all things to all people stems from international carriers' duty to serve all user requirements, including nonprofitable but socially desirable services. By restricting access to discounted services and perhaps also limiting the number of options available, incumbent carriers have greater control over traffic and revenue streams. Ostensibly to prevent bypass and lost revenues, limitations also provide incumbents with greater control over access to markets. That reflects the view that the opportunity to have traffic terminated in a foreign locale and to receive foreign-originated traffic is a privilege conferred on carriers by a sovereign government or its duly designated franchisee. Accordingly, the foreign government and its carrier may balk at speedy deployment of new services, insisting on the need for study to determine whether the new service complies with ITU regulations and recommendations. While compatibility with ITU edicts is important, incumbent carriers may stall for time to ensure that a new service does not cannibalize existing, more profitable services through customer migration to the new offering. Technological innovations increasingly prevent incumbent carriers from successfully implementing stalling tactics. For example, entrepreneurs can offer imported foreign dial tone cheaply and without detection. Such callback or call-reorigination services largely prevent incumbent carriers from overcharging and mandating a monopolist perspective. The ability to use the Internet as a medium for even cheaper or free international calling provides an even greater check on incumbent carrier behavior.

3.1.8 Commingled Costs and Cross-Subsidies

Government-owned or franchised incumbent carriers traditionally have dominated international telecommunications markets. Despite growing interest in privatization and deregulation, most nations have only begun to divest incumbents of their monopolies. Government PTTs or privatized monopolies continue to operate partially insulated from unfettered competition, but in exchange they must bear significant service duties. Those duties include the deliberate underpricing of certain services like some domestic telecommunications services to achieve universal service objectives (e.g., affordable service to rural residents and the disadvantaged).

In the transition to competition, regulators have to calibrate closely the pace of liberalizing incumbent carrier regulation versus the scope of regulation applicable to market entrants. Too great a spread creates regulatory asymmetry, that is, a dichotomy in regulatory burdens borne by competitors offering similar services but subject to different regulatory classifications. When two categories of service providers offer functionally the same service, the potential exists for an unlevel competitive playing field because one type of venture has to bear more burdensome and costly regulatory requirements. Regulators should not deny incumbents the opportunity to compete by retaining unnecessary legacy regulations. On the other hand, incumbents possess significant market power and the ability to frustrate market entry and competition. Regulators also have to create sufficient incentives for market entrants to invest in facilities rather than continue to resell incumbents' services.

Advocates for cost-based pricing and competition must recognize the threat presented by initiatives that would eliminate high profits from some international services. The typical international telecommunications regulatory regime requires carriers to cross-subsidize basic, domestic telephone services using revenues derived by charging excessive rates for services used by international users. For example, international switched telephone service may generate triple-digit returns so that incumbent carriers can underprice domestic telephone service and undertake major construction projects. Such pricing dampens demand for international calling, and the matter grows worse when incumbents cannot use the revenues to improve service, but must contribute some or all to the general treasury. Carriers cannot price all services on the basis of cost, because public policies typically require that pricing help achieve certain socially desired goals, for example, widespread availability of basic telephone service even in sparsely populated areas where service is costly to provide.

3.1.9 Closed or Conditional Market Access

During the reign of the old world order, many nations significantly restricted access to domestic markets, including the opportunity for foreign manufacturers to bid on equipment procurements by the incumbent carrier. Sovereignty and national security concerns often served as a thinly veiled effort to blunt comparative advantages enjoyed by foreign carriers and manufacturers. Nations may have industrial policies designed to favor a national hero, that is, a domestic manufacturer who might lose in a competitive tender. Likewise, nations sought to insulate carriers from competition that might foreclose below-cost pricing of socially desired services.

Most nations restricted market access by foreign carriers and manufacturers until pro–market access philosophies reached critical mass. Governments first embraced procompetitive views when it became clear to them that a majority of the electorate demanded more choice, better service, and lower rates and when they concluded that incumbents could not or would not reform. Consumers drove change and turned against incumbents when technological innovations and market convergence provided the opportunity. Having been captive to incumbents, who may have treated them cavalierly, consumers embraced new and different options. The proliferation of services and operators, including ruthlessly competitive information and entertainment ventures, provided ways to exploit technological innovations to aggregate traffic and to route it across political borders without detection.

3.2 The PTT Industrial and Regulatory Model

Until the late 1980s and early 1990s, few nations had adopted procompetitive policies in telecommunications. The procompetitive model, defined by law and codified regulations supporting competition and implemented by an independent expert agency, significantly contrasted with the rest of the world that vested both policy oversight and operational management within a state bureaucracy.[9] While many developed and developing nations alike have privatized the incumbent carrier and taken other market-liberalizing steps, most markets remain substantially regulated and many lack robust competition in at least some market segments (e.g., local exchange services).

3.2.1 The Traditional PTT Model

To understand why nations have embraced conditional competition and have changed the industrial structure, this section examines the traditional PTT model. The PTT model represents a monopoly postal and telecommunications enterprise owned, operated, and regulated by government. Advocates for this model believed that only centralized, public management can temper profit-maximizing strategies with the duty to achieve the following social policy goals:

- Universal service or at least affirmative efforts to provide affordable basic telecommunications services to rural areas and low-income users;
- Rate making that prevents price gouging, except where socially desired, (e.g., overpricing outbound international telephone rates to generate a source of revenues for cross-subsidizing local telephone services);
- Price averaging that blends high- and low-cost routes (urban versus rural and dense versus sparse) and transmission technologies (cable versus satellite) into a single, composite rate;
- Rate setting that serves social goals (e.g., subsidized lifeline rates to low income users);

---

9. See Communications Act of 1934, *as amended,* 47 U.S.C. Sec. 151 *et seq.* (1999); Communications Satellite Act of 1962, *as amended,* Pub. L. 87-624, 76 Stat. 419 (1962), *codified as amended* at 47 U.S.C. Sec. 701-744 (1999); 47 C.F.R. Telecommunications, Parts 0-100.

- Long-range planning that achieves development objectives (e.g., deploying advanced broadband digital facilities and services to rural locales even in the absence of demand and the likelihood of fully recouping investment);

- Political brokering that balances, on one hand, the interests in weaning the telecommunications sector from government subsidies and moving it toward global competitiveness and, on the other hand, securing political support from the large, often unionized body of telecommunications employees whose numbers may decline as a result of streamlining initiatives.

Nations unwilling to pursue the difficult and possibly painful migration from monopoly-based service to competition remain convinced that only a single network service provider can serve public policy objectives. That rationale was supported by now-disputed economic analysis that competition would result in wasteful duplication of facilities and would deprive the incumbent of the ability to generate the lowest service costs. Status quo advocates believe centralized planning can optimize investment in infrastructure improvements and best serve the national interest. Even if a nation pursues structural modifications aimed at eliminating impediments to efficiency, like spinning off postal operations, hard-liners in the old world order remain unconvinced that the marketplace can support multiple service providers and networks.

PTTs justify monopoly service on philosophical and economic grounds. Despite reevaluation and increasingly aggressive experiments with competition, many foreign governments still consider telecommunications a matter of such social importance that they cannot rely entirely on private enterprise, open markets, and competition. A declining number of governments still believe competition cannot be allowed to usurp from the incumbent carrier the opportunity to achieve economies of scale by operating most efficiently, that is, at the lowest cost. But even now it is surprising to see how many nations lack robust competition as measured by, for example, market share.

The PTT model proved resilient in the face of technological change, consumer desires, and the aspirations of prospective market entrants, because many governments believed that fundamental economic principles and political pragmatism supported the status quo. PTTs, particularly in developing countries, have invested in system improvements with the expectation of having the opportunity to serve all users for the long term. PTTs often affiliate with or own the single "national-hero" telecommunications equipment manufacturer. The PTT serves as a virtual captive customer of the domestic equipment manufacturer, and users typically have no service alternative.

Monopoly status and the ability to charge high rates provide an attractive source of hard currency. Historically, many PTTs recouped a significant percentage of their total facility investment through favorable toll revenue division arrangements for terminating international calls. Additionally, they exploit the relative inelasticity of demand for outbound international services, that is, charging business users of international message telephone service (IMTS) a steep premium for immediate access to the rest of the world.[10] Multiple facilities-based carriers (enterprises with their own transmission networks) compete on price and willingly reduce the rate they charge foreign carriers with an eye toward stimulating demand and capturing a larger share of calls.

Many telephone administrations in Africa, Central America, and the Caribbean still have telecommunications monopolies, despite growing consumer dissatisfaction and technological innovations that make it possible to circumvent restrictive rules and avoid excessive charges. The exclusive carrier model became unsustainable and unacceptable when a growing body of empirical evidence proved that competition does not jeopardize the ability of the incumbent to survive, even as it stimulates lower rates, service diversity, efficiency gains, and infrastructure development as measured in such quantitative indices as number of telephone lines per 100 inhabitants.[11] While accepting that the old world order could not remain, governments have wrestled with finding ways to continue working toward achieving public policy objectives, like universal service, while unleashing competitive forces. Governments work incrementally and cautiously for fear that competition will prove harmful to consumers and incumbents alike.

10. See *International Accounting Rates and the Balance of Payments Deficit in Telecommunications Services,* Report of the Common Carrier Bureau to the Federal Communications Commission, Dec. 12, 1988.

11. See International Telecommunication Union and Telegeography, Inc., *Direction of Traffic Trading Telecom Minutes* (Geneva: ITU, 1999); Edward R. Leahy and Michael O'Brien,

3.2.2 Incentives to Maintain the Status Quo

The old world order emphasized continuity and certainty enforced by regulators, who kept carriers on a tight rein, and vice versa. In fact, many national governments used to combine the functions of service provider and regulator. As an important government institution, employer and carrier of last resort for underpriced postal and telecommunications services, PTTs asserted that both economic and public policy rationales supported monopolization.

- The economic rationalization was that only one enterprise could reach the size and operational wingspan needed to accrue economies of scale and scope. In other words, telecommunications supported a natural monopoly, because only one enterprise would operate most efficiently and offer consumers the lowest costs. Put another way, this view rationalized that market entry would result in wasteful duplication of resources that could be allocated to more efficient and useful purposes.
- On public policy grounds, PTTs maintained that only a single entity could manage to offer below-cost services to favored constituencies, including rural residents, the poor, and the elderly. Competition would prevent the PTT from using profits generated from monopoly services to subsidize other services that no venture would ever offer voluntarily.
- Incumbent PTTs presented persuasive arguments or at least generated enough apprehension among legislators and regulators alike so that privatization, liberalization, and deregulation occurred at a measured pace. Such reluctance to experiment can be explained by the service mandate of the typical PTT, its institutional heritage, and notions of national sovereignty. Telecommunications in most nations is not merely a technical system, but a social, political, and economic institution [3]. Having merged the PTT's telecommunications function (which possibly can tolerate competition) with postal operations (which more clearly qualify for monopoly status),

---

"Telecommunications Law and Technology in the Developing World," *Boston College International and Comparative Law Review*, Vol. 1, Winter 1999, p. 22; International Telecommunication Union, *Trends in Telecommunication Reform—Convergence and Regulation* (Geneva: ITU, 1999).

> nations seeking greater efficiency may separate the P from PTT. But divestiture of postal operations is unlikely when a government would rather tap a somewhat hidden source of subsidies to keep the post office solvent. Certain telephone services, particularly business lines and international long-distance service, generate substantial revenue surpluses that are used to keep local service and postal rates artificially low.

We should not underestimate the caution that national governments exercise when they are considering policy and regulatory changes. Legislatures typically must take the lead in fostering change, because PTTs have two countervailing preoccupations:

- How to keep ratepayers captive, that is, preserve a telecommunications equipment and service monopoly in the face of technological innovations that provide ways to "bypass" the PTT or to avoid the most expensive services;
- How to maintain exclusive and lucrative revenue streams, that is, prevent market entry.

PTTs will avoid any procompetitive initiative that threatens profitable revenue streams used, in part, to subsidize postal operations and domestic telecommunications services. Concern over undesirable competition or bypass creates incentives for PTTs collectively to preserve conservative, if not collusive and anticompetitive, policies.

3.3 Revenge of the Phoneheads: The Power of Incumbents

We should not underestimate the power of incumbents to thwart and delay change even if the PTT model looks illogical and unsustainable. Stakeholders in the old world order in telecommunications have every incentive and plenty of resources to forestall the transition to the new world order. Few ventures willingly give up market share and industry dominance even with the promise of greater operational flexibility, more market niches to serve, and a liberalized regulatory environment. As inevitable as it may be, the onset of new world order characteristics will arise incrementally and inconsistently, and nations will not achieve the migration at the same time and in the same way.

Incumbents have had up to 100 years (and longer in some cases) to develop the linkages, connections, and resources for weathering any sort of political, financial, regulatory, or consumer-driven reform campaign. Incumbents have extraordinary expertise in exploiting the regulatory system and in stifling progress. They have established long-standing relationships with consumers, many of whom have no interest in acquiring knowledge about the advantages and cost savings in developing new business relations. Likewise, incumbents have installed the infrastructure needed to provide and bill for services.

Market entrants by definition start with zero market share and the duty to prove they can improve on what the incumbent offers, based on the criteria consumers use to evaluate performance. While they may have access to new technology, which can offer better and cheaper services, market entrants bear the financial burden of deploying an infrastructure that incumbents have installed over the course of their tenure. Even well-capitalized entrants take time and work incrementally when they install facilities. On one hand, that makes it possible for them to conserve capital and target the most lucrative consumers and service territories. But on the other hand, it means that market entrants must rely on incumbents for access to customers not directly linked to the entrant's network.

Incumbents have four major advantages: control over bottleneck facilities, superior access to political and regulatory systems, employer leverage, and the ineffectiveness of the ITU and the World Trade Organization (WTO) to trigger speedy changes in policy.

3.3.1 Persistent Bottlenecks

The need to conserve capital by gradual deployment of network facilities coupled with regulatory policies typically limiting first market entry to niches places newcomers in a precarious position. Market entrants typically do not completely duplicate the network facilities of incumbents, which means that, absent effective regulation, incumbents have the power to limit the effectiveness and impact of competition. Market entrants may first operate as resellers of capacity leased from incumbents. Alternatively, they may have installed a limited backbone network that directly serves some, but not all, customers. As well, some market entrants receive regulatory authority to serve one type of user (e.g., the mobile wireless market), thereby requiring the market entrant to seek interconnection of its network with the far more widespread network operated by the incumbent.

Even to this day, many incumbent carriers continue to maintain control over essential bottleneck facilities needed by market entrants to secure access to consumers. To receive traffic from a call originator, or to deliver traffic to an intended call recipient, market entrants typically have to use the existing network facilities of the incumbent. Incumbents have every incentive to refuse to interconnect their network facilities or to do so at great expense, inconvenience, and only after time-consuming negotiations. Because market entrants need access to such bottlenecks, incumbents have both the incentive and possibly the power to engage in a price squeeze, raising the cost of an intermediate service element (e.g., the wireline portion of a call originating or terminating over a wireless network), with an eye toward making it more difficult for competitors to operate efficiently and effectively.

3.3.2 Access to Political and Regulatory Systems

Incumbents have established a long-standing, mutually beneficial relationship with legislators and regulators. Critics consider the arrangement as one in which the service provider captures the regulator. Even if policy makers retain control over the process, they may very well grow reliant on the incumbent telecommunications carrier for information and support vis-à-vis the legislature. Market entrants may have the benefit of new and refreshing political and economic philosophies, but incumbents typically enjoy superior access to decision-makers.

3.3.3 Employer Leverage

Incumbent carriers and equipment manufacturers also benefit by employing a large number of people. Despite evidence that competition generates more employment in the information and telecommunications sector, incumbents have generated fears of near-term massive layoffs. Incumbents may delay the inevitable onset of the new world order by bargaining for more time to adjust and respond to changed circumstances.

Incumbents may bargain for less drastic procompetitive initiatives by galvanizing employee and union opposition to change. At the onset of procompetitive initiatives, such tactics achieved generally favorable results. Later, employees grew to realize that a privatized venture typically triggered greater valuation, particularly when shares in the venture became publicly traded on a stock exchange. Government owners of incumbent carriers helped secure employee support for privatization by agreeing to allocate a portion of the shares in the commercial venture directly to employees and their pension funds. Over time, incumbents lost leverage with their employees, who recognized that while some workers might lose their jobs, everyone would receive a near-term monetary bonus as a result of privatization.

3.3.4 Ineffectiveness of the ITU and WTO Multilateral Forums

Forums that set multilateral standards, allocate spectrum, and make trade policy contribute uniform rules of the road in telecommunications. However, as will be examined in Chapter 7, such organizations lack effective enforcement tools, may avoid tackling particularly troublesome issues, and increasingly fail to achieve global consensus. One key reason for declining success lies in the broadening array of stakeholders and the expanding range of issues the organizations must address.

Incumbents have the same incentives to target international forums to support the status quo as they have with domestic legislative and regulatory bodies. In both arenas, incumbents have established a long-standing and credible presence, having had representatives on countless delegations and having performed extensive work on the variety of committees and working groups that generate the recommendations and other work products for consideration.

Multilateral telecommunications and trade forums have not fully adjusted to technological and marketplace convergence and the diversification of issues they generate. The ITU management may want to address Internet governance issues, in view of the potential for Internet mediation and a datacentric infrastructure to dominate the future. However, the ITU governance process was formulated by government representatives who expected to work primarily in an intergovernmental, treaty-based forum. Netheads have limited, if any, exposure and experience with such a process and probably will have limited, if any, patience with it.

The WTO has successfully formulated a baseline frame of reference for market access and trade policy in the new world order. However, not all nations have embraced it, and fewer still have fully implemented the market access commitments they have made. The WTO also has had difficulty in applying a trade policy mechanism to telecommunications issues that integrate trade and market access issues with what appears to involve carrier-to-carrier operating procedures.

3.4 Teaching Old Dogs New Tricks

While they may resist change, few old world order incumbents persist in thinking they can control the international telecommunications marketplace. Technological innovations, market entrants, and the infusion of information-processing applications and culture absolutely prevent that. Incumbent carriers will lose market share, particularly when they have to relinquish a monopoly. But technological innovations and regulatory liberalization create opportunities and trigger lost market share. Furthermore, most incumbents have generated substantial retained earnings that can provide the finances for buying market share.

Incumbents have to reshape themselves if they are to survive. They must have the will to do so, which for the most part will arise with the infusion of new personnel and a new culture or when lost market share results in declining earnings and share prices. The key to a successful renewal lies in a cultural change in which incumbents embrace change and the need to compete, rather than expect to exploit the legislative and regulatory process.

References

[1] American Telephone and Telegraph Co., et al., 73 FCC 2d 248 (1979), *on recon.*, 83 FCC 2d 1 (1980).

[2] ITT World Communications, Inc., 77 FCC 2d 877 (1980), *reversed and remanded sub nom.,* ITT World Communications, Inc. v. FCC, 699 F.2d 1219 (D.C. Cir. 1983).

[3] Noam, E., "International Telecommunications in Transition," in R. Crandall and K. Flamm, Eds., *Changing the Rules: Technical Change, International Competition, and Regulation in Communications*, Washington, D.C.: Brookings Institute, 1989, p. 258.

4

The New World Order: Developing Models in International Telecommunications and Information Processing

4.1 The Challenge to Sovereignty and Control

Technological innovations in telecommunications, liberalizing policies, and the Internet's ascendancy threaten the ability of national governments to control the flow of communication and centrally manage that sector of the economy.[1] Such control can help bolster a political regime and insulate it from outside, destabilizing influences. But in the new world order no degree of government control or centralized management can maintain the status quo, continue to hold consumers captive, and avert the impact of outside cultural and commercial forces.

Some nations have been reluctant to liberalize telecommunications, because of concerns for stability of the political status quo. Historically, managing access to telecommunications facilities and limiting or prohibiting market entry succeeded in bolstering a governmental regime, particularly

1. See Henry H. Perritt, Jr., "The Internet as a Threat to Sovereignty? Thoughts on the Internet's Role in Strengthening National and Global Governance," *Indiana Journal of Global Legal Studies*, Vol. 5, Spring 1998, p. 423.

when it could provide employment and below-cost services by enforcing policies that required cross-subsidization from services particularly used by visitors and expatriates. Governments and carriers no longer can deliver such benefits and demand favors in return, because technological and market forces largely preclude inefficiency. Governments cannot command and control the telecommunications and information sectors when liberating technologies offer just about anyone the opportunity—legal or illegal—to bypass restrictive policies and excessive rates. Similarly, carriers cannot force consumers to underwrite inefficiency when more options exist for consumers to vote with their pocketbooks and use the services of alternative carriers.

The new world order marks the end of government and carrier control over telecommunications. Incumbent carriers no longer can leverage their still considerable political power and major employer status to maintain the status quo, including service pricing requirements that burden large-volume users and international callers with excessive rates. Technological innovations, market access opportunities, and aggressive marketing initiatives of some carriers have made it easier for those burdened users to access cheaper alternatives. Users who previously underwrote cheap domestic phone and postal service now have more options. Even some of the beneficiaries of cross-subsidies have begun to view centralized planning and monopolies as contributing to a lag in development compared to the level of progress mandated by a more business-oriented and competitive environment.

Integrated telecommunications networks, comprising geographically dispersed access points called *nodes* and traffic aggregation points called *hubs*, challenge national sovereignty and conventional concepts of control over territory. Broadcast signals and telecommunications networks traverse borders with the consequence that individuals in one nation have increasingly easy and cheap access to databases, news, and entertainment from other nations. While economic theory endorses the view that the utility and value of such networks increase with the number of users, simple politics confirms the potential destabilizing force such access creates.[2]

2. Geoff J. Mulgan, in *Communications and Control: Networks and the New Economies of Communication* (New York: Gilford Press, 1991), presents a different view of control. The ubiquity, number, and pervasiveness of new telecommunications technologies accrue greater strength in the state to monitor the populace and to determine who should have access and who should not. However, with increasing distribution of network control to end users and with proliferating access points, the ability to monitor and control may, in fact, decline. The number of successful penetrations of secure computer networks by hackers corroborates that view.

4.2 New Models and Descriptive Terms

The PTT model no longer constitutes the primary way nations structure their telecommunications industries. A number of factors explain this change:

- Successful and progressively more expansive experimentation with private initiatives in former PTT model nations;
- Greater user sophistication and ability to bypass inflexible and expensive carriers;
- The overlap of telecommunications with information-processing technology;
- A recognition by incumbents that they must operate in a more efficient and businesslike manner regardless of the domestic marketplace and regulatory environment.[3]

With increasing frequency, nations have considered ways to stimulate widespread, market-driven competition. Most governments now seek to enhance user welfare without threatening the incumbent's ability to provide subsidized service and to serve as the carrier of last resort for rural users. Nations typically pursue incremental changes to the PTT such as:

Mulgan also views control in the narrow context of telecommunications law, regulation, and policy in which governments can sponsor control by vesting power in a PTT monopoly, restrict the monopoly through regulation, and arbitrate on matters of access to frequency spectrum, networks, and markets. For those types of controls, technological innovations present two kinds of pressure: (1) political pressure for government to exit or to reduce the scope of its control and (2) economic pressure for government to avoid sponsoring monopolies and to eliminate restrictions on the scope of activities former monopolies can pursue after entrants have made the marketplace more competitive (Mulgan, pp. 141–160).

3. "Against this 'conservative' (PTT) perspective, we have to compare the North American approach, which has generally favoured the use of private enterprise for the provision of telecommunications services.... It is true that, in part under the pressure of new technology and capital users, we are now witnessing a 'commercialization' of telecommunications together with a weakening of the 'territorial' perspective of how monopolies see their rights. This phenomena will continue." R. Butler, "The Changing Telecommunications Environment," *Telecommunication Journal*, Vol. 55, No. 2, 1988, pp. 130, 132–33.

- Spinning off the postal operation into a separate money-losing enterprise;
- Separating regulation from operations by creating an independent regulator or by spinning off operations from the ministry of posts and telecommunications;
- Liberalizing the environment in which the PTT operates by changing its service mandate and charter, typically as a first step toward privatization;
- Privatizing the PTT, for example, through the issuance of publicly held stock, as has occurred in over 50 nations;
- Pursuing deregulatory initiatives, for example, authorizing market entry of at least one more basic service carrier and of value-adding carriers who either construct new facilities or lease lines.

4.3 More Descriptive "C" Terms

The international telecommunications marketplace is changing at an increasingly speedy rate.[4] While it will not match the pace in some domestic markets, we can anticipate change and an eclectic blend of initiatives often derived from domestic experiments throughout the world.[5] The vision of a global village has become increasingly possible as nations grow more interdependent through trade, facilitated by high-speed, efficient, and ubiquitous international telecommunications and the connectivity offered by the Internet.[6] Few would dispute the view that telecommunications has become a critical factor in the global information economy and a vehicle for nations and individual companies to develop a comparative advantage. Different

4. See Aileen A. Pisciotta, "Regulation of International Communications in the Age of the Internet: Lagging Behind the Future," *International Lawyer*, Vol. 33, Summer 1999, p. 367; Peter Cukor, Raul Katz, Lee W. McKnight, and Paul M. Vaaler, "Creative Destruction—or Just Destruction? Telecoms in Transition: Survival and Success in the Global Internet Economy," *Fletcher Forum of World Affairs*, Vol. 24, Spring 2000, p. 103.

5. "The increasing trend towards regulatory liberalization has created a new breed of transnational telecommunications enterprise and has opened some doors for them that would have been closed previously." Tim Logue, "Who's Holding the Phone? Policy Issues in an Age of Transnational Communications Enterprises," in D. Wedemeyer and M. Lofstrom, eds., *Proceedings of PTC '90* (Honolulu: Pacific Telecommunications Council, 1990), p. 95.

nations, however, continue to have vastly divergent views on how to maximize benefits with the least amount of dislocation, unemployment, and other adverse consequences.

Recent policy liberalization has occurred primarily to spur efficient operations of incumbents and to respond to the improved competitiveness achieved by foreign carriers with plans for global market expansion. Despite market-opening commitments made under the auspices of the WTO, nations still try to stimulate their economies and promote ease in commercial transactions while not necessarily supporting market access for foreign carriers and equipment manufacturers. Accordingly, many of the opportunities for strategic alliances and foreign investment are possible only if an individual company's expertise supports evolving governmental policies, or if an incumbent realizes that to serve foreign markets it will have to establish teaming arrangements with other enterprises to tap needed expertise and an existing marketing presence.

Particularly important and relatively new policies for many nations include the following:

- Pursuit of a conditionally open marketplace;
- Reduction of the scope of government ownership in telecommunications carriers and the degree of government oversight;
- Support of a unified and globally competitive region;
- Mandating a more businesslike, global, and consumer-oriented approach by incumbents.

Most policy makers recognize how an efficient and ubiquitous telecommunications infrastructure can stimulate a national economy and the fact that a global enterprise requires fast, efficient, and reliable telecommunications. With growing concurrence, nations are considering alternatives to a single PTT administration owned by government. The willingness to consider alternatives to an entrenched model stems from growing recognition that nations cannot afford to have their telecommunications administration behave as monopolists "shelving (or feigning) avoidable innovation, ducking

6. See, for example, John H. Harwood II, William T. Lake, and David M. Sohn, "Competition in International Telecommunications Services," *Columbia Law Review*, Vol. 97, May 1997, p. 874.

investment wherever possible and keeping a beady eye on their own convenience rather than the customer's" [1].

Multinational enterprises can exploit the opportunities that are created when consumers demand more and better service and policy makers prefer to spur innovation rather than insulate incumbents from having to compete. On the other hand, to achieve a goal of becoming "a leading worldwide provider of telecommunications and information management services," an enterprise must structure a procompetitive proposal that "complements the national telecommunications strategies of … [targeted] nations" [2]. In other words, market entry occurs only when national governments can envision public benefits that more than offset the costs borne by incumbents.

New government policies that address competition, liberalization of policies, and partial deregulation constitute the fundamental predicates for expanded and more diversified telecommunications services. In a growing number of nations, the incumbent carrier has lost the ideological and functional argument that it alone should provide all telecommunications services and equipment as a natural monopoly,[7] because it can operate most efficiently and maximize consumer welfare by achieving economies of scale.[8]

---

7. "Traditionally telecommunications has been seen by many people as a classic example of a 'natural monopoly'—said to exist when a single firm is technically able to provide goods or services at a significantly lower cost than two or more competing firms. Such natural monopolies benefit from economies of scale." "Is There Life After Monopolies?" *Public Network Europe*, Vol. 1, No. 10, October 1991, pp. 27–28. Professor Gerald Faulhaber presents a more comprehensive definition: "A natural monopoly occurs in an industry in which the production technology is such that one producer can supply the entire market more cheaply than two or more producers.... Until recently, it was thought that a natural monopoly occurs when the average costs of production decline with increasing volume.... This allows the resulting monopolist to charge prices substantially above costs and cut back the amount supplied and consumed to the profit-maximizing level. More recent work suggests that the mere presence of cost advantage is not sufficient to enable the monopolist to raise its prices above cost; it appears that only the presence of sunk costs, investments that once made cannot be retrieved for purposes other than original intent, can deter entry and permit above-cost pricing and monopoly profits." G. D. Faulhaber, *Telecommunications in Turmoil: Technology and Public Policy* (Cambridge, MA: Ballinger, 1987), p. 106.

8. "Economies of scale refer to the situation where an increase in inputs leads to a proportionally greater increase in outputs (for example, a doubling of inputs would lead to more than a doubling of outputs)." K. Wilson, "Deregulating Telecommunications and the Problem of Natural Monopoly: A Critique of Economics in Telecommunications Policy," *Media Culture and Society*, Vol. 14, 1992, pp. 343, 345.

Deregulation results when policy makers consider telecommunications in terms of a number of submarkets in which niche competitors can operate successfully without handicapping the incumbent carrier's ability to provide core services. While initially preserving a monopoly for basic services, many governments have grown to believe that many areas of telecommunications now have low barriers to entry and do not require a substantial degree of up-front, not easily salvaged (i.e., "sunk") investment in physical plants. For example, a nation may continue to reserve to the incumbent a monopoly for basic, switched voice and data services, but authorize competition in mobile communications (e.g., cellular radio and paging) and for value-added services that enhance lines leased from the incumbent. Competition in some market segments can flourish if the regulator does not abdicate its responsibility to ensure that the incumbent carrier offers cost-based interconnection to remaining monopoly components of an integrated telecommunications system, for example, the public switched telecommunications network (PSTN) that provides the "first and last miles" of a communications link to users. Likewise, to maximize public benefits, the regulator should ensure that at least some market entrants construct new facilities rather than simply resell the incumbent's lines and services.

With some support from old-line economic theory, the incumbent monopolist responds to the campaign for market access by claiming that market entry will foster unfair and inefficient competition, that is, newcomers will thrive only by "cherry-picking" the most profitable customers for discounted service and by "cream-skimming" (serving the densest routes), leaving the incumbent with fewer resources and revenues to serve remaining users and to upgrade the network. Opponents to competition claim that it will result in wasteful duplication of facilities and bypassing of incumbent facilities not on efficiency grounds, but simply because regulators have imposed financial burdens on the incumbent, in effect requiring it to raise its rates to underwrite the below-cost provision of other services. "The dilemma for policy-makers is to ensure an adequate contribution to the local network infrastructure without increasing incentives for corporate bypass" [3]. Deregulation and competition should generate consumer benefits and cost savings through better, cheaper, and more efficient services and not by avoiding regulator-imposed fees or encouraging large-volume users to build their own networks.

Despite the potential for adverse financial consequences, many governments and most customers demand change:

> Opening the telecomms operating environment to competition is no longer a purely ideological concept: advances in technology now make it

> practical, policy changes make it possible, and moreover, customers want it. [4]

Tables 4.1 and 4.2, compiled by the U.S. Federal Communications Commission, provide helpful statistics for tracking the growth in international telecommunications traffic from the United States, an outcome that has triggered a substantial transfer payment from U.S. carriers to other carriers that terminate this traffic.

4.3.1 Change

With each passing year, the new world order gathers momentum as more nations embrace various procompetitive models. The models operate under the assumptions that telecommunications and information-processing markets do not support natural monopolies and that incumbent carriers have to improve their performance to survive. Measures of efficiency include statistical indices such as number of telephone lines per 100 inhabitants, number of employees per telephone line, average waiting time for service installation, and probability of securing dial tone. Competition has become less a theoretical possibility and more a baseline model with perhaps implicit expectations that telecommunications sectors can become more like information-processing markets characterized by relentless efforts to stimulate efficiency, responsiveness to consumers, innovativeness, and ruthless competition.

Procompetitive nations reject the belief that telecommunications supports a natural monopoly and that the status quo best serves social and political policies like universal service and the deliberate underpricing of some services. Nations reluctant to abandon old world order views persist in their belief that centralized planning can optimize investment without wasteful duplication of resources, and without abandoning quality and availability of service concerns.

4.3.2 Corporatization

Even without the spur of privatization, just about every incumbent telecommunications carrier has recognized the need to become more businesslike, less bureaucratic, and more efficient. That orientation may result from external prodding by a revised charter or through legislation. Internal changes result from a new strategic assessment that change and competition are inevitable. The proactive incumbent carrier recognizes the need to target and serve new, often foreign, markets, because other carriers will attempt to "poach" on domestic customers previously deemed off-limits. While some

Table 4.1

International Telephone Service by Region

U.S. Billed Minutes[A] (in Thousands)											
Year	Western Europe[B]	Africa	Middle East	Caribbean	N. and Central America	South America	Asia	Oceania	Eastern Europe[B]	Antarctica & Maritime	Total
1980	408,823	11,190	44,215	98,387	803,076	86,728	99,186	17,544	—	1	1,569,150
1981	481,598	17,095	57,533	123,080	919,224	108,378	128,124	22,171	—	1	1,857,207
1982	582,579	24,349	82,724	143,787	1,003,628	145,102	176,717	27,677	—	2	2,186,565
1983	694,985	36,265	121,939	191,536	1,135,563	193,266	241,546	34,582	—	1	2,649,683
1984	840,358	45,902	136,539	205,588	1,224,101	216,446	325,591	42,495	—	1	3,037,022
1985	992,183	54,367	134,923	211,521	1,381,581	232,223	389,001	50,464	—	1	3,446,264
1986	1,172,136	62,354	147,807	273,737	1,623,497	286,470	505,057	55,000	—	2	4,126,060
1987	1,406,047	74,331	168,998	324,737	1,827,508	326,413	617,945	73,237	—	2	4,819,217
1988	1,655,338	90,253	178,574	396,705	2,105,751	387,251	775,394	90,142	—	3	5,679,412
1989	1,903,188	113,122	213,521	463,909	25,42,156	465,831	939,636	109,310	—	3	6,750,675
1990	2,227,075	139,730	284,846	530,267	3,062,478	534,269	1,126,524	124,547	—	4	8,029,740
1991	2,273,028	147,210	461,787	610,967	3,326,063	580,745	1,262,571	129,045	164,112	3,268	8,985,796
1992	2,506,566	204,900	347,902	681,267	3,882,931	715,660	1,473,809	146,942	192,533	3,701	10,156,244
1993	2,713,102	235,272	400,363	716,139	4,308,664	849,390	1,745,312	165,900	246,782	11,892	11,392,816
1994	3,017,507	303,369	474,164	870,610	4,918,310	1,028,605	2,226,637	206,223	337,457	10,308	13,393,190
1995	3,408,026	408,084	555,730	1,075,390	5,577,100	1,282,233	2,850,979	261,952	408,769	8,869	15,837,132
1996	4,073,314	522,324	654,529	1,221,275	6,399,135	1,582,822	3,756,276	370,367	534,644	4,365	19,119,051
1997	5,076,850	619,062	654,528	1,352,773	7,274,808	1,815,218	4,652,910	607,443	592,177	11,057	22,656,826
Annual growth since 1992[C]	14.9%	26.5%	14.7%	16.2%	13.6%	21.2%	26.7%	32.1%	26.1%	6.8%	17.8%

A. International MTS data are derived from the annual FCC publication Statistics of Communications Common Carriers (SOCC). The SOCC includes data for Canada and Mexico starting in 1991. AT&T data for Canada and Mexico have been added to the SOCC data for 1980 through 1990. In addition, the SOCC data for 1980 through 1984 were adjusted to remove data for Alaska, Hawaii, and Puerto Rico. Starting in 1991, the SOCC shows totals with and without nondomestic traffic between U.S. points. This table uses the totals for foreign points, which exclude nondomestic traffic between U.S. points.

B. Data for countries in Eastern Europe and Western Europe were published together through 1990. The combined data are shown in the table as Western Europe data for those years.

C. The annual rates of growth were estimated by fitting a log linear regression to traffic data for 1992 through 1997.

Source: FCC.

carriers may not welcome a rousing from the previously comfortable monopoly world, most players recognize the inevitability of having to operate more efficiently and commercially.

Table 4.2
U.S. Traffic Compared with World Traffic

Message Telephone Service (measured in millions of minutes)[A]								
	1991	1992	1993	1994	1995	1996	1997	Growth 1992 to 1997[B]
Total United States	8,986	10,156	11,393	13,393	15,837	19,119	22,586	17.7%
Total world	37,900	43,000	48,200	53,900	60,300	70,000	81,800	13.5%
U.S. as percentage of world	23.7%	23.6%	23.6%	24.8%	26.3%	27.3%	27.6%	—
AT&T	6,596	7,039	7,201	8,040	8,831	9,546	10,331	8.5%
MCI	1,600	2,101	2,857	3,529	4,486	5,372	5,913	23.2%
Sprint	728	946	1,181	1,490	1,772	2,745	2,794	26.1%
BT	2,105	2,188	2,310	2,489	2,909	3,158	3,735	11.4%
Deutsche Telekom[C]	3,605	4,087	4,680	5,147	5,244	5,100	5,333	4.7%
France Telecom	2,295	2,449	2,576	2,603	2,805	3,116	3,545	7.4%
HongKong Telecom	913	1,137	1,377	1,578	1,692	1,739	1,718	8.4%
Stentor	1,425	1,520	1,552	1,525	1,467	1,650	1,778	2.7%
Swisscom	1,429	1,551	1,572	1,649	1,778	1,936	2,164	7.0%
Telecom Italia	1,220	1,473	1,610	1,708	1,908	2,184	2,352	10.1%

A. U.S. data are taken from 43.61 International Traffic Data reports and represent U.S. billed minutes of telephone service for all international points. Estimates for world total and for selected foreign carriers are taken from TeleGeography, Inc. (www.telegeography.com). In some instances, data for previous years have been revised.

B. Estimated using log linear regressions.

C. DBP Telecom in 1991.

Source: FCC.

Corporatization means that, to varying degrees, carriers have migrated from a mind-set appropriate for government-owned or franchised monopolies, to one embraced by business enterprises obligated to operate in a competitive world. PTTs traditionally have been regulated as public utilities and considered natural monopolies. They now transform into a number of different enterprises running the gamut from a slightly more business-minded government monopoly, perhaps with workers hired outside the civil service, to a shareholder-owned business like any other publicly traded corporation.[9]

---

9. See Gregory C. Staple, ed., *Telegeography 2000* (Washington, D.C.: Telegeography, 1999).

4.3.3 Commercialization

Advocates of the new world order point to empirical evidence that a commercial environment enhances consumer welfare without adversely affecting market countervailing policies like mandatory cross-subsidies to support universal service. Cost-based competition leads to more rational pricing based on commercial imperatives like users' demand elasticities, that is, the intensity of consumers' service preferences, their sensitivity to price increases and decreases, and the extent of service alternatives. Pricing on the basis of users' elasticity of demand and supply typically means that large-volume users will receive service at the lowest per-unit rates. If those users previously paid significantly higher rates, then competition will present them with a major financial benefit at the expense of other users with fewer options who secured service at averaged rates.

Users in some countries with monopoly service have complained that half of the population has yet to secure telephone service while the other half cannot get dial tone from their telephones. A commercial orientation prompts incumbent carriers to become more responsive in terms of price, service packages, and speed in resolving service problems and outages. The Canadian Radio-television and Telecommunications Commission endorsed this view:

> No matter how benignly disposed a single supplier is, or how technologically advanced it strives to be, it cannot differentiate adequately enough or quickly enough to service the particular requirements of a variety of customers. [5]

Likewise, competition can spur improvements in service quality and the speed with which carriers introduce new services and exploit technological innovations.

National governments want to stimulate a future-thinking, progressive orientation in telecommunications by both operators and regulators so that the country can accrue a comparative advantage in the information economy and secure or maintain a reputation for having carriers that offer world-class service, particularly to multinational enterprises. Carriers that offer "best-practices" efficiency and prices generate a comparative advantage in two ways:

1. They help showcase the nation as an attractive location for conducting business, particularly for enterprises that require instantaneous, reliable, feature-rich, and inexpensive telecommunications.

2. They evidence the efficiency, responsiveness, and skills necessary to win global, one-stop-shopping tenders for turnkey (ready to use) global or regional networks.

A globally known player in telecommunications and information processing also may show how its home market can serve as a hub for routing regional traffic, because of low prices and operating efficiencies in one nation. Large-volume users may elect to entrust a carrier from that nation with responsibility to manage traffic originating and terminating from several countries within a region.

Procompetitive nations conduct cost-benefit analyses and conclude that the gains from competition outweigh the costs. That outcome differs from previous evaluations in which regulators concluded that competition would ruin incumbent carriers by stranding their facility investments as high volume and the most profitable users migrated to newcomer services or bypassed the incumbent carrier's most expensive services and facilities. Decision-makers in procompetitive nations recognize the potential for traffic and revenue migration from the incumbent but offset that adverse outcome with the expectation that competition will stimulate demand. A growing market compensates for lost market share, making it possible that—despite market entry incumbent—carriers will not suffer a downturn in either revenues or facilities utilization.

4.4 Competition Now!

Nations that embrace competition appear to have an increasing sense that they must act quickly and definitively. The failure or delay in seeking to capture the benefits of competition, which include service diversity, lower rates, and greater responsiveness to consumer requirements, can generate the risk that other nations in the region will do so. Technological innovations have created a greater premium on timely action. New opportunities to aggregate traffic at hubs mean that some of the traffic that otherwise would have directly originated and terminated in-country might in the future follow a circuitous routing via another carrier's hub in another country to capture cost savings and performance. As technological innovations have reduced the distance sensitivity in telecommunications service costs, such routing has no price handicap. The lack of a technological wherewithal to determine the origination point of traffic means that traffic routed via a hub may be injected into other countries' public switched telecommunications network without being designated international traffic. The inability to classify traffic as international may foreclose the imposition of surcharges and other fees. Accordingly, nations inclined to support market entry may perceive the need to act so that national carriers can start the process of protecting their vulnerable markets and entering the vulnerable markets of others.

4.4.1 Variations on a Theme: The Three Phases of Competition Policy

Competition, in one form or the other, has contributed to the substantial upheaval currently underway in telecommunications. In the last few years:

- The Internet has reached a critical mass with credible forecasts that data communications soon will predominate.[10]
- Internet-mediated telephone service challenges the traditional toll revenue–sharing arrangements and a pricing system based on voice services.[11]
- Regional and global market opening trade initiatives have become a reality.[12]

---

10. See Grant Lenahan, "Next Generation Networks: A Practical View of Network Evolution," available at http://www.telcordia.com/newsroom/knowledgebase/index.html; T. M. Denton Consultants, *Netheads Versus Bellheads: Research into Emerging Issues in the Development and Deployment of Internet Protocols,* Final Report (1999), available at http://www.tmdenton.com/netheads.htm.

11. See Robert M. Frieden, "Dialing for Dollars: Will the FCC Regulate Internet Telephony?" *Rutgers Computer and Technology Law Journal*, Vol. 23, 1997, pp. 47–79.

12. See John J. Alissi, "Revolutionizing the Telephone Industry: The World Trade Organization Agreement on Basic Telecommunications and the Federal Communications Commission Order," *Connecticut Journal of International Law*, Vol. 13, Spring 1999, p. 485; Taunya L. McLarty, "Liberalized Telecommunications Trade in the WTO: Implications for Universal Service Policy," *Federal Communications Law Journal*, Vol. 51, Dec. 1998, p. 1; Laura B. Sherman, "'Wildly Enthusiastic' About the First Multilateral Agreement on Trade in Telecommunications Services," *Federal Communications Law Journal*, Vol. 51, Dec. 1998, p. 61; Arthur E. Appleton, "Telecommunications Trade: Reach Out and Touch Someone?" in Symposium Linkage as Phenomenon: An Interdisciplinary Approach, *University of Pennsylvania Journal of International Economic Law*, Vol. 19, Summer 1998, p. 209; Paula Barnes Sours, "The Impact of U.S. Regulatory Activity on Prospects for Implementation of the WTO Agreement on Basic Telecommunications," *North Carolina Journal of International Law and Commercial Regulation*, Vol. 23, Summer 1998, p. 645; Stefan M. Meisner, "Global Telecommunications Competition a Re-

- Strategic alliances grow in importance as carriers strive to exploit new market-access opportunities.[13]

- Deregulation and market liberalization have become more widespread in developed and developing nations alike.[14]

However, the scope and nature of competition vary widely across nations. Such deviation does not track the level of national development, or gross domestic product. In large part, the type of competition occurring in a nation's telecommunications sector depends less on the size of market and more on whether and how the factors that sustain competition operate in a particular nation.

This section examines telecommunications competition in the context of three phases, with an eye toward constructing a map for tracking the progress of achieving maximum sustainable competition. Not all nations may

---

ality: United States Complies with WTO Pact," *American University International Law Review*, Vol. 13, 1998, p. 1345.

13. See John H. Harwood II, William T. Lake, and David M. Sohn, "Competition in International Telecommunications Services," in "Symposium: Telecommunications Law: Unscrambling the Signals, Unbundling the Law," *Columbia Law Review*, Vol. 97, May 1997, p. 874; Doug Galbi and Chris Keating, Federal Communications Commission, International Bureau, *Global Communications Alliances*, Feb. 1996, available at http://www.fcc.gov/ib/; Peter Drahos and Richard A. Joseph, "Telecommunications and Investment in the Great Supranational Regulatory Game," *Telecommunications Policy*, Vol. 19, No. 8, 1995, pp. 619–635.

14. See Mark Naftel, "Status of Telecommunications Competition in Europe and Comparison with the United States," *Journal of Transnational Law and Policy Fall*, Vol. 7, 1997, p. 1; Ingo Vogelsang, "Micro-Economic Effects of Privatizing Telecommunications Enterprises," *Boston University International Law Journal*, Vol. 13, Fall 1995; Jeffery Atik, "Technology and Distribution as Organizational Elements Within International Strategic Alliances," *University of Pennsylvania Journal of International Business Law*, Vol. 14, Fall 1993; Meheroo Jussawalla, "Is the Communications Link Still Missing?" *Telecommunications Policy*, Vol. 16, No. 6, pp. 485–503; Muhammad I. Ayish, "International Communication in the 1990s: Implications for the Third World," *International Affairs*, Vol. 68, No. 3, 1992, pp. 487–510; Peng Hwa Ang, "The Causal Relationship Between Telecommunications and Economic Development: Cause for Re-Analysis," paper presented at *44th Annual Conference of the International Communication Assn.*; Heather E. Hudson, "Telecommunications and Development Ten Years After the Maitland Report: The Need for New Policies and Strategies," *Proceedings of PTC '94*, pp. 69–78; Sam Pitroda, "Development, Democracy, and the Village Telephone," *Harvard Business Review*, Nov.–Dec. 1993, pp. 66–79.

have the economic and political conditions conducive to reaching the end point identified here: full facilities-based competition in all sectors with limited government regulation targeted to serve narrow and defined functions (e.g., enforcement of antitrust and consumer protection laws). However, the competition road map provides a three-phased trip that many nations have taken or will take over time.

4.4.2 How Did Competition Appear?

Not too long ago, economists, regulators, legislators, judges, and most stakeholders considered many telecommunications sectors as unsuitable for competition. Terms like *natural monopoly*, *market failure*, and *national security* articulated a sense that marketplace forces somehow could not trigger market entry, or that should market entry occur, it would generate few consumer dividends. Worse yet, the prevailing view considered competition potentially harmful as market entrants would cream-skim and cherry-pick their way to profitability by serving large-volume customers, leaving the incumbent to serve less attractive customers and satisfy nonremunerative social policies like universal access. In essence, government had to manage, if not own, the sector and operate the carriers providing service competition, lest consumer welfare decline and the nation suffer threats to its sovereignty and security.

In a noncompetitive environment, government must ensure the availability of essential services, typically subsidized for rural and high-cost areas, the poor, and the elderly. Even where government does not provide telecommunications services, it perceives the need for heavy-handed, command-and-control regulation to remedy carrier and market shortcomings. Government has to act in an intrusive and countervailing manner to offset what economists deem market failure.

Under such circumstances, we could properly infer that governments have a stake in maintaining the status quo and a political economy dominated by government as the central manager. It would take a revolution in government, economic theory, or politics to trigger change, coupled with technological innovations that make competition sustainable. Over time, such monumental changes have constituted the initial catalyst for change in the telecommunications sector: Thatcherism, Reaganomics, the economic theory of market contestibility, unbundling, new legislation, and treaty commitments have triggered dramatic change and unleashed the power and creativity of entrepreneurialism to serve pent-up demand. In the early 1980s, the Thatcher government in the United Kingdom embraced deregulatory initiatives previously undertaken in the United States, leading to such initial procompetitive steps as unbundling telephone service from the lease of telephone handsets, which allowed competition to evolve. What a radical concept: the right of sovereign consumers to own their telephones and choose from a wide variety rather than have to settle for one-size-fits-all black rotary units. Economists articulated theories that supported competition on the grounds that the market would evolve and support competition even if none had previously existed.

4.4.3 Phase 1 Competition: The Grand Experiment[15]

Procompetition advocates in the 1980s had few supporters across the globe. Policy makers viewed initiatives in the United States and the United Kingdom as isolated exceptions to the prevailing wisdom. Such reticence to embrace change stemmed, in part, from the stakes involved and real or perceived fears. The stakes involved meddling with a major employer and a large group of often unionized employees. The fears included concerns about destructive competition, the sense that the public would suffer in the long run if competition proved unsustainable even if in the short run it generated price reductions. Other fears included apprehension about whether a competitive environment would support public policy objectives, the potential for large numbers of telecommunications workers to lose their jobs with the incumbent carrier as it streamlined to meet competition, and open-ended concerns about the unknown.

Phase 1 competition has an experimental character, because nations have little empirical evidence available to support deregulation of the telecommunications sector, including the liberalization of rules applicable to the incumbent carrier. The trigger for phase 1 competition lies in political and economic philosophy rather than grassroots advocacy for change. Even in the United States, where private ventures predominated, many objected to the divestiture of AT&T and other procompetitive initiatives. Additionally, incumbent stakeholders had an obvious interest in maintaining the status quo and significant political clout to block change.

Government officials, on their own initiative, initiate phase 1 competition. They may base such action on a real or perceived public mandate for change, perhaps corroborated by the fact that such initiatives typically ensue when control of the executive or legislative branch of government changes.

15. For background on this phase, see Robert M. Frieden, *International Telecommunications Handbook* (Norwood, MA: Artech House, 1996).

The new administration or majority party may have captured votes on a platform promising less government.

The less-government party platform leads to initiatives for government to exit ownership and control of industries like telecommunications, transportation, natural resources, and utilities. Such privatization generates funds for reducing a nation's deficit and tax burdens. It provides new opportunities for the private sector to execute alternative strategies that often achieve greater profit and efficiencies. In response to privatization, governments have to erect new regulatory mechanisms, typically resulting in the creation of new legislation that, among other things, creates an independent regulatory authority. Because no legislation drafter can anticipate every contingency and because some stakeholders have incentives to thwart or delay change, phase 1 competition may manifest itself first in the courtroom rather than the marketplace. Actual competition may not occur for months or years after legislative approval, because incumbents may not have to comply with interconnection and facilities access requirements until they have exhausted all judicial appeals.

4.4.4 Phase 2 Competition

The next phase in telecommunications competition features the unleashing of entrepreneurial creativity and actual market entry in some sectors. Phase 2 competition typically occurs in market niches with low barriers to entry. Such sectors include resale of the incumbent carrier's facilities and basic services, adding value to and enhancing leased lines and marketing (but not manufacturing) of telecommunications equipment like handsets and private branch exchanges. The advocates for competition may perceive the need to tilt the competitive playing field in favor of market entrants by creating rules and regulations that obligate the incumbent carrier to facilitate competition. Such requirements may extend the concept of common carrier duties to include a widespread duty to interconnect facilities, an inability to foreclose resale, and a requirement that the incumbent carrier subdivide network functionality into an à la carte menu of service elements available to newcomers on cost-based prices significantly below the retail rate. The legislature or a regulatory agency may have to create inducements for the incumbent carrier to accommodate market entrants, because doing so results in lost market share. In the United States, the Congress authorized the Regional Bell Operating Companies (RBOCs) to provide long-distance services contingent on full compliance with a 14-point competitive checklist designed to achieve full and sustainable competition in both local and long-distance markets. For its part, the incumbent carrier may execute a globalization strategy to seek new investment and strategic partnering opportunities for acquiring market share and revenues abroad as an offset to what it will lose at home.

Phase 2 competition requires substantial investment in new facilities by market entrants. Such investment occurs when a business case identifies pent-up and increasing demand for services, making such additional capacity easily absorbed rather than duplicative and redundant to embedded capacity. The viability of phase 2 competition depends on the nature and efficacy of regulatory oversight. Whether newly created or not, the regulatory agency must devise mechanisms for effective intervention and enforcement of rules when marketplace abuses or violations occur. Light-handed regulators operate as referees, much like referees in competitive sports. Light-handed regulation emphasizes forbearance, compared to traditional, heavy-handed regulation, which emphasizes frequent and direct intervention.

Much of the telecommunications sector will operate in flux in this phase. As a result, the regulator runs the risk of lagging behind current marketplace conditions and micromanaging the sector based on the mistaken view that imperfect competition needs its intervention to become closer to a perfect economic model. Asymmetric regulation occurs when a different regulatory template applies to the incumbent vis-à-vis market entrants.[16] In the transition to full and robust competition occurring in the final phase, asymmetric regulation may have to occur with the incumbent more heavily regulated than market entrants. However, absent a definitive timetable for "sunsetting" and eliminating the imbalance, inconsistent regulation may unfairly handicap the incumbent carrier with no justification and no public welfare benefits.

Asymmetric regulation generates the biggest risk for delay in the migration to phase 3 competition. It stimulates the already keen instinct to litigate and game the system, thereby triggering counterproductive outcomes. Here are three current examples of asymmetric regulation in the United States:

- Imposing common carrier obligations on telephone companies, requiring the unbundling of network service elements with à la carte pricing, versus exempting cable television operators from similar

16. See Antonio Perrucci and Michela Cimatoribus, "Competition Convergence and Asymmetry in Telecommunications Regulation," *Telecommunications Policy*, Vol. 21, No. 6, 1997, pp. 493–512.

duties even when the cable television network provides functionally the same telecommunications services;[17]

- Requiring interexchange (long-distance) carriers to pay above-cost local exchange access fees, coupled with a duty to help fund universal service subsidies while exempting ventures that use the same type of access to originate and terminate services that qualify for a different regulatory classification (e.g., Internet telephone services);
- Allowing an ISP to form a local exchange carrier that exclusively handles the ISP's dial-up access and qualifies for compensation from the incumbent local exchange carrier that handed off the traffic, but who will not receive traffic and compensation from the ISP's affiliated carrier.

4.4.5 Phase 3 Competition

Phase 3 competition represents the theoretical endpoint where robust facilities-based competition thrives in all sectors of the industry. Currently no nation has reached this phase, either because marketplace conditions do not necessarily support competition everywhere (as in New Zealand), or because national deregulatory initiatives remain incomplete (as in the United States and the United Kingdom). Accordingly, the examination here involves a prospective identification of what changes would ensue as a nation transitions from phase 2 to phase 3.

While the clear trend toward competition exists globally, the nature and scope of competition will vary between nations. The three-phased model presented here identifies a trajectory that many nations may follow over time. However, not all nations will follow this model, nor will phase 3 constitute the endpoint or competitive solution everywhere.

Regardless of whether and how national government promotes telecommunications competition, the nature and scope of government regulation must change. The infusion of private capital and the involvement of newcomers all but guarantees the onset of a new perspective and perhaps significant cultural change. That new perspective emphasizes

17. For extensive analysis on this topic, see Barbara Esbin, "Internet over Cable: Defining the Future in Terms of the Past," Federal Communications Commission, OPP Working Paper Series No. 30, Aug. 1998, available at http://www.fcc.gov/opp/workingp.html.

business goals and discipline and may contrast with an incumbent's orientation that blends business with public service. An incumbent operator may accept government regulation as a necessary evil and may execute strategies to capture the regulator. The incumbent may find ways to exploit its regulated status with an eye toward thwarting or blunting competition. A newcomer, particularly one already operating in other unregulated environments, may bristle at any government oversight and lack patience with the regulatory process and expertise in operating within it.

Under any future scenario, government must change its mission and its functions.[18] Regardless of the competitive phase underway in any nation, government does not completely exit the regulatory arena. Libertarian visions of laissez-faire fail to appreciate the public benefits that accrue from a specific, narrowly drawn set of governmental functions substantially different from previous versions. The future government role operates from a specific set of enforceable rules and requirements rather than an open-ended mandate to serve the public interest. Such an open-ended mission has led even the most reluctant regulator into a quagmire of functions that on balance provide little public benefit with great potential to handicap or favor various categories of stakeholders.[19]

The pace of change in telecommunications will continue fast and furious. Predictably, governments cannot revise and revamp their functions on a timely basis. "Regulatory lag" refers to situations in which governments fail to respond to changed circumstances and public requirements as they continue applying unmodified regulations. Absent a revised regulatory platform that addresses current requirements, governments may doubly harm the telecommunications sector by micromanaging aspects that could operate freely and by failing to safeguard consumers and ensure full and fair competition.

---

18. See International Telecommunication Union, *Chairman's Report of the Sixth Regulatory Colloquium, Regulatory Implications of Telecommunications Convergence*, Dec. 11–13, 1996, available at http://www.itu.int/itudoc/osg/colloq/briefrep/; see also European Commission, "Green Paper on the Convergence of the Telecommunications Media and Information Technology Sectors, and the Implications for Regulation Toward an Information Society Approach," COM(97) 623, Dec. 3, 1997, available at http://www.ispo.cec.be/convergencegp/97623.html.

19. See Eli M. Noam and Anjali Singhal, "Supra-National Regulation for Supra-National Telecommunication Carrier?" *Telecommunications Policy*, Vol. 20, No. 10, 1996, pp. 769–787.

4.5 Convergence

The Internet provides a medium for the delivery of both telecommunications and information processing often on a faster, better, cheaper, and smarter basis. The Internet's rise in importance has triggered a blending of technologies and markets, thereby challenging many of the assumptions underlying the current legal and regulatory models. That convergence also has the potential to affect the stability of preexisting international telecommunications conventions and protocols. For example, technological innovations coupled with the desire to exempt the Internet and ISPs from conventional telecommunications regulation make it possible for the delivery of long-distance telephone services at a fraction of what carriers previously charged. Convergence has the potential for disrupting long-standing rules of the road. Application of traditional telecommunications regulation may unnecessarily burden operators under such changed circumstances. Complete unregulation, however, would leave consumers at risk and eliminate the benefits that accrue when governments can resolve disputes and create reasonable rules of the road (e.g., deciding who has the right to use a particular Web address).

Convergence also creates opportunities for entrepreneurs to carve out new and profitable market niches. As never before, barriers to market entry have receded such that new ventures, with plans for offering the next best thing, can receive a fair marketplace test. Figure 4.1 graphically sets out the current, broad telecommunications/information-processing topology. Figure 4.2 narrows the broad macro view in Figure 4.1 by focusing on the three major access technologies that consumers will increasingly use for Internet access: digital subscriber links, a broadband, digital retrofit to the dial-up public switched telephone network; cable modems using the broadband cable television plant; and wireless networks, including third-generation broadband services.

4.6 The New Mandate: Privatize, Liberalize, Deregulate, and Globalize

The new world order does not lend itself to a uniform collection of descriptive key terms or to a single definitive regulatory or industrial model. A variety of new descriptive terms, representing the entire alphabet, have evolved, including *privatization*, *liberalization*, *deregulation*, and *globalization*.

Such commonly used terms belie the controversy that arose when the first initiatives were announced. In the early 1980s, the U.K. government

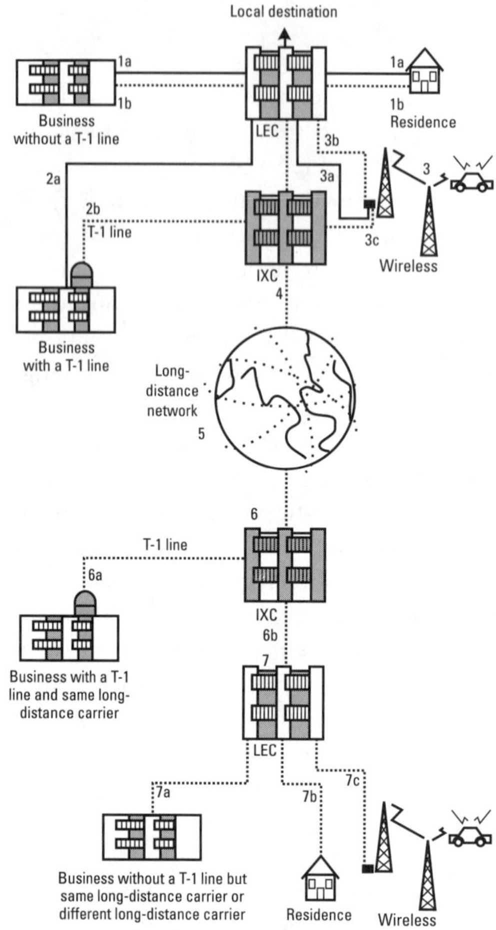

1a: For local calls originating from a home or business, the call routes to the Local Exchange Company (LEC), such as GTE, Bell South, or Ameritech. The LEC switches the call to its destination.
1b: For long-distance calls, the signal is routed by the LEC to the Inter Exchange Carrier (IXC). The IXC is the long-distance company (for example, AT&T, MCI, or Sprint) to which the caller subscribes.

2a: For a business equipped with a high capacity T-1 line, local calls are routed by the business's PBX (Private Branch Exchange—an internal switching unit) to the LEC. The LEC routes local calls to their destination.
2b: For long-distance calls from a business equipped with a T-1 line, calls are routed by the business's PBX directly to the IXC, bypassing the LEC.

3: Wireless calls are transmitted from a mobile unit to a transceiver tower. Signals are passed through repeater towers, and are switched at a base station as follows:
3a: Local calls are routed directly to the LEC, which switches the call to its destination.
3b: Long-distance calls follow the same path to the LEC, and are then switched to the caller's IXC, where it follows the process detailed in Step 4 below.
3c: In a case where the wireless operator also provides long-distance service, such as AT&T, the call bypasses the LEC and is switched directly to the IXC.

4. Calls coming from T-1, LEC, and wireless sources arrive at the IXC's switching computer. Each of the major long-distance carriers owns and operates its own switching systems; minor carriers often rent capacity from major carriers.

5. Calls are then routed around the country (or around the world) through a network of computers/switchers, utilizing the most efficient path available at the time the call is placed.

6: The call is forwarded to the recipient's local IXC.
6a: If the call's recipient is a business equipped with a T-1 line, and the recipient uses the same long-distance carrier, the call is routed directly from the IXC to its final destination.
6b: Otherwise, the call proceeds to the destination LEC.

7: The LEC directs the call to the recipient, which may be:
7a: A business that does not have T-1 access, but uses the same long-distance carrier; or a business that uses a different long-distance carrier (IXC).
7b: A residence.
7c: A mobile unit, in which case the call is routed through the recipient's wireless network.

Figure 4.1 Converge promotes network integration. (*Source:* Telecommunications®.)

decided to split postal operations from telecommunications and then to privatize the telecommunications enterprise, subject to both independent regulatory scrutiny and competition. The British initiative stemmed from a largely unpopular "conviction that private ownership would be more efficient than public ownership … [and] better able to control the trade unions" [6].

Now, eliminating or limiting the incumbent's monopoly has expanded support from economists, policy makers, and the public, despite efforts by

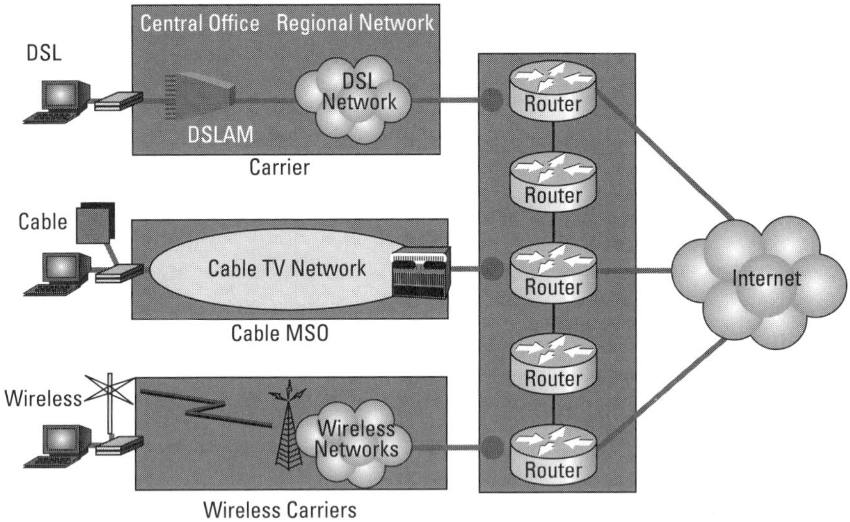

Figure 4.2 Separate networks for each access technology. (*Source:* Telecommunications®.)

incumbents[20] to generate concerns about lost jobs, reduction in service availability, and rate increases.[21] Since 1984, privatizations have occurred in such diverse nations as Argentina, Australia, Belgium, Bolivia, Brazil, Canada, Chile, Denmark, France, Germany, Hungary, Ireland, Israel, Italy, Japan,

20. "For almost a century the key institutional feature of traditional telephony around the world has been a ubiquitous network operated by a monopolist .... Public telecommunications were not merely a technical system, but social, political and economic institutions....The PTTs were supported by a broad political coalition, a 'postal-industrial complex.' It included the PTT itself and the equipment industry as its supplier, together with residential and rural users, trade unions, the political left, the newspaper industry (whose postal and telegraph rates were heavily subsidized), and affiliated experts. The system worked in no small measure to the benefit of the [domestic] equipment industry." E. Noam, "International Telecommunications in Transition," in R. Crandall and K. Flamm, eds., *Changing the Rules: Technological Change, International Competition and Regulation in Communications* (Washington, D.C.: Brookings Institution, 1989), pp. 257, 258.

21. See Kristina Driscoll Carey, "Competition Law and the Privatization of Telecommunications Markets in the European Union," *Suffolk Transnational Law Review*, Vol. 22, Summer 1999, p. 747.

Mexico, the Netherlands, New Zealand, Norway, Peru, Portugal, Russia, Singapore, South Africa, Spain, Sweden, Switzerland, and Venezuela [7].

It is important to understand the key terms that describe the new world order. *Privatization* involves the change in legal status of an incumbent carrier from public to private ownership. However, the newly privatized company may hold a private (as opposed to public) monopoly, the government may retain some degree of stock ownership in the new company, and all preexisting regulations may remain in force. Typically, but not always, privatization is coupled with deregulation: Government streamlines regulatory requirements of carriers and reduces structural safeguards that insulate the incumbent from competition in selected markets. Even in a deregulatory climate, government regulators must ensure that newcomers can interconnect directly with the incumbent carrier's networks on fair terms and conditions.

Liberalization also may occur around the time the incumbent carrier is privatized. The government typically affords the incumbent greater flexibility in identifying and serving new profit centers, thereby offsetting market share lost to new carriers. In many instances, the incumbent carrier seeks opportunities in foreign markets. The changes that have occurred in the last few years contrast with the dozens of preceding years, when change occurred at a glacial pace, because a broad "postal-industrial complex" [8] had a vested interest in maintaining the comfortable status quo.

Deregulation refers to the reduction and streamlining in regulation of telecommunications. Given the typical absence or limited nature of information-processing services regulation, governments cannot maintain heavy-handed command-and-control regulation of telecommunications services, particularly now that Internet-mediated services increasingly offer a substitute for traditional telecommunications services (e.g., Internet telephony). A wide degree of asymmetry between regulatory treatment of the two categories would destabilize the competitive playing field. Accordingly, the telecommunications sector has become, or soon will become, less pervasively regulated in most nations, with the nature of ongoing government involvement more light-handed and oriented toward sustaining full and fair competition and protecting consumers from marketplace abuses.

Globalization refers to the need for incumbents to expand their operations and market perspective to include international opportunities. Globalizing incumbents can recover the reduced revenues and profits occurring as a result of domestic competition. However, an ill-conceived globalization strategy can exacerbate losses. Globalization does not simply involve the export of techniques and expertise that seem to work fine domestically. Instead it requires a keen understanding of how the markets an incumbent seeks to enter differ from home and a willingness to develop or acquire the skills and resources needed to excel in that market. Globalization surely does not mean that an incumbent can tap into its retained earnings and accrue ample dividends simply by deciding to invest in markets abroad. See Figure 4.3.

4.7 Revamping Regulatory Oversight

Many nations, including Australia, Canada, Mexico, Peru, and the United Kingdom, also have revamped the structure and scope of telecommunications regulation by creating independent regulatory agencies. A freestanding, expert regulatory agency enhances the possibility that government can establish rules and regulations that promote full and fair competition. This referee function is particularly important when a nation authorizes facilities-based competition and the incumbent carrier seeks freedom from prior regulations to compete even as it provides facilities needed by new competitors to originate and terminate service. An independent, expert regulatory agency can balance the interests of incumbents and newcomers, while achieving procedural and substantive fairness, by:

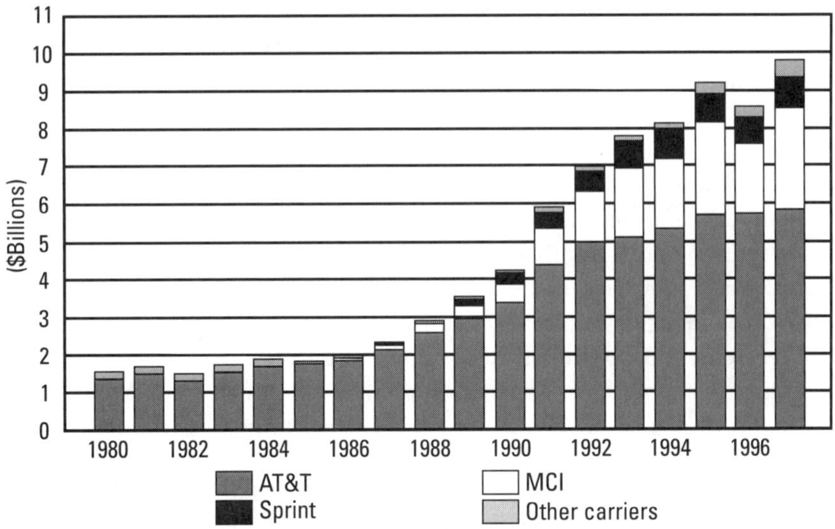

Figure 4.3 Market shares for the total international services market. (*Source:* FCC.)

- Affording the public an opportunity to participate in the decision-making process by filing comments and data;
- Generating a factual record and using transparent and fair procedures;
- Explaining how the record supports the final decision or policy;
- Standing ready to defend its decisions before an appellate court.

Regulation often requires carriers to reduce profits and to comply with rules designed to serve the "public interest, convenience and necessity" [9]. Regulatory agencies continually have to establish policies and rules that flesh out how best to serve the public interest.

4.7.1 Regulated Competition

In nations that adopt the competitive model, international telecommunications carriers typically operate with somewhat less government oversight than the PTT model, but the industry remains subject to a still significant degree of government oversight. Market entry eliminates the need for much of the traditional regulation, which was designed to safeguard the public from monopoly abuses like price gouging, because, theoretically at least, no competitor can affect the price or supply of equipment and service. Advocates for the procompetitive model seek to limit regulation, because it artificially segments the marketplace, creates barriers to competition, and generates market distortions.

The international telecommunications marketplaces in the United States, Canada, the United Kingdom, Australia, Chile, and New Zealand represent the regulated competition model. Significant telecommunications competition exists in those nations with carriers subject to streamlined regulation that government officials hope will balance the need to maintain safeguards without curbing the scope of competition.[22] Despite over 20 years of progressive deregulatory initiatives, many international telecommunications markets remain concentrated and dominated by a few major carriers, even in most progressive nations. That condition may reflect defects in the scope and the nature of regulation. But it also may evidence limitations on the extent to

22. See Lawrence J. Spiwak, "From International Competitive Carrier to the WTO: A Survey of the FCC's International Telecommunications Policy Initiatives 1985–1998," *Federal Communications Law Journal*, Vol. 51, Dec. 1998, p. 111.

which some telecommunications markets can sustain competition, particularly when mergers and acquisitions concentrate market shares.

Fostering facilities-based competition constitutes a major step even for nations inclined to support competition. Many nations first increase the number of carriers to two, thereby establishing a duopoly. The United Kingdom, Australia, Korea, and some other nations initially adopted such a duopoly. In the satellite facilities marketplace, even fewer nations have supported full, "open-skies" competition, that is, unlimited market entry by qualified applicants. Not until 2000 did U.S. law support multiple domestic investors in the world's primary international satellite facilities operator: INTELSAT. Previously, Comsat Corporation, recently acquired by Lockheed-Martin Corporation, served as the sole U.S. investor in INTELSAT. In 2000, the FCC signaled its intent to permit carriers other than Comsat to secure direct access to INTELSAT satellites without first having to coordinate with Comsat and take service through a Comsat tariff.[23] For its part, INTELSAT privatized only recently, having previously operated as a cooperative of more than 135 nations designed to serve the world's international "fixed satellite requirements," that is, using satellites to transmit voice, data, video, and other services required by users in different stationary locations. As the still-dominant operator of international satellites, INTELSAT provides a case study in the challenges that occur in the transition from government-sanctioned monopolies to full competition. Because of the limitation in the number of satellites that can park in geosynchronous orbits, even now domestic and international satellites markets are concentrated, and incumbents enjoy competitive advantages based in part on their preexisting access to choice orbital slots.

Even now, telecommunications legislation and regulation can result in market segmentation and concentration. For example, privatization of the incumbent carrier may not occur at the same time as market entry. Many nations prohibit or limit facilities-based competition, that is, competition from carriers that own and operate transmission facilities rather than lease and resell the capacity of incumbent carriers. Even the most progressive nations act with caution in opening up markets to competition, lest they inadvertently handicap the ability of incumbent firms to pursue social policies like universal service. In seeking to provide ubiquitous service, carriers

23. See "Availability of Intelsat Space Segment Capacity to Users and Service Providers Seeking to Access Intelsat Directly," IB Docket No. 00-91, Notice of Proposed Rule Making FCC 00-186, 2000 WL 668960 (F.C.C.) (rel. May 24, 2000).

typically average costs among profitable dense routes and money-losing sparse routes. Likewise, they implicitly or explicitly engage in cross-subsidization among classes of service and types of users.

In view of the typical claim by incumbent carriers that competition will handicap or prevent them from achieving the universal service mission, newcomers must convince both policy makers and incumbents of the benefits accruing from competition. As well, they must seek authorization to interconnect facilities with the incumbent carrier, which may claim such access will provide a way for users to "leak" traffic into the PSTN without proper payment to the incumbent carrier for access. Local exchange carriers fear that new long-distance carriers will find ways to receive traffic from callers and have it delivered to call recipients without fully compensating the local exchange carrier. Porous network access from long-distance or local networks means that local carriers have legitimate concerns.

Procompetitive initiatives first target nonessential niche markets to demonstrate the merits in competition without the risk of harm to essential core services. Even when nations permit core-market competition, incumbent carriers tend not to lose significant market share.

4.8 Sustaining Competition and Protecting Consumers

Consumers and technological innovation are the primary drivers that foster change in telecommunications and information processing. Few countries can tolerate the financial and political fallout from failing to establish a mechanism for guaranteeing reliable telephone service. High-volume corporate users increasingly can resort to strategies that make it possible to access services that the incumbent carrier cannot or will not offer. If the incumbent is so unable to finance infrastructure improvements, then large-volume users may resort to self-help by securing the legislated right to install and operate their own equipment. Technological innovations make that option both cost-effective and executable by nontelephone company personnel. If the incumbent is unwilling to innovate, perhaps because of fear that users will migrate to cheaper services, individual corporate users may attempt to bypass the uncooperative or incompetent incumbent carrier. Other users, including small businesses and individuals, have access to new illegal or gray-market options that exploit technological innovations and the inability of the incumbent to detect and prevent such bypass. Such self-help efforts include the use of callback services, which allow callers in nations with high international call rates to secure dial tone in nations with lower rates. In fact, any satellite Earth station can provide the basis for accessing a cheaper (but not necessarily legal) transmission option.

Slowly, many governments have grown less willing to accept unconditionally incumbent carrier claims that migration of large-volume users to new carriers or to private options (i.e., bypass) would jeopardize the ability to improve service. Governments increasingly side with consumers who demand the right to choose whether to operate their own facilities or to select from a number of carriers and services. Users increasingly view telecommunications as both a key cost center and a vehicle to achieve a comparative advantage in the global marketplace. In turn, governments have come to believe that a more efficient telecommunications system can stimulate the national economy and grow impatient with incumbent claims that they are improving the situation with all deliberate speed.

Nations typically adopt change and procompetitive initiatives in an incremental manner. Even with changed marketplace conditions, national governments do not expect to eliminate their regulatory function. It may change in scope and pervasiveness, but "[e]xperience suggests . . . that particularly during the early period of any transition from a noncompetitive (or minimally competitive) market to a competitive (or more competitive) one, the government will have to oversee and manage change" [10].

Governments must closely oversee the behavior of incumbents during the transition from a closed, often monopolized environment to one in which market entry exists in an increasing number of market sectors. Existing equipment and service providers recognize the likelihood for significant loss of market share and the potential for reduced earnings absent market growth. Incumbents have every incentive to delay, through litigation and regulatory challenges, the onset of competition and its impact. In the marketplace, they may stifle competition by anticompetitive practices, including selective price cuts, in which the incumbent sets prices below costs to drive out competition. Incumbents also may try to deny competitors access to their facilities or provide inferior and overpriced interconnection to facilities needed by market entrants who have installed only limited networks or who operate using leased lines.

The history of interconnection disputes in the United States points to the potential for incumbents to abuse their control over essential bottleneck facilities. Interconnection between the facilities of different carriers is essential to maximize network access and promote the goals of universal service. Market entrants typically do not initially erect networks that fully duplicate what the incumbent carrier has installed. To avoid waste and to concentrate on productive deployments of network facilities, some newcomers install only long-haul facilities, relying on the existing local-loop facilities of incumbent carriers to originate and terminate service.

Newcomers often rely on incumbents' carriers to provide call aggregation functions and to carry traffic over the "first and last miles," that is, closest to the caller and the call recipient. Such facilities are both capital and labor intensive. Because few newcomers have the wherewithal or inclination to invest in local distribution facilities, incumbent carriers have the ability and the incentive to restrict access to these potential choke points, or bottlenecks. Competitors cannot survive in the marketplace without efficient and cost-based access to incumbent carrier facilities. Governments must ensure that the incumbent carrier provides equal access to such facilities, as is required of common carriers and public utilities. The migration from phase 2 or phase 3 competition requires an effective regulatory regime over the terms and conditions of interconnection between carriers.

The installed base of network equipment and a large subscriber base provide incumbents with the capacity to respond effectively to competition. Governments must ensure a fair competitive response, rather than anticompetitive practices that tilt the newly competitive playing field even more in the direction of the incumbent. The potential for market manipulation, disputes between newcomers and the incumbent carrier, reductions in service, and immediate and substantial rate hikes means that governments have to maintain a role as regulator and referee.

Regulation does not end in the new world order. Instead, changes occur as to the emphasis of regulation and perhaps even which government agency performs the still necessary regulatory oversight. Having removed many, but not necessarily all, barriers to market entry, governments have every incentive to ensure that competition develops. Because competitors may want to secure artificial marketplace advantages by handicapping others, government regulators need to enforce rules of engagement. For example, the goal of sustaining competition means that governments must remain vigilant against pricing strategies by the incumbent carrier that are designed to drive out competition. An incumbent with deep pockets can endure a period of destructive competition, in which prices fail to meet costs. Such predatory pricing may confer a short-term consumer benefit, but in the longer term, rob consumers of the benefits accruing from ongoing competition. Despite the political gains in helping to achieve drastic price cuts, government regulators must ensure that prices do not serve any particular carrier's strategy to drive out the competition, after which the carrier can recoup its losses with above-market rates made possible by the lack of competition.

Likewise, the new world order requires greater attention to protecting consumers from sharp dealing, fraud, and deception. In the United States, the FCC has expanded its scrutiny of long-distance carrier tactics to shift consumers' choice of carrier without informed consent. The Commission has increased the penalties for *slamming*, as the practice is called, and as well had to launch an educational campaign to inform consumers of the risk. The FCC recently has addressed another unsavory carrier strategy called *cramming*, loading consumers' bills with charges for services they did not order.

4.9 Gray Market Ascendancy: Entrepreneurs Push the Legal Envelope

As governments attempt to manage change and ease the transition to competition, some consumers and entrepreneurs grow impatient. Technological innovations provide more opportunities to remedy such impatience through self-help. A greater tendency to pursue suspect (gray market) and clearly illegal (black market) tactics results when new technology makes it easier to evade detection and the risk taker feels justified in doing something others can do legally in other, more liberal jurisdictions.

Technological innovations work to make markets more porous. Put another way, innovations prevent the effective enforcement of rules that reserve markets to select carriers and foreclose carriers and users from doing something they could do more easily without incurring significant cost. For example, a private branch exchange (PBX) lawfully enables even small businesses to conserve leased telecommunications line expenses by switching the traffic generated by dozens of telephone users over a smaller number of trunk lines based on the statistical probability that not every telephone would be in use at the same time. The switching function of the PBX and other types of switches provide traffic aggregation functions that make it possible for a venture to hold itself as a long-distance telephone company even when it starts off small with access to a few leased lines.

Additionally, switches provide users the ability to evade artificial market and territorial limitations. For example, telephone company tariffs typically limit the use of a private line to a single user, who typically uses the line to access one or more distant points. A law firm with offices in several cities might link all offices via private lines, making it possible for attorneys and staff in each office to contact others without incurring a metered long-distance charge. Tariffs typically prohibit or require additional payments for the practice of private-line subscribers using PBXs and other switches to link private lines with local trunk lines. In our law firm example, such linkage would make it possible for calls to reach both the law firm office in a distant city and any other telephone within the local calling area without incurring a toll charge. Leaky PBXs and other switches provide users opportunities to evade toll charges and other restrictions imposed by the incumbent telephone company.

4.9.1 Accounting Rate Arbitrage

The excessively high rates charged by many incumbent telephone companies have encouraged creative use of leaky PBXs and other devices to avoid triggering a toll charge. Users and entrepreneurial service providers alike seek to configure a service that provides voice dialing capability but does not trigger what is known as an accounting rate settlement. International carriers negotiate a fixed division of toll revenues. That rate substantially exceeds carrier costs, justifying high retail rates and encouraging users and new ventures to secure call routings that do not fall within the accounting rate settlement process.

Because of high accounting rates, individuals and companies have devised ways to configure private line service that is exempt from the settlement process, but that provides internal calling capability. Other options include the importation of dial tone from nations with significantly lower international long-distance rates. That callback or call-reorigination process involves the use of relatively simple signaling from a high-cost area to a switch in a low-cost area that responds by calling back with outbound calling capability. (Chapter 9 examines accounting rate and arbitrage opportunities at greater length.) See Figure 4.4.

4.9.2 Internet Telephony

Internet-mediated long-distance telephone calling greatly expands the opportunity for accounting rate arbitrage. Should this type of calling reach critical mass, the existing voice-oriented pricing methodology of carriers must change, because they no longer will have the ability to accrue significant profits from such services. Internet telephony provides a vehicle for avoiding an accounting rate settlement. Additionally, using the Internet to make telephone calls exempts the ISP from having to pay for access to the traffic delivery facilities of the local exchange carrier that are needed to originate and terminate calls. ISPs also do not have to make contributions to funds used to subsidize universal service.

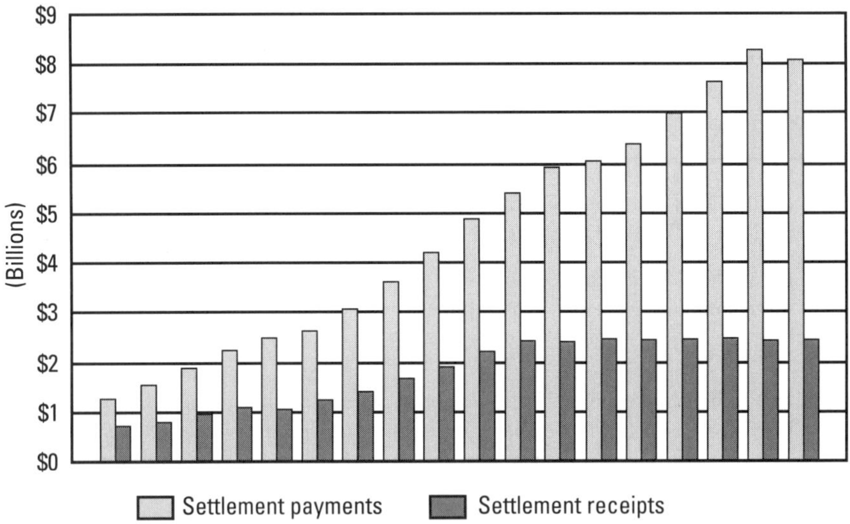

Figure 4.4 Gross settlement amounts: settlement payments and settlement receipts. (*Source:* FCC.)

Internet telephony has evolved quickly from something of limited appeal available only via personal computers to something of widespread appeal available via almost any telephone. Companies like Net-to-Phone, Ibasis, and Delta3 offer Internet-mediated international calling at levels in the pennies per minute range in contrast to service costing over $1.00 from incumbent carriers. (The case studies examined in Chapter 9 include a comprehensive look at Internet telephony.)

References

[1] "Politics on the Line: A Survey of Telecommunications," *The Economist,* Oct. 5, 1991, p. 1.

[2] Frost, D., Vice President, Bell Atlantic International, "Pacific Rim Telecom Status Report: Privatization and Competition Strategies in Mexico and the Lands Down Under," in M. Lofstrom and D. Wedemeyer, eds., *Proceedings of the 14th Annual Conference of the Pacific Telecommunications Council,* Honolulu: PTC, 1992, p. 176; see also Christopher J. Sozzi, "Project Finance and Facilitating Telecommunications Infrastructure Development in Newly-Industrializing Countries," *Santa Clara Computer and High Technology Law Journal,* Vol. 12, Aug. 1996, p. 435.

[3] Wilson, K., "Deregulating Telecommunications and the Problem of Natural Monopoly: A Critique of Economics in Telecommunications Policy," *Media Culture and Society*, Vol. 14, 1992, p. 362.

[4] "Is There Life After Monopolies?" *Public Network Europe*, Vol. 1, No. 10, p. 27.

[5] Canadian Radio-television and Telecommunications Commission, "Competition in the Provision of Public Long Distance Telephone Service and Related Resale and Sharing Issues," Telecom Decision CRTC-92-12, June 12, 1992, p. 57.

[6] Dixon, H., "The Sleeping Giants Awaken," *Financial Times Survey*, Section III World Telecommunications, Oct. 7, 1991, p. 1.

[7] *Ibid.*, p. 1.

[8] Noam, E., "International Telecommunications in Transition," in R. Crandall and K. Flamm, eds., *Changing the Rules: Technological Change, International Competition and Regulation in Communications*, Washington, D.C.: The Brookings Institution, 1989, p. 258.

[9] Communications Act of 1934, as amended, 47 U.S.C. Sec. 302 (1999).

[10] Director, M. D., *Restructuring and Expanding National Telecommunications Markets: A Primer on Competition, Regulation and Development for East and Central European Regulators*, Washington, D.C.: Annenberg Washington Program, 1992, p. 31.

5

The Technologies in Modern International Telecommunications

The recent and dramatic changes in international telecommunications result in large part from the creative and entrepreneurial application of technological innovations. Regulatory barriers to market entry have fallen in many nations, making it possible for new ventures to compete. On the other hand, the extremely high cost of providing broadband telecommunications over a vast geographic area favors large, well-capitalized ventures that more likely can accrue economies of scale and scope, that is, the lowest cost per unit of capacity for all service types offered.

The technological innovations that diversify services and expand consumer choice also create financial and marketing pressure on firms to merge, acquire, or forge strategic alliances. Ironically, a marketplace first considered a natural monopoly may evolve into a somewhat concentrated industry notwithstanding the removal of regulatory barriers to entry. Although some localized market niches can support robust competition, the potential exists for domination by relatively few multinational carriers for many global markets. Telecommunications and information-processing technology (e.g., long-distance telephone service resellers, ISPs, and equipment retailers) can support robust competition at some levels. But other markets (e.g., providing broadband transcontinental and transoceanic transmission facilities) require massive size, a regional or global reach, and multibillion-dollar capitalization.

The necessity to acquire substantial funding and broad geographic reach does not necessarily condemn the telecommunications marketplace to monopolization or excessive concentration. As never before, investors and bankers seem willing to fund new telecommunications and Internet ventures. Likewise, new companies can start off small and grow into multinational powers. For example, companies like MCI and Worldcom started as tiny regional players in the United States. MCI entered the switched long-distance telephone business with a single microwave radio backbone between Chicago and St. Louis. Worldcom started by reselling long-distance services in the southern United States.

Thanks to regulatory and technological developments, new international telephone companies can operate without owning transmission facilities. Innovations in telecommunications switches and interfaces with customers make it possible to enter the long-distance business with a relatively minor investment. Some newcomers lease transmission capacity from facilities-based carriers and use devices that make it possible to resell to many users lines initially intended for the private use of a single customer.[1]

Other entrepreneurial companies stretch domestic law and recommendations of the ITU by installing switching devices that offer callback services: importing dial tone from nations with cheap domestic or international calling to users physically located in nations with high rates. The callback service operator makes it possible for callers to establish a virtual presence in another country simply by placing an international call and hanging up as the connection is made.[2] The simplest device used by callback operators, sometimes called a boomerang box, can identify the caller from stored information, dial out to the caller at a preregistered telephone number, and then provide dial tone for outbound calls. The circuitous routing of calls via a boomerang box

1. Gregory C. Staple provides an insightful analysis of how converging technologies and regulatory changes provide consumers with new opportunities for cheaper and diverse services. See G. C. Staple, "International Telecommunications: The Challenge of Convergence," in G. C. Staple, ed., *TeleGeography 1994* (Washington, D.C.: International Institute of Communications, 1994), pp. 11–33.

2. For an outline of how callback services operate, see Rob Frieden, "The Impact of Call-Back and Arbitrage on the Accounting Rate Regime," *Telecommunications Policy*, Vol. 21, No. 9/10, 1997, pp. 819–827; Kenneth R. Propp, "The Eroding Structure of International Telecommunications Regulation: The Challenge of Call-Back Services," *Harvard International Law Journal*, Vol. 28, Spring 1996, p. 493; D. Briere and M. Lagner, "Automated Call-Back Services: An Update," in G. C. Staple, ed., *TeleGeography 1993* (Washington, D.C.: International Institute of Communications, 1993), pp. 38–39.

makes it possible for callers in nations with high rates to avoid being captive to an incumbent monopoly carrier. The callback service operator, physically located in a nation with cheap rates and private line resale opportunities, can provide a profitable service that largely bypasses the incumbent in high-cost nations.

Callback services and use of the Internet to provide the functional equivalent of telephone service enable small ventures operating on a shoestring to offer substantially lower rates than what incumbents charge. Such technological applications demonstrate how the combination of innovation, deregulation, and entrepreneurialism in some nations increasingly pressures the status quo. Previously, the cost and scale of international telecommunications technology supported aggregating traffic and consolidating investment through cooperatives and consortia. Now, the per-unit cost of telecommunications technologies favors market entry, proliferating services and putting pressure on incumbent organizations. For example, new private satellite companies have so challenged the INTELSAT and Inmarsat cooperatives that both incumbents have spun off private, commercial ventures and privatized all but core activities. New submarine cable private and joint ventures challenge the centralized management achieved when a consortium of many incumbent carriers pool financial resources and traffic requirements.

Technological innovations enhance consumer options by making it possible for individuals and small businesses to use services previously available only to large-volume corporate users. For example, when satellites were first commercialized, nations universally authorized a monopoly that initially operated a single Earth station with antennas 30m or more in diameter. Decision-makers agreed that monopoly status was necessary to ensure that the operator could recoup a sizable and risky investment in the international satellite gateway. Technological innovation now has shrunk the size of international Earth stations to 1m or less, and the number and the location of Earth stations have vastly increased as the cost dropped from millions to thousands of dollars per installation. Rooftop antennae and very small aperture terminals (VSATs) do not require monopoly control, despite efforts by the incumbent operator to perpetuate the status quo.

5.1 Satellites

Satellites traditionally have provided a "bent pipe" in space for receiving signals and retransmitting them back to Earth.[3] In this basic mode, satellites receive a traffic flow and redirect it back to Earth much as the ionosphere reflects short-wave radio signals beamed upward from Earth. The conventional term used as a measurement of satellite capacity is the *transponder*, typically a unit equivalent to 36 MHz of bandwidth. A transponder receives a signal, amplifies it, and may convert it to another frequency for retransmission to antennas on Earth.

To optimize the transponding and retransmission of uplinked signals, satellite operators have positioned stations in an orbit that maximizes the potential for signal coverage on Earth, commonly known as the *footprint*. Operators position satellites in an orbit in which the satellite appears stationary relative to Earth, making its easier to point transmitting antennas that are called Earth stations. Only a small sliver of space, approximately 22,300 miles above the Earth, has the physical properties of holding an object in synchronicity with the Earth's orbital speed. While the speed of a satellite and the speed of the Earth differ, the satellite hovers above the same Earth location, creating the appearance of being *geostationary*. The orbital arc constitutes a scarce, shared resource, because only in that relatively narrow sliver of space do satellites and the Earth travel at the same speed relative to each other, making the satellite a stable target for transmitting signals upward (also known as *uplinking*).[4] Figure 5.1 shows the manner by which satellites reach orbit, and Figure 5.2 shows the location where satellites operate in an orbit synchronized with the Earth's orbit.

Recent innovations in satellite technology add intelligence and versatility to the processing and retransmission capabilities of satellites. Additional new satellites can operate at several orbital locations closer to Earth. New satellites have onboard signal processing that enables operators to transmit on one frequency and receive signals on another frequency. Such processing capability also can enable users to change the beam size or location of the signal footprint. For example, a video programming company seeking to

---

3. For further examination of the technology and economics of satellite telecommunications, see Elke A. Hofmann, "The World Telecommunications Policy Forum: Globalization, Liberalization, and Privatization in the Provision of Satellite Services," *Law and Policy in International Business*, Vol. 28, Spring 1997, p. 929; M. L. Smith III, "The Orbit/Spectrum Resource and the Technology of Satellite Telecommunications: An Overview," *Rutgers Computer and Technology Law Journal*, Vol. 12, 1987, pp. 285–304; H. Hudson, *Communications Satellites: Their Development and Impact* (New York: Free Press, 1990); M. Giget, "Economics of Satellite Communications in the Context of Intermodal Competition," *Telecommunications Policy*, Vol. 18, No. 6, 1994, pp. 478–492.

4. See R. F. G. Hart, "Orbit Spectrum Policy-Evaluating Proposals and Regimes for Outer Space," *Telecommunications Policy*, Vol. 15, No. 1, 1991, pp. 63–74.

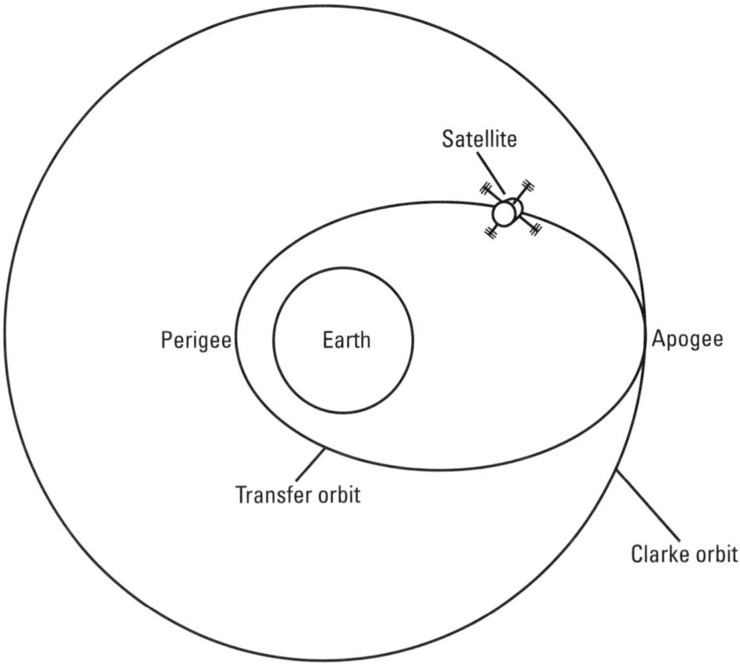

Figure 5.1 Satellite orbits.

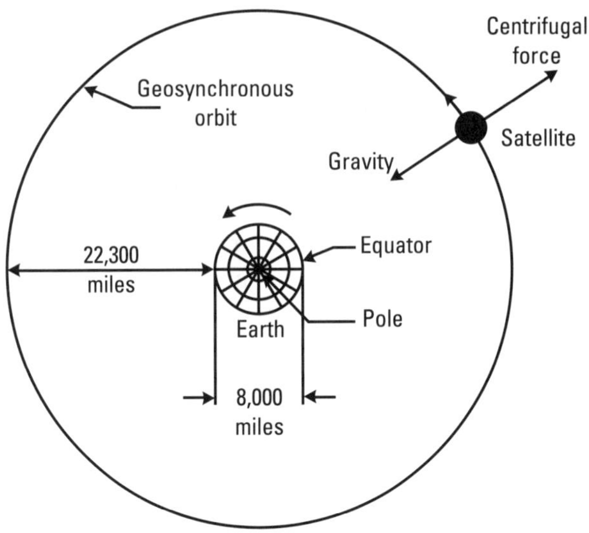

Figure 5.2 Geosynchronous orbital location.

distribute a cable television channel might use a large *global, hemispheric,* or *zone* beam to distribute the programming throughout a wide geographic region in which cable facility operators have installed large satellite dishes at their Earth stations. The video programmer might want the satellite carrier to transmit the programming via a different frequency or to narrow the footprint to a beam to permit reception of the concentrated signal by smaller dishes that end users might install at their homes and offices. Figure 5.3 graphically shows the reltionship between a geostationary satellite and its signal illumination, or footprint, on Earth.

Another satellite innovation, the intersatellite link (ISL), enables carriers to connect satellites by signals transmitted to and from the satellites rather than via Earth stations. ISLs reduce expense by eliminating the need to route some long-haul traffic via multiple satellite hops, that is, transmitting down to an Earth station, back up to another satellite, then back down a second time. ISLs eliminate the delay of approximately one-half second that a user would incur in such an 89,200-mile routing, A carrier seeking to establish global or multiple region coverage can link the satellites in orbit, thereby reducing the cost and the echo that results when signals must take a double hop up and down via two Earth stations to link two satellites in different regions of the world.

New satellite service markets may support the use of orbits closer to the Earth's surface. For services that require the use of portable, preferably handheld transceivers like those used for terrestrial cellular radio, satellite operators must deploy their stations closer to Earth. Satellites orbiting closer to Earth need less power to receive and transmit signals, and the transceivers used to transmit and receive satellite signals can be lightweight and have acceptable usage times before they need a battery recharge. On the other hand, their closer proximity to Earth means that individual footprints will be smaller, and their orbital speed relative to Earth will increase. The combination of smaller footprints and nongeostationary satellite operation requires a constellation of satellites whose number grows as the orbits get closer to Earth. One proposed mobile satellite service to handheld transceivers will require 288 refrigerator-sized satellites (Teledesic) operating in low Earth orbit (LEO). The Globalstar satellite network, operating in a somewhat higher orbit, uses 48 satellites.[5] The commercially unsuccessful Iridium

5. See R. Frieden, "Satellites in the Wireless Revolution: The Need for Realistic Perspectives," *Telecommunications,* Vol. 18, No. 6, June 1994, pp. 33–36; R. Frieden, "Satellite-Based Personal Communication Services," *Telecommunications,* Vol. 17, No. 12, Dec. 1993,

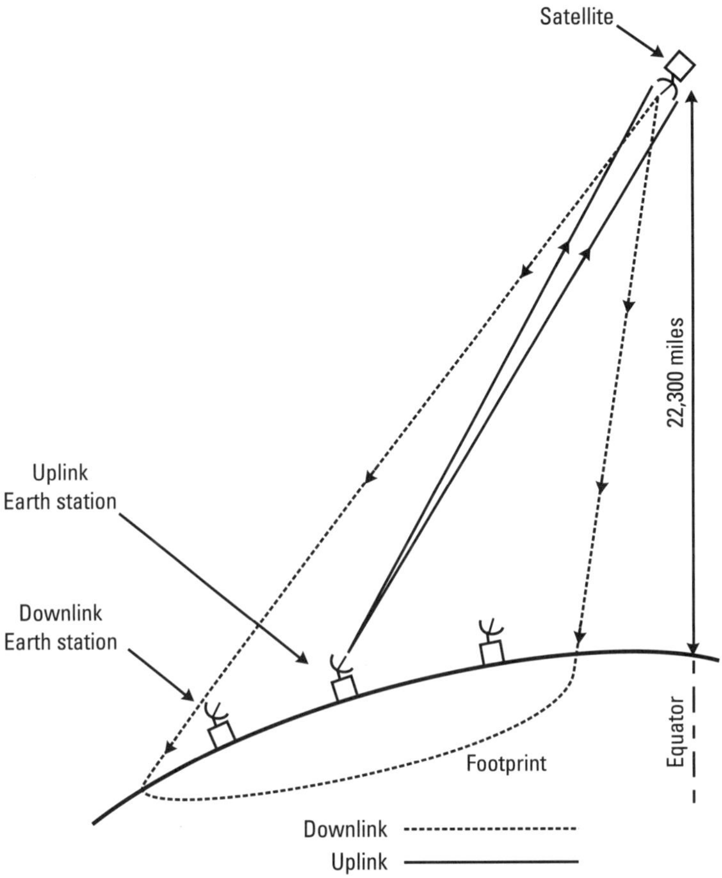

Figure 5.3 Satellite footprint.

satellite network nevertheless successfully proved a number of technological innovations including ISLs to transfer a telephone call from a satellite near the caller to a satellite near the gateway Earth station closest to the intended call recipient. Additionally, it created the need for assembly-line manufacture of satellites at the rate of one per month, in contrast to the 18- to 24-month lead time for one larger communications satellite.

---

pp. 25–28; "WARC-92 and Low Earth Orbiting Satellites: A Case Study for the Process for Accommodating Spectrum Requirements for New Technologies," *Proceedings of PTC '93,* (Honolulu: Pacific Telecommunications Council, 1993), pp. 271–287.

For some services, satellites have achieved a technological comparative advantage over other transmission media, like submarine cables and terrestrial microwave facilities. The broad geographic coverage of a satellite footprint favors point-to-multipoint services, for example, the delivery of video programming to a number of terrestrial broadcast or cable television distribution facilities. Additionally, satellites can efficiently and economically deliver traffic to interior locales far from coastal points where submarine cables make a landfall. Satellites may lack a comparative advantage for high-volume traffic routes, particularly ones with low-cost access to submarine cables.

The cost of launching satellites and the limits on fuel used to keep a satellite in the proper orbit are two major factors that affect the economic viability of satellite technology. On average, it costs approximately $75 million to $100 million to launch a satellite. Approximately one in four launches fails to place a satellite in the proper orbit. Satellites typically reach their end of usable life in 10 years, primarily because they have exhausted station-keeping fuel, which is used to keep the satellite in proper orbit. Expanded launch options and new satellite station-keeping technologies promise improvements in the life expectancy and financial viability of satellites.

Until the mid-1980s, only three nations or national alliances had commercial satellite launching capabilities: the United States, the European Space Agency, and the former Soviet Union. In recent years, China has entered the commercial launch marketplace, and other nations, including Japan, Norway, and Israel, have developed launching capabilities that may have commercial applications. A new station-keeping technology that involves ion thrusters rather than gaseous fuel promises to help maintain satellite orbits for 15 or more years. As long as the satellite power generation and electronic equipment remain usable, a station-kept satellite can continue to provide service well beyond current lifetime expectations.

5.1.1 Satellite Fundamentals

Communications satellites receive and retransmit signals much as very tall radio towers do. In 1945, science writer Arthur C. Clarke predicted that three strategically located space stations could provide service to most of the world. Clarke speculated that there existed a particular orbital location in which satellites would appear stationary relative to Earth, thereby presenting a fixed, geostationary location for sending and receiving signals. If the satellite hovered over a particular point on Earth, television and other signals could travel up to that known location occupied by a stationary receiver/transmitter and then downward to Earth. The satellite could operate like a bent pipe: receiving signals and bending them back to Earth, much as what portions of the ionosphere do to "skywave" radio signals.

Clarke correctly predicted that objects sent into a particular orbital location could operate in synchronicity with the Earth's orbit. In other words, both communications satellites and the Earth rotate at a velocity of one revolution every 24 hours. At 22,300 miles (35,800 km) above the Earth, satellites appear to hover in a stable location, even though they are traveling at 6,879 mph. Satellites operating in geostationary orbits above the equator can maximize the illumination of Earth because the signal is divided equally between the northern and southern hemispheres. Put another way, a satellite with no tilt in its footprint can maximize coverage (one designed to serve North America exclusively would have its footprint tilted northward). Another way to maximize coverage of a preferred geographic area involves a noncircular, elliptical orbit with a high point (apogee) farthest from Earth and a low point (perigee) closest to Earth. Some early communications satellites, particularly ones from the former Soviet Union, operated in an elliptical orbit to maximize availability in preferred service regions.

5.1.1.1 Large Footprints and Point-to-Multipoint Services

Satellites have made such a substantial impact on cable television because of two fundamental characteristics:

- With an orbital location so far from Earth, satellites transmit a weak but usable signal over a broad footprint.
- Large geographic coverage makes it possible to serve thousands, if not millions, of different broadcaster and cable television facilities from the same satellite, called point-to-multipoint service.

You can visualize the concept of a satellite footprint and point-to-multipoint service by using a flashlight or a globe of even a straight surface. A flashlight quite close to the globe illuminates a small area, but provides a strong (bright) signal. Consider the illuminated area to be the footprint. As you pull the flashlight farther from the globe, the coverage area (footprint) increases, but the signal strength (light intensity) decreases. Point-to-multipoint service means that a single flashlight can provide signals to any location it illuminates. A single signal transmitted upward (uplinked) to a satellite can be received (downlinked) by users anywhere within the footprint. Hence, in the video program distribution marketplace, a single satellite can, for example, receive an uplinked movie originating at a Home Box Office's operations facility on Long Island, New York, and downlink it to cable television head-ends equipped with satellite receiving dishes throughout the continental United States.

Point-to-multipoint transmission capabilities change the calculus and economics of video program distribution. In lieu of a single cable system distributing a movie over an unused channel, a company can package a series of movies and other sorts of premium entertainment for distribution over the same unused channel. The programmer can aggregate audiences throughout the nation, while the cable operator can serve a new market niche, without the labor and logistic effort involved in physically receiving tape or celluloid. Once the cable television operator invests in satellite receiving technology, a number of satellite-delivered program options become available.

Because satellite footprints typically cover a wide geographic area, carriers can provide service throughout a region with only the minor investment of adding an additional point of communication. Such widespread coverage also means that the cost of satellite service can be spread over a number of different routes of different length and traffic density. The insensitivity to distance and traffic density means that, with proper coordination by governments, carriers, and users, satellites have the potential to achieve two important outcomes affecting the distribution of video and other kinds of content:

- Satellites can provide point-to-multipoint service, for example, widespread distribution of a video program to a number of broadcast and cable television outlets, at roughly the same cost as a single point-to-point transmission.
- Satellites can achieve networking economies of scale by aggregating audiences that singularly would not generate the demand and revenue sufficient to support a programming venture but that collectively do.

Figure 5.4 shows the transmission contours of satellite footprints.

5.1.1.2 Physics and Electronics of Satellite Transmission and Reception

The great distance traversed by satellite signals requires both electronic amplification and the natural concentration of signal strength achieved by application of the principles of physics. Satellites operate at extremely high frequencies where the wavelength approaches the infrared and visible light

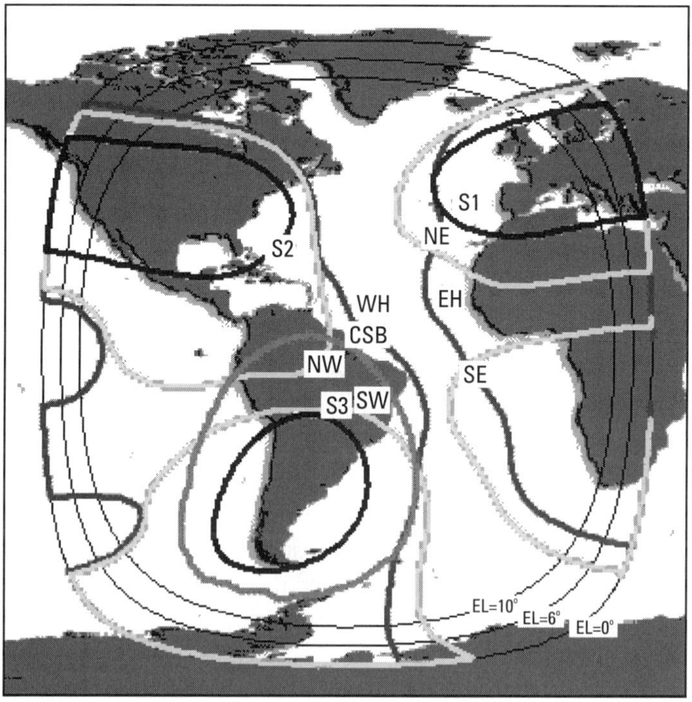

Figure 5.4 Transmission contours of satellite footprints.

spectrum. At those frequencies, signals have many of the same characteristics as lightwaves. Satellite beams travel in a straight line-of-sight manner. Scientists exploit that characteristic by using physics to concentrate beams and thereby increase signal strength, commonly referred to as *gain*.[6] Satellite transmitting and receiving dishes have a parabolic design that aggregates and

---

6. Typically, satellite signal strength is measured in decibels, units of measure that are 10 times the logarithm to the base 10 of a number. Using decibels makes it easier to measure very small and very large values associated with measured signal strength and the associated gain or loss resulting from amplification and the use of signal-concentrating technologies as well as the loss resulting from transmitting a signal through the atmosphere. For comprehensive coverage on satellite technology topics, see Gordon and Morgan, *Principles of Communications Satellites* (Boston: Wiley, 1988), and Morgan and Gordon, *Communications Satellite Handbook* (Boston: Wiley, 1989).

concentrates signals. An elementary school science project provides a visual analogy. With nothing more than a magnifying lens, a sunny day, and a piece of paper or wood, students can carve their initials or even ignite a fire. The glass in the magnifying lens concentrates solar energy onto a single fixed point, thereby increasing the temperature of the solar beam.

Parabolic antennas provide a similar beam concentration function. Manufacturers can achieve further beam concentration by configuring the parabolic antennas on the satellite to concentrate signal strength. That reduces the size of the signal footprint but increases the signal strength in the smaller area served. Satellites typically have multiple signal beams that run the gamut from unfocused to fully concentrated footprints. An unfocused, "unshrunk" footprint is commonly called a global beam and has maximum geographic coverage. Beam concentration runs a continuum from global to zone to hemispheric to spot beams. Spot beams, which offer the most limited geographic coverage, have become increasingly important, because of the physical tradeoff between beam size and signal strength: the more concentrated the beam, the smaller the size of the Earth station receiving dish needed to receive an adequate signal. Geographic areas served by spot beams may receive an adequate signal using dishes the size of a large pizza, that is, less than 1m in diameter. Such small Earth station dish size makes it possible for installation of unobtrusive, lightweight terminals in homes and businesses.

The developing direct broadcast satellite/direct to home (DBS/DTH) industry needs receiving dishes that are low-cost, lightweight, and easy to install. Those three characteristics have enabled DBS/DTH satellite technology to become a consumer electronics option. Previously costly and heavy components limited satellites to relatively few hobbyists, rural residents desperate for any source of video programming, and operators of cable television and satellite master antenna television (SMATV) services.

Scientists have solved the problem of achieving adequate signal strength, despite the signal attenuation resulting from the long transmission path through a dusty, ionized, often rainy or snowy, and sometimes obscured atmosphere. However, they cannot make the signals travel faster than the 250 ms it takes for uplinking and downlinking. Despite advances in echo suppression (i.e., reducing the potential for a receiver to process signals at slightly different times), consumers do not like using links to geostationary orbiting satellites for voice communications. Without the uncharacteristic discipline of sequential and uninterrupted speaking, satellite-carried calls result in breakups in the transition from one speaker to the other.

Accordingly, satellite operators have lost most of the voice communications traffic that can be loaded onto submarine cables.

5.1.1.3 Interference-Free Operation

Interference-free operation of satellites requires coordination on the use of radio frequencies and the geostationary orbital arc. Nations must agree on which frequencies different types of satellites will operate and how they will register orbital slot usage. Likewise, the satellite and associated receiving Earth stations must operate with sufficient signal strength to override potential interference resulting from other terrestrial transmissions like microwave relays.

Again physics plays a primary role. The receiving Earth station must have an unobstructed "look angle," that is, a direct link to the satellite unblocked by trees, terrain, and buildings. Additionally, the Earth station must aim to the satellite at an angle sufficiently above the plane of the Earth so that the link extends above the horizon. The greater the distance a signal has to travel through the atmosphere, the greater the difficulty a satellite receiver will have in distinguishing between a desired signal, and noise or interfering signals.

The physics of satellite telecommunications also affects the vulnerability of Earth stations to interference. As Earth stations locate farther from the center point of a transmission, where signal strength is highest, signal quality from the intended source weakens as other satellite signals increase, particularly if they operate on the same frequency. Signal strength typically degrades in concentric circles or contours. The farther from the satellite's boresight, the weaker the signal becomes with signal rolloff (degradation) accelerating as the distance from the boresight increases.

5.1.1.4 Signal-Compression Technologies

Innovations that digitize and compress signals expand the number of channels for the same amount of radio bandwidth. Video compression can squeeze in as many as six digital video signals into the same amount of spectrum that used to carry only one channel. The squeezing process involves the use of computer processing that makes it possible to examine video frames, which change at the rate of 30 frames per second, and allocate bandwidth for processing only the changes that have occurred from the preceding frame. A voluntary compression standard established by the Motion Pictures Expert Group (MPEG) seeks to support interoperability among computers and other devices, including cable television-settop converters.

5.1.1.5 Comparative Advantages of Satellites Versus Terrestrial Options

Given the emphasis on fiber optic cables as the preferred medium for deploying broadband information superhighways, satellites seem to have been relegated to subordinate status. Many consider wireless options inferior to wireline services in view of the broader bandwidth and echo- and interference-free operations available from wireline. However, wireless options will play a role in information-age infrastructure development, and they possess a comparative advantage for some applications.

The broad geographic coverage of a satellite footprint promotes the use of satellites for service throughout a large geographic region. Likewise, it makes it possible for satellite operators to provide a comparatively cheaper, more complete service than what operators of terrestrial and submarine fiber optic cable networks would charge. Once a satellite-delivered application illuminates a region, access from an additional point within the footprint typically requires lower investments than if a terrestrial network had to be extended outward to serve that additional point of communication. Similarly, the distance insensitivity of satellite footprints means that it costs no more to serve a sparse user population in a remote locale than to serve a point of communication closer to the content source. This feature may prove especially helpful for delivery of video streaming and other broadband Internet services to remote locales and to users far from the server hosting the content. Once successfully launched and inserted into orbit, satellites can provide service without outages that might beset terrestrial facilities as a result of natural catastrophes (e.g., storms and earthquakes) and man-made disasters (e.g., war and environmental hazards).

5.1.1.6 Positive Networking Externalities

Economists use the term *positive networking externalities* to reflect the enhanced consumer welfare generated when satellites serve increasing numbers of users and points of communication.[7] Satellites have the technological and economic wherewithal to serve additional points of communication at minimal additional (incremental) cost, while increasing the overall utility

7. Most telecommunications and information networks, such as the Internet, also reflect positive networking externalities. As more subscribers and sources of information come on-line, presumably the overall value of the network grows even if any single user has no interest in accessing the additional users or information sources.

(value) to users.[8] The unconcentrated signal from a geostationary orbiting satellite can illuminate as much as one-third of the Earth's surface.[9] Once a carrier incurs the substantial sunk cost to make that footprint available, the incremental cost for it to serve an additional point of communication and additional users via another Earth station approaches zero. An additional point of access requires users to install or interconnect with an Earth station, acquire domestic facilities to link their premises with the Earth station, and pay space segment charges.

The value of satellite service accruing to users can increase as the satellite serves more Earth stations and more users, with no degradation in service quality and often without higher charges to reflect the increased utility.[10] The concept of direct network externalities reflects the enhanced value of service accruing to users as more points of communication and more users come on-line.[11] The benefit is considered an externality, because standard economic analysis and the pricing of service may not take into account that favorable outcome. Indirect network externalities result when increasing coverage and market penetration result in more plentiful, lower-cost

---

8. See Michael L. Katz and Carl Shapiro, "Technology Adoption in the Presence of Network Externalities," *Journal of Political Economy*, Vol. 94, 1986, p. 822; Michael L. Katz and Carl Shapiro, "Network Externalities, Competition, and Compatibility," *American Economics Review*, Vol. 75, 1985, p. 424.

9. For an introduction on satellite technology, see Andrew F. Inglis, *Satellite Technology: An Introduction* (Boston: Focal Press, 1991); Donald M. Jansky and Michel C. Jeruchim, *Communications Satellites in the Geostationary Orbit* (Norwood, MA: Artech House, 1987).

10. INTELSAT typically does not engage in price discrimination on the basis of demand elasticity and user desire to lease capacity on a particular satellite. Private operators typically do. The so-called "hot bird" concept reflects the added value and commensurately higher lease prices for satellites that become home to the most desirable video programs and networks. Because users have a financial incentive in limiting the number of Earth stations they need to install and maintain, they prefer to access only a few satellites for their complete inventory of video programming. Private satellite operators who have executed transponder leases with programmers having the most desirable video product find that other programmers, perhaps offering less attractive fare, want to lease capacity and possibly exploit the benefits of being more widely accessible.

11. "There are many products for which the utility that a user derives from consumption of the good increases with the number of other agents consuming the good." Michael L. Katz and Carl Shapiro, "Network Externalities, Competition, and Compatibility," *American Economics Review*, Vol. 75, 1985, p. 424.

complementary goods, that is, goods often consumed at the same time. For example, the greater the use of satellites to deliver telecommunications services, the greater the incentive for reaching consensus on technical standards for Earth stations accessing the proliferating satellites. Industry-wide equipment compatibility helps manufacturers achieve economies of scale by having to support fewer product lines with different technical standards.[12]

Satellite cooperatives, like INTELSAT and Inmarsat, generated positive network externalities simply by succeeding in commercially exploiting satellite technology previously used primarily for defense, space exploration, and intelligence-gathering applications. They increased the likelihood for and extent of such positive network externalities by enacting a governance document that deliberately priced space segment on an averaged cost basis, thereby enabling developing nations strapped for hard currency to participate in the cooperative by investing in an ownership share of as low as 0.05%.

5.1.1.7 Satellite Multicasting

Because satellites operate at a substantial distance above the Earth, the transmission contour greatly exceeds the coverage available from an Earth-based tower. The economics of satellite operation favor point-to-multipoint multicasting, because the incremental cost to serve an additional location is largely a function of the costs incurred to transmit and receive traffic to and from the satellite, that is, the construction and operation of the Earth station. In other words, the vast majority of satellite costs are sunk in the construction and launch of the satellite such that activating an additional point of communication imposes no additional operational costs with little if any impact on existing users.

Satellites promote ubiquitous telecommunications access to the rest of the world by overlaying infrastructure throughout a large geographic region. Nations that lack the resources to invest in their own satellite assets can tap into the network launched by another nation or by a private company. Telecommunications development has improved with efforts to provide global connectivity from virtually anywhere in the world, including the creation of cooperatives with a global service mission, a mandate to average costs among dense and sparse routes, and an ownership structure that allows lesser developed nations to acquire a small investment share in the venture.

---

12. See Carmen Matutes and Pierre Regibeau, "'Mix and Match: Product Compatibility Without Network Externalities," *Rand Journal of Economics*, Vol. 10, 1988, p. 221.

Heretofore, the amount of spectrum allocated for satellite service and the technical limitations on the bandwidth that a single satellite can support have constrained the benefits of global connectivity and point-to-multipoint multicasting. Operators of satellite services must vie with operators of other spectrum-using services for allocations at the ITU and from domestic regulatory authorities. Additionally, satellites have capacity restraints insofar as the number of transponders that can be installed on a single frame. The typical geostationary orbiting satellite has approximately 24 transponders, each representing 36 MHz in bandwidth. The ability to reuse the same bandwidth over different geographic regions and by polarizing the signal in two different formats (vertical and horizontal or right circular and left circular) helps abate bandwidth limitations. Frequency reuse means that satellites serving two different geographic regions can use the same frequency, because antennas on the satellite can point to two separate regions. Polarization offers the ability to use the same transponder twice by transmitting the signal in two different patterns.[13] Satellites' primary function in Internet transmission involves the delivery of content to sites in closer proximity to users. This delivery to "caches" conserves Internet resources by locating frequently accessed content closer to consumers. (See Figure 5.5.)

5.1.1.8 Uniform Quality

Satellites deliver large bandwidth with virtually the same signal quality throughout a broad geographic footprint. Terrestrial facilities, particularly the cascading tree and branch cable television infrastructure, lack uniform quality in view of the multiple connections and amplifications that occur in the route from head-end to individual television sets. Ironically, DBS marketers herald the superior signal quality of a digital signal that has traversed 44,600 miles relative to the cable television signal that may have run fewer than 10 miles. Until such time as fiber optic technology provides a closer or complete link to the consumer terminal, analog transmissions, multiple connections, and sequential amplification inject noise and degrade signal quality.

13. A way to understand satellite polarization is to consider the function of polarizing sunglasses. Instead of duplicating the use of frequency range, polarizing lenses block reception of harmful lightwaves.

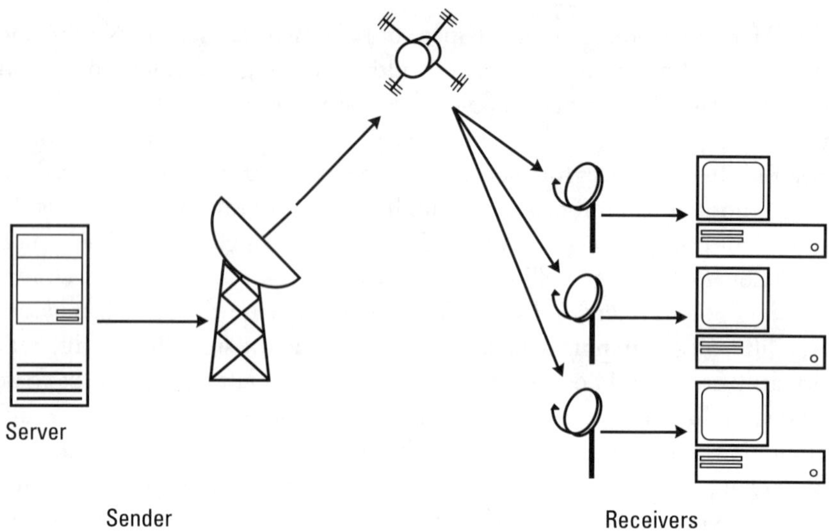

Figure 5.5 Internet via satellite. (*Source:* Telecommunications®.)

5.1.1.9 Distance Insensitivity

Satellite transmissions throughout a large geographic footprint mean that operators incur no additional incremental costs to serve an additional point regardless of its distance from the program source. In terrestrial point-to-point networks, operators can attribute direct costs in extending a network to an additional point. Accordingly, satellites may possess a comparative advantage relative to wireline networks for applications that require a large number of distribution points in diverse locations.

5.1.1.10 Efficient Point-to-Multipoint Distribution

The distance insensitivity in satellite transmission means that operators incur little if any additional cost in serving an additional point of communication within the footprint. Terrestrial networks typically use a separate line to deliver programming to each destination. Accordingly, satellites possess a comparative advantage in terms of both cost and logistics for distributing programming to geographically diverse cable head-ends.

5.1.1.11 Mobility and Ease in Reconfiguration

Electronic component miniaturization on integrated circuits and microprocessors, higher-powered satellites, more sensitive receiving terminals, and orbiting satellites closer to Earth make it possible to support diverse mobile, wireless applications. Service providers cannot use fiber optic cables access to users in vehicles, ships, aircraft, and sparsely populated locales. Likewise, they cannot easily add and subtract service areas and users by reconfiguring their networks. A wide satellite footprint makes it possible to add another service point simply by installing an Earth station, some of which are portable and configured in a few minutes.

5.1.1.12 One-Stop Shopping

Satellite carriers recognize the potential for footprints to traverse national boundaries and the service territories of different cable television operators and broadcasters. They can package programming in such a manner that the retailer simply plucks programming from the satellite for distribution to end users. Cable television networks, which use satellites to distribute their programs, offer an example of this characteristic.

5.1.2 Challenges to the Satellites' Comparative Advantage

Recent developments in terrestrial networking challenge satellites' multicasting comparative advantage. A single satellite cannot keep pace with bandwidth expansion in terrestrial and submarine options. Recent developments in fiber optic cable include throughput expansion to multiple-gigabit-per-second transmission rates and dense wave division multiplexing, which expands capacity by using different laser beam frequencies. Additionally, cable operators can install plenty of reserve "dark fiber" cables, which can be activated (lit) in the future at comparatively much lower costs. Satellites are launched with a finite amount of transponders onboard.

However, new spectrum reallocations, lower-than-geostationary orbits, and networks involving a constellation of dozens if not hundreds of satellites create the likely prospect that satellite technology can continue to compete with terrestrial or submarine cables for broadband consumers. In the context of the information age and the global information infrastructure (GII), a global grid of LEO satellite constellations can provide "Internet-in-the-sky" services. The first generation of LEO satellites provided narrowband, mobile telecommunications services, for example, Iridium,[14] Globalstar,[15] and ICO, Ltd.,[16] with less commercial appeal than anticipated. Future constellations

14. Iridium operated a $5-billion global constellation of 66 LEO satellites, about 400 miles above Earth, able to provide voice, data facsimile, and position determination services to

will provide wideband fixed and mobile services, for example, Teledesic,[17] Expressway,[18] Celestri,[19] Astrolink,[20] and Cyberstar/Skybridge.[21]

---

handheld transceivers. Motorola created the Iridium concept and spent almost five years and several hundred million dollars developing the technology and soliciting investors to form a global consortium. In 1993, an initial private placement of $700 million created a consortium that comprised a geographically diverse set of investors representing different aspects of space, telecommunications, and venture capital industries: Khrunichev Enterprise, builder of the Russian Proton launch vehicle; China's Great Wall Industry Corp., operator of the Long March launch vehicle; Iridium Nippon, a Japanese investment group led by Daini Denden, a major cellular radio operator; Kyocera Corp., a diversified manufacturer; two venture capital groups, Mawarid Group of Saudi Arabia and Muidiri Investments BVI, Ltd. of Venezuela; incumbent carriers BCE, Inc. of Canada, Sprint of the United States, and STET of Italy; manufacturers of Iridium network equipment, Lockheed and Raytheon; and telecommunications enterprises in developing nations, including United Communications Industry of Thailand. In 1997, the venture raised $223 million through initial public offering of Iridium World Communications and sold S. Pacific gateway rights for $100 million. By 1999, the venture failed commercially and Iridium filed for bankruptcy.

15. Globalstar operates an LEO constellation of 48 satellites in 8 orbital planes. With fewer satellites operating in orbits about 800 miles above Earth, the $2.5-billion Globalstar network provided service at a wholesale cost of less than $0.50 a minute plus a monthly service charge of $60 to $70. However, less in-orbit resources means that the network will rely heavily on the widespread availability of gateway Earth stations to route calls. Globalstar initially was organized by Space Systems/Loral, a satellite manufacturer, and Qualcomm, Inc., an innovator in code division multiple access mobile radio technology and provider of vehicle location services via GSO satellites. In 1994, the venture announced that it had secured the infusion of several hundred million dollars from new investors, including Alcatel, a major French aerospace and telecommunications equipment manufacturer; France Telecom, the country's monopoly telecommunications carrier; Alenia, an Italian aerospace manufacturer; Deutsche Aerospace and DASA, aerospace manufacturers in Germany; Hundai, a major diversified Korean manufacturer; DACOM, Korea's second telecommunications carrier; AirTouch, the cellular radio spinoff of Pacific Telesis; and Vodafone, a British cellular radio operator.

16. ICO Global Communications was spun off from the Inmarsat global cooperative in January 1995. In 1999, having run out of operating funds and investor confidence, it filed for bankruptcy protection.

17. Rather than provide ubiquitous narrowband (less than 4,800 bps) capacity to mobile users, Teledesic will offer a global overlay of wideband functionality using Ka-band frequencies (20–30 GHz). The system will offer throughput rates in excess of 2 Mbps initially from 288 refrigerator-sized satellites at a total cost of $9 billion.

Like its visionary backers, who include Bill Gates and Craig McCaw, Teledesic pushes the envelope with an eye toward providing a ubiquitous, broadband GII. The system will

Heretofore, video program distribution has generated most of the multicasting opportunities for satellite operators. In the future, the opportunities should expand as multimedia and Internet services proliferate. On the other hand, technological innovations in terrestrial program distribution should generate keener intermodal, terrestrial versus satellite competition. Terrestrial wireless options like local multipoint distribution systems and other extremely high-frequency, microwave radio distribution technologies will provide cheap and quickly deployed competitive options. Some terrestrial wireless delivery options already exist (e.g., multichannel, multipoint distribution systems), while even higher-capacity systems, with more available allocated frequencies, soon will enter urban markets.[22]

---

commercialize a technology developed as part of the Strategic Defense Initiative, which deploys observer ("Brilliant Eyes") and interrupter ("Brilliant Pebbles") satellites into a seamless, global array. In 1997, Boeing agreed to invest up to $100 million in the venture for a 10% share.

18. Hughes Communications, Inc. submitted the Spaceway broadband Ka-band application to the FCC in December 1993. The company expected the first launch to culminate in an 11-satellite constellation. Initially the Spaceway network will provide coverage to approximately 90% of the world's population segmented into four overlapping regions, each with two geostationary satellites. In 1997, Hughes proposed the $3.85-billion Expressway satellite network, the first commercial proposal to use the 40-GHz to 50-GHz frequency band. The Expressway proposal calls for 10 orbital slots, with two satellites each. Hughes plans to launch the first Expressway satellite 50 months after FCC approval.

19. In 1997, Motorola proposed a $13-billion, Ka-band satellite network comprising both geostationary and LEO satellites. Until Motorola appeared to back off the venture and ally with Teledesic, the Celestri system was designed to provide high-speed data and video transmissions to 99% of the globe beginning in 2002. The proposed system was designed to integrate two previously proposed Motorola satellite projects, Millennium and M-Star. The initial backbone of the system will be one GEO and 63 LEO satellites. While the LEO spacecraft would provide high-speed interactive communications around the globe, the GEO component would broadcast to users in a send-only mode. The total Celestri system was designed to have the capacity to transmit 80 Gbps.

20. Lockheed Martin Telecommunications has proposed a network of nine satellites operating from five orbital locations to provide global coverage.

21. Loral Space and Communications Ltd. has proposed a comparatively modest three-satellite Ka-band system. The company teamed up with Alcatel in 1997 to propose a $3.9-billion video and data venture called Sky Bridge, comprising 64 LEO satellites.

Satellite operators can remain competitive and retain comparative advantages through service diversification. Mobile satellite systems (MSSs) offer user-friendly, seamless access to the rest of the world, a desirable option where terrestrial options do not exist. Direct-to-the-consumer access to the rest of the world, anytime and anywhere, provides a powerful tool for commerce and personal safety. However, as the Iridium commercial failure attests, no matter how spectacular the technology, consumers must consider the price reasonable.

Another promising example of service diversification lies in providing high-speed, wideband Internet applications anytime and anywhere. Currently, the on-ramp used by most users to access the Internet involves the retrofitting of a narrowband, analog system configured for short-term, dedicated switched connections between two telephone callers. Even for Web users in high-income nations with the greatest telecommunications line density (also known as teledensity),[23] the mode of access involves the use of modems operating over twisted-wire pairs with throughput limited to 56,000 bps. Satellites already offer throughput rates in excess of 400 Kbps (400,000 bps), but at significantly higher cost than currently available terrestrial options (e.g., integrated services digital networks, digital subscriber links, and cable modems).

A new generation of wideband, fixed satellite service ventures promise Internet-ubiquitous access at throughput rates of multimegabits per second (i.e., millions of bits per second). By providing extraordinarily broad bandwidth, satellite operators can provide access to the information superhigh-

22. See Rulemaking To Amend Parts 1, 2, 21, and 25 of the Commission's Rules To Redesignate the 27.5-29.5 GHz Frequency Band, To Reallocate the 29.5-30.0 GHz Frequency Band, To Establish Rules and Policies for Local Multipoint Distribution Service and for Fixed Satellite Services, Petitions for Reconsideration of the Denial of Applications for Waiver of the Commission's Common Carrier Point-to-Point Microwave Radio Service Rules, CC Docket No. 92-297, Suite 12 Group Petition for Pioneer Preference, PP-22, Second Report and Order, Order on Reconsideration, and Fifth Notice of Proposed Rulemaking, 12 FCC Rcd. 12545, 62 Fed. Reg. 16514 and 23148 (1997) adopting Subpart L of Part 101 of the Commission's Rules, 47 C.F.R. §§ 101.1001-1112; Order on Reconsideration, 12 FCC Rcd. 6424 (1997); Second Order on Reconsideration, 12 FCC Rcd. 15082, Third Report and Order, 12 FCC Rcd. 22310 (1997) Third Order on Reconsideration, FCC 98-15, 1998 WL 54591 (F.C.C.) (rel. February 11, 1998); Fourth Report and Order, 13 FCC Rcd. 11655 (1998).

23. For a collection of Internet use and penetration statistics and a helpful overview of Internet issues, see International Telecommunication Union, *Challenges to the Network Internet,* For Development (ITU, Geneva, 1999).

way from localities that lack access to terrestrial broadband cables. Because a business case cannot be made for ubiquitous deployment of fiber optic cables, it remains likely that satellites will have a key role in provisioning the GII.

5.1.3 New Letters in the Satellite Spectrum Alphabet

Proliferating services, increasing demand for basic services like video program distribution, and growing interference potential from terrestrial microwave facilities have caused satellite operators to expand the frequencies used. In the 1980s, carriers began to use the Ku-band (14-GHz uplink and 11-GHz downlink) in addition to the C-band (6-GHz uplink and 4-GHz downlink). Operators will continue up the spectrum to the Ka-band (30-GHz uplink and 20-GHz downlink), frequencies currently used for terrestrial microwave services and satellite links primarily oriented to tracking and controlling satellite orbits. As well, some carriers may stake claims to other less cluttered frequency bands, for example, the L-band (1.5 to 2 GHz), S-band (2 GHz), and X-band (7 to 8 GHz).

Expanding spectrum options provides opportunities for satellites to become broadband friendly. Ironically, satellite capacity has been measured in terms of 36 MHz–equivalent transponders, even as most users consider satellites' capacity constrained and narrowband oriented, except for video program distribution. That perception may have resulted from carrier pricing policies and marketing strategies that appear to have targeted video to the exclusion of other broadband applications.

5.1.3.1 Conversion from Designer to Off-the-Rack Satellite Manufacturing

The information age will stimulate a proliferation of services, frequencies, orbits, and operators. Satellite manufacturers must respond to changing market conditions by migrating to an assembly-line manufacturing process capable of turning out more satellites in less time. Satellites now require as much as a two-year lead time, even though any particular satellite in a product line contains many common parts.

The commercial communications satellite marketplace will continue to diversify in terms of products and markets. The industry no longer supports just the one-by-one manufacture of a few dozen designer satellites per year, with services provided by a few predominately government-owned carriers participating in global or regional cooperatives. Instead, a variety of developing markets present the prospect for service diversity and perhaps even greater price competition. The option of procuring satellites available in orbit or for quick launch has accelerated the development of a more diversified and maturing marketplace.

What an off-the-rack manufacturing process loses in terms of design flexibility, it gains in speedy deployment and economies of scale. Diverse markets and satellite roles require quicker turnaround times in the manufacturing process, particularly when market opportunities occur in a window caused by short-term capacity shortages. Likewise, market opportunities will require different types of satellites, some containing less than the standard of 24 to 36 transponders and some operating in nongeostationary orbits. Some satellites that offer Internet and mobile telephone services will operate at LEOs and will number in the dozens to make up a constellation providing global access.

5.1.3.2 Cheaper by the Dozen: New Launch Options

Launch operators also will need to respond to diversifying satellite product lines. Most launches now insert one or two satellites into orbit. Satellite constellations, operating in LEO, require launches of six or more small, light satellites at a time and the quick deployment of replacements. Those requirements mean that launch operators must redesign rockets that can hold several satellites whose weights collectively reach the total of two conventional satellites. Likewise, the LEO orbital location can stimulate competition, because a larger set of operators might participate, including companies adapting missile technology to commercial applications.

5.1.4 New Orbits

New, primarily mobile telecommunications markets favor the use of satellites operating in LEO, middle Earth orbit (MEO), and inclined circular orbit (ICO). From those locations, satellites can concentrate a satellite signal and avoid the transmission delays and echoes that vex callers using geostationary options. Lower Earth orbital locations enable users to access satellites using handheld, low-powered terminals weighing only slightly more than cellular radio telephones. However, the stronger signal achieved by closer proximity to Earth also reduces the scope of geographic coverage and necessitates a larger number of satellites to operate in a global constellation. In nongeostationary orbital locations, satellites will speed across the horizon rather than appear stationary. Such moving targets will require complex tracking and coordination.

5.1.5 Alternatives to "Cradle-to-Grave" Satellite Procurements

The proliferation of satellite service options and carrier types will lead to alternative product lines to the one-size-fits-all high-capacity geostationary satellite model. It also may result in lower market shares held by any single carrier. The proliferation of satellite types should stimulate the provision of many new services and market segments. It also makes it possible for a larger number of nations, both developed and developing, to acquire their own satellite networks. New procurement options include the following:

- Satellites relocated from one orbital slot to another and perhaps from one owner to another;
- Satellites retrieved from a useless orbit caused by launch failure or at a time when key components have failed, including the exhaustion of all station-keeping fuel;
- Satellites previously used but still capable of providing adequate service typically for several more years;
- Satellites streamlined in size, weight, and capacity for shared launches;
- Renovated and relaunched satellites;
- Satellites in inclined orbits that require tracking, to conserve station-keeping fuel;
- Satellites privatized and transferred from government to private ownership;
- Satellites available on short notice and able to satisfy pressing and immediate requirements, particularly for capacity less than the conventional inventory of 36 transponders.

5.1.6 Basic Components in Satellite Systems

Satellite systems divide into five basic components:

- The transmitting Earth station(s);
- Uplink;
- Satellite;
- Downlink;
- Receiving Earth station(s).

While simply identified and listed, these components involve complex and costly technology. Scientists from a number of different disciplines have needed to find solutions to difficult problems like finding ways to:

- Propel satellites weighing thousands of pounds into orbits 22,300 miles above Earth;
- Transmit and receive signals of adequate strength, despite the distance to and from the satellite;
- Keep satellites operational and free of interference from other satellites and terrestrial systems for 10 years or more;
- Load satellites with more video channels and Internet content to accommodate demand and to compete with terrestrial options.

5.1.7 Improvements in Satellite Design

Recently launched satellites use the same digital transmission capabilities as used by fiber optic undersea and terrestrial networks. The satellites operate with higher power, thereby reducing the size of receiving dishes to that of a pizza. They generate digital signals that can be transmitted over the Internet, compressed, coded, packetized, and distributed simultaneously to multiple users. More selective Earth stations promote higher performance and the ability to derive more channels.

New satellite networks will convert all traffic into a digital bit stream and next generation satellites will likely have more sophisticated onboard processing capabilities to switch traffic to the appropriate downlink beam, or link to another satellite in an operating constellation. Satellites operating as part of the information superhighway look less like simple, unintelligent bent pipes that simply relay signals, and more like complex airborne alternatives to terrestrial, fiber optic networks. Traffic on such networks is digital, divided into packets for efficient processing and transmission, switched and routed by the satellite down to Earth stations and/or onward to another satellite via intersatellite links.

Satellite designers have pursued improvements in the materials that make up the satellite chassis, antennas, and electronic components, as well as the station-keeping propulsion primarily to reduce weight and to extend the satellite's usable life. New satellites combine aluminum with lighter metals like lithium and beryllium along with graphite, the material used to make tennis rackets lighter and more durable. The fuel used to keep satellites in the proper orbit and orientation to Earth now combine fuel with electricity to generate more thrust and longer life. Ion thrusters use an electrified metal grid to ionize xenon gas and generate a highly efficient gas plasma that results in thrust 10 times more efficient than previously used chemical thrusters. New satellite antenna designs make it possible to use larger but lighter one-piece dishes that flexibly spring into shape rather than require a mechanical activation.

New satellite generations also take advantage of advances in lasers, component design, and amplification. ISLs, which are just being implemented to link multiple, in-orbit satellites, in the future will use higher-frequency laser beams. Because of the narrowness of laser beams, satellites will have to maintain a more stable location in orbit. New designs for solar cells, amplifiers, and integrated circuits will result in more efficient circuitry that will last longer, use less power, generate less heat, and weigh less.

5.2 Submarine Cables

Undersea cables provided the first telecommunications links between nations separated by large bodies of water. The first cables supported only a few telegraph channels but made it possible for social and commercial transactions to occur on a far speedier basis, compared to multiweek transoceanic crossings by ship. After a period of time during which innovations in high-frequency radio made it possible to transmit voice conversations, submarine cables expanded capacity and again became the least-cost routing option for many country pairs. In fact, for many years after the commercial debut of satellites, regulators ordered the "balanced loading" of facilities that in application required carriers to activate satellite circuits, regardless of whether the medium provided service at a lower cost. That policy bolstered the use of satellites to ensure the redundancy of traffic routing and to support the commercial viability of satellite operators.

Most international telecommunications carriers and customers view satellites and submarine cables as complementary media. Satellites provide cost-effective, optimal service for point-to-multipoint applications, such as the distribution of video programming to a number of broadcast stations and cable television head-ends. Cables possess a comparative advantage for echo-free voice transmissions and especially for high-volume, point-to-point routes, particularly between countries close to the points where cables make landfall, thereby reducing the cost of terrestrial "tail-circuit" links to the cable. The use of glass fibers instead of copper wire and the associated migration from direct current amplification to low-powered laser transmission have increased the bandwidth and reliability of submarine cables.

Fiber optic cables are the preferred medium for broadband telecommunications. New generations of such cables can space signal amplifiers, known as regenerators, hundreds of miles apart, thereby reducing installation labor and risk of outages. Increasingly, fiber optic cables are suitable for some types of video transmissions, in addition to high-speed data and voice traffic. When installing fiber optic cables, ventures can include extra fiber strains, known as dark fiber, that can be activated as demand warrants. A developing technology known as dense wave division multiplexing substantially increases the throughput of fiber optic cables by using laser beams of different color and frequency transmitted over a single strand. Soon optical switching will obviate the need to convert laser-based transmissions into electronic traffic streams for delivery to consumers.

Recently installed submarine cables provide increased reliability and lower costs, because they are configured in a "self-healing" network. Instead of having to resort to satellites for circuit restoration, carriers can achieve the near instantaneous ability to route around a cable outage via another operational cable. An intelligent submarine cable network can sense a break in a cable and route traffic via a parallel cable on the fly. Such functionality means that users will perceive no significant disruption in service because traffic can continue to the destination via another, perhaps circuitous routing. Table 5.1 offers background on the growth in submarine cable capacity and the commensurate reduction in per-channel costs.

5.3 Enabling Technologies

Rarely does a week go by without some mass market newspaper or magazine heralding the convergence of technologies that will generate a GII.[24] A multimedia GII, or information superhighway, results from the merger of

24. Vice President Albert Gore envisioned the construction of information superhighways on the basis of five principles: (1) primarily private investment; (2) competition; (3) flexible regulation; (4) open access to the network for all information providers; and (5) universal service. Remarks of Vice President Gore at the International Telecommunication Union, *World Telecommunications Development Conference*, Buenos Aires, March 21, 1994; see also U.S. Dept. of Commerce, Information Infrastructure Task Force, *The National Information Infrastructure: Agenda for Action*, Progress Report, Sept. 1993–1994, Sept. 15, 1993.

Table 5.1
Transatlantic Cable Systems[A]

System[B]	Year	Technology	Estimated Cost ($ Million)	Total 64-Kbps Circuits	Usable 64-Kbps Circuits[C]	Annual Investment Cost per Usable Circuit[D]	Investment Cost per Minute[E]
TAT-1	1956	Coax cable	49.6	44.5	40.1	21,996	2.443
TAT-2	1959	Coax cable	42.7	49.0	44.1	167,308	1.910
TAT-3	1963	Coax cable	50.6	87.5	78.8	111,027	1.267
TAT-4	1965	Coax cable	50.4	69.0	62.1	140,238	1.601
TAT-5	1970	Coax cable	70.4	720.0	648.0	18,773	0.214
TAT-6	1976	Coax cable	197.0	4,000.0	3,200.0	10,638	0.121
TAT-7	1983	Coax cable	180.0	4,246.0	3,821.4	8,139	0.093
TAT-8	1988	Fiber optic	360.0	7,560.0	6,048.0	10,285	0.117
TAT-9	1992	Fiber optic	406.0	15,120.0	10,584.0	6,628	0.076
TAT-10	1992	Fiber optic	300.0	22,680.0	18,144.0	2,857	0.033
TAT-11	1993	Fiber optic	280.0	22,680.0	18,144.0	2,667	0.030
TAT-12	1996	Fiber optic	378.0	60,480	60,480	1,080	0.012
TAT-13	1996	Fiber optic	378.0	60,480	60,480	1,080	0.012
Gemini	1998	Fiber optic	520.0	241,920.0	241,920.0	371	0.004
AC-1	1998	Fiber optic	850.0	483,840.0	483,840.0	304	0.003

A. Including systems removed from service and planned systems.

B. Carriers have announced two additional transatlantic cable systems. TAT-14 would link the United States with five European countries, have a capacity of 640 Gbps, and cost $1.5 billion. OXYGEN (USA) would link the United States with more than 75 other countries. The system is projected to have a capacity of 640 Gbps.

C. With allowance for redundancy, restoration, etc.

D. The annual investment cost per usable circuit is the annual payment rate for the life of the asset that produces a present value equal to the initial investment cost. This calculation assumes a 25-year cable life and a discount rate equal to the average cost of capital for the firm. For purposes of comparison, the discount rate is based on 40% debt at an embedded cost of debt of 9% and 60% equity at a 14% rate of return, with the latter increased to reflect a 37% income tax rate. These assumptions translate to a discount rate of 16.9333%.

E. The investment cost per minute assumes that average activated circuits are used 8 hours a day, 365 days a year and that 50% of circuits are idle (not activated). These assumptions are consistent with current utilization rates but probably overstate the utilization rates for early cable systems. This calculation is intended to illustrate relative costs of new versus old cable capacity. The Gemini and AC-1 systems are designed primarily to handle private-line circuits.

Source: FCC.

previously discrete technologies and the erosion of regulatory policies that supported such mutual exclusivity. In a digital environment, "a bit is a bit is a bit," which means that single transmission media can support a multimedia environment that bundles voice, data, facsimile, video, text, information processing, financial services, and electronic commerce, including catalog sales, news, and entertainment.

The migration from analog to digital transmission technologies, in conjunction with increasing reliance on software-based networks, challenges the status quo. Carriers must acquire a new nimbleness in responding to user requirements, because they no longer can expect any customer base to remain captive. Digitization and software network management make political boundaries more easily penetrated and accessible to services provided by outsiders. In a software-defined networking environment, service providers can avoid the unwieldiness of hardware and older electromechanical switching devices that used high-maintenance, movable parts. Easy network reconfiguration and flexible service arrangements facilitate traffic aggregation at regional hubs, much as airlines route feeder traffic into and out of airport hubs.

Facilities-based carriers aim to exploit market access opportunities singularly where possible. However, global alliances with other carriers have become a major business vehicle for securing market access opportunities and providing one-stop-shopping solutions for the requirements of multinational enterprises. Likewise, regulatory initiatives have made it possible for market entry by nonfacilities-based resellers of leased lines. Even if some regulatory authorities want to insulate the national carrier from competition, the porousness of networks makes competition all but unavoidable. Sophisticated users have found ways to migrate some traffic streams away from high-cost, inflexible carriers and onto low-cost, software-defined, virtual networks created by low-cost innovative carriers or service providers.

Single facilities-based carriers, global strategic alliances of such carriers, and new private-line resellers seek to engineer a patchwork network of lines throughout the world or a region. The software used to configure switches can also manage ad hoc networks that can be reconfigured as a function of user requirements. The lines themselves have grown speedier in terms of bit rate (also known as *throughput*) and are able to handle larger traffic capacity. Digital transmission technologies like asynchronous transfer mode (ATM) provide a standardized vehicle to carry various kinds of high-volume traffic over information superhighways. ATM provides a more flexible, software-defined protocol for switching and routing even larger streams of voice, data, text, and video. It uses packet switching, the subdivision of a digital bit stream into uniform packets, or cells, that can be routed over different physical routes and reassembled in the proper order.

Digitization, software engineering, and new switching and routing technologies work to convert the telecommunications infrastructure from hardwired physical lines to ad hoc, software-engineered links. Visions of a global village seem more plausible when the telecommunications infrastructure begins to look more like the Internet, which has millions of users who can communicate using a protocol that makes millions of nodes and users accessible on a global basis.

5.4 The Internet

The Internet, because of its versatility and the manner in which operators charge for service and recover costs, presents the potential to change just about every prior assumption about telecommunications. First and foremost, consistent triple-digit annual growth in the Internet and other data applications now largely drive operators' financial, network development, and operational assumptions. Depending on which traffic analysis we examine, data volumes soon will exceed, if they have not already done so, voice traffic. A datacentric, Internet-driven telecommunications infrastructure requires different thinking than what prevailed when voice services predominated.

Data services, including Internet-mediated communications, work well in a packet-switched, digital environment. While voices can and will fit into that datacentric atmosphere, the previous paradigm involved analog, circuit-switching, that is, networks designed to set up dedicated pathways for voice conversations. Data can be transmitted on a software-defined, virtual-circuit basis, in which intelligent networks route traffic on a space-available basis shifting between usable links. The Internet provides global connectivity among thousands of different networks, thanks to the flexibility of data communications. Because data services can travel over diverse routes, any available network capacity can provide transport, thanks to the interconnection and integration of numerous discrete networks.[25]

---

25. For a comprehensive and helpful introduction to the Internet, see Adam Gaffin, EFF's Guide to the Internet, available at http://www.eff.org/pub/Net_info/ EFF_Net_Guide/netguide.eff (viewed April 23, 1999). A concise and straightforward jurist's summary is available at Reno v. American Civil Liberties Union, 521 U.S. 844, 117 S.Ct. 2329 (1997) available at

The baseline technical standards for the Internet promote global access and connectivity among thousands of different networks. One can think of the Internet as a network of networks in the sense that a digital bitstream can travel across any available network, not just one dedicated link owned or leased by the carrier providing service. Additionally, the packet-switched nature of the Internet and many data applications mean that a digital bitstream can divide into smaller portions and secure delivery via different routings. Circuit switching generally sets up one routing for the duration of a call.

5.4.1 Features of the Internet

The features of the Internet's routing flexibility are documented in technical standards known as the Transmission Control Protocol and the Internet Protocol (TCP/IP). Those protocols provide the foundation for coordinating the delivery of traffic via multiple networks, a process that requires an addressing scheme, and a fast, seamless, and user-friendly way for intelligent devices to look up and identify traffic originators and destinations. The TCP/IP standards provide a way for users to access any network resource for the Internet simply by keying in an address. IP offers users the opportunity to key in the letters of an electronic mail address or a World Wide Web site and have those letters translated behind the scenes into a sequence of numbers that represents a specific computer situated somewhere in the Internet cloud but accessible in seconds. TCP provides a traffic cop function for network operators to coordinate the receipt and delivery of traffic that will traverse many different networks.

The Internet accrues positive networking externalities,[26] because its value and utility to users grow as the number of points served increases. The

---

http://www2.epic.org/cda/cda_decision.html (viewed April 23, 1999), 844; see also the lower court's equally helpful summary at 929 F.Supp. 824, 64 USLW 2794 (E.D. Pa. 1996).

26. "[N]etwork externalities arise when increasing the number of users of a network benefits the existing network. The Internet, whose primary function is unfettered communication, is clearly an example of such a network, as are fax machines, telephones, modems, and the like. A 'low tech' example is the VHS VCR standard which defeated Beta and enabled video stores to invest exclusively in VHS tapes." Carole E. Handler and Julian Brew, "The Application of Antitrust Rules to Standards in the Information Industries—Anomaly or Necessity?" *Computer Lawyer*, Vol. 14, No. 11, Nov. 1997, p. 1; see also Mark A. Lemley and David McGowan, "Legal Implications of Network Economic

proliferation of Web sites enhances consumer welfare in terms of economic theory and personal experience. Internet users benefit when they have access to more content sources, points of communications, and potential e-mail addresses even if they visit only a tiny fraction of what is available. Absent network congestion and other impediments to accommodating growing demand, the value of the Web rises with the number of subscribers linked via the Internet. Heretofore, ISPs have conferred expanding benefits to consumers without using such enhancements as justification for raising rates. On the other hand, many telecommunications carriers historically have used expanding value accruing from broadened service opportunities to support a rate hike. For example, a domestic regulatory authority might require flat-rated, usage-insensitive local telephone service rates, but accept the carrier's view that value-of-service ratemaking justifies an automatic rate increase once the carrier reaches specific benchmarks (e.g., number of subscribers or the substitution of dedicated lines for shared, or party, lines).

Some Internet and telecommunications subscribers have grown to expect fixed, "all you can eat" (AYCE), usage-insensitive rates. Even though consumers pay on a usage-sensitive, metered basis for some services (e.g., long-distance telephone service), they nevertheless may expect to pay a monthly flat rate for Internet access.[27] Such users assume that the ISP incurs no additional expense in furnishing another minute of access. In view of all the electronic commerce and advertisements on the Web, Internet subscribers also may expect their initial monthly subscription rate for Internet service to remain the same for the foreseeable future. That expectation becomes increasingly unrealistic, because most ISPs have made substantial additional investments to support access by a larger number of subscribers to an increasingly diversified array of Internet services, some of which require comparatively larger, broadband capacity.

Effects," *California Law Review*, Vol. 86, 1998, p. 479; Teague I. Donahey, "Terminal Railroad Revisited: Using the Essential Facilities Doctrine to Ensure Accessibility to Internet Software Standards," *American Intellectual Property Law Association Quarterly Journal*, Vol. 25, Spring 1997, p. 277; Mark A. Lemley, "Antitrust and the Internet Standardization Problem," *Connecticut Law Review*, Vol. 28, 1996, pp. 1041, 1043–54 (examining the potential for positive network externalities, accrued by Internet standards); David S. Evans and Richard Schmalensee, "A Guide to the Antitrust Economics of Networks," *Antitrust*, Vol. 10, Spring 1996, p. 36.

27. See T. B. Fowler, "Internet Access and Pricing: Sorting Out the Options," *Telecommunications Policy*, Vol. 21, Feb. 1997, pp. 44–52.

For example, most Internet consumers previously used the Internet primarily for e-mail, a narrowband application that need not be transmitted and received instantaneously. Now many users have requirements for immediate, real-time delivery (streaming) of video and other broadband applications. In both the Internet and telecommunications worlds, service providers incur real costs when they add long-haul trunk lines to new servers and routers as well as short-haul local loops to end users.[28] In the Internet arena, however, the cost-recovery process might not ensure that the persons and companies responsible for such costs fully and fairly compensate the service provider.[29]

5.4.2 Impact of an Internet-Centric Environment

If and when the Internet becomes a primary medium for telecommunications, network architectures and operational expectations will have changed from fragmentation to consolidation. While digitization makes bitstreams appear interchangeable and fungible, not all bits are equal in value. The best-efforts routing characteristics of the Internet will not satisfy some users who would willingly pay for premium service with greater reliability, the kind of 99.99% availability telecommunications carriers typically offer.

---

28. "Considering that the end client is paying the local ISP for comprehensive Internet connectivity, when a client's packet is passed from one ISP to another at an interconnection point, where is the revenue for the packet? Is the [appropriate] revenue model one where the packet sender pays or one the packet receiver pays?" Geoff Huston, "Interconnection, Peering and Settlements," Sec. 4, *Settlement Models for the Internet*, available at http://www.telstra.net/gih/peerdocs/peer.html (viewed March 18, 1999). Huston gives this answer: "The retail base of the Internet is not an end-to-end tariff base. The sender of traffic does not fund the first-hop ISP for the total costs of carriage through the Internet to the traffic's destination, nor does the ultimate receiver pay the last-hop ISP for the costs. The ISP retail pricing structure reflects an implicit [rough justice] division of cost between the two parties, and there is no consequent structural requirement for interprovider financial balancing between the originating ISP and the terminating ISP" (Sec. 6).

29. "Congestion is also made worse by the absence of a satisfactory way for firms running the Internet to charge for carrying each other's traffic. These firms tend to use a simple barter system based on the roughest approximations of real cost, characteristic of an informal process developed years ago. But while only a few firms existed then, today there are thousands, with many grudgingly carrying more noncustomer traffic than customer traffic." Laurie Thomas Lee and Preeti Sharma, "The Internot? Understanding the Problem of Internet Congestion," *Journal of Media Economics*, Vol. 11, No. 1, 1998, pp. 13, 21.

An Internet-centric environment emphasizes the versatility of the Internet in terms of its ability to provide a medium for a wealth of different services and functions. But the Internet as we know it now will have to evolve and diversify, because a uniform, one-size-fits-all system cannot satisfy all particular user requirements. The Internet grows in importance because more users will resort to Internet-mediation for more services, including a variety of commercial applications. That will require Internet carriers and service providers to address and resolve a host of problems (e.g., quality of service, responding to consumer requirements, elasticity of demand-based pricing, customer care, network reliability, handling peak demand conditions) that perpetually have challenged telecommunications carriers.

We should think of an Internet-centric environment as one in which network planners use the TCP/IP standards to create a medium capable of providing voice, data, graphics, text, real-time audio and video, and diverse electronic commerce applications. This orientation requires customization and the ability to provide guaranteed quality-of-service routing instead of "best efforts." Future networks will become even more versatile and reliable in terms of the services they can support.

5.5 What the Internet Changes

The Internet's ascendancy highlights the importance of digital networking, data communications, and the convergence of telecommunications and information-processing markets and technologies. While it might not change everything, the Internet already has made a substantial impact on how we do business, receive news and entertainment, and communicate. As a network of networks, the Internet provides user-friendly interconnectivity for a variety of services and applications. Previously we could expect such connectivity only among networks offering a relatively narrow range of services and functions. For example, international direct distance dialing and the ITU established a global numbering system that enables one telephone caller to reach just about any other telephone, regardless of the type and vintage of equipment used to switch and route the call. The Internet provides such connectivity in an efficient digital mode for a variety of services and applications.

We have not yet reached a point where the networks that provide Internet services constitute the primary links for global telecommunications as well. That day will come, however, as digitization, packet switching, and

TCP/IP support a datacentric and Internet-centric infrastructure. With digitization, most services can be transmitted and processed in a digital, computerized format. The Internet can simultaneously switch, route, and transmit text, voice, audio, graphics, and video traffic. Of course, not all bits have the same value, time sensitivity, and quality of service requirements. While the Internet may become the generic telecommunications and information network, ironically ISPs will have to diversify their functionality and prices in response to consumer requirements. Currently, the Internet offers best-efforts routing with limited ability to differentiate quality and price of service. (Chapter 9 addresses Internet pricing and other issues relating to growing reliance on Internet-mediation.)

5.5.1 The Internet Comes of Age

At some point in the late 1990s, the Internet became the primary driver for the changes necessary to hasten the onset of an information-age economy. Those changes include investment by telecommunications carriers in more transmission capacity and by consumers in personal computers and other information appliances.[30] Even as blue-sky prognosticators continue to extrapolate and project future accomplishments, the current Internet actually has delivered on the mantra of faster, better, cheaper, smarter, and more convenient. It has reached what Andrew Grove terms the "strategic inflection point,"[31] or critical mass, for a number of reasons.

---

30. For background on the technology, business, and economics of the Internet, see generally Kevin Werbach, Federal Communications Commission, Office of Plans and Policies Series, No. 29, *Digital Tornado: The Internet and Telecommunications Policy* (Washington, D.C., March 1997), available at http://www.fcc.gov/Bureaus/OPP/working_papers/oppwp29pdf.html (viewed July 26, 1999); Thomas M. Siebel and Pat House, *Cyber Rules* (New York: Doubleday, 1999); Kevin Kelly, *New Rules for the New Economy: 10 Radical Strategies for a Connected World* (New York: Penguin USA, 1999); Carl Shapiro and Hal R. Varian, *Information Rules: A Strategic Guide to the Network Economy* (Cambridge, MA: Harvard Business School, 1998).

31. "A strategic inflection point is a break in a curve where the old way of doing business gives way to a new way of doing business, where the old concepts and the old constructs transition into new concepts and new constructs and, at this point, a business, a society—or, for that matter, individual—can make that adjustment to the new or continue to work harder and harder with greater determination in the old way. The first one will rise to new prominence, new prosperity; the second one will lead to decline." Dr. Andrew S. Grove, Plenary Speech at the *World Economic Forum*, Feb. 3, 1997, Davos, Swit-

In a nutshell, the Internet has developed into a major communications and commercial medium, because other long-touted concepts also became real and provided the foundation on which to build a thriving international network of networks.[32] The Internet could not become a vibrant and credible medium without the following factors:

- Proliferation of high bandwidth capacity to house, deliver, and route desirable content to a large and geographically diverse population;
- Technological innovations that promote the convergence of previously discrete media and services;
- Wise government decisions that encourage new technologies without regulating and managing Internet applications;
- Creative entrepreneurs, activists, and citizens of cyberspace who trigger innovations and "buzz," attracting more users at all levels.

5.5.2 Distance Insensitivity, the Rising Importance of Packet Switching, and IP

Telecommunications carriers and information service providers increasingly find that they can recover facilities investments on the basis of fixed rather than mileage-based rates. Distance insensitivity means that information service providers need not meter traffic, price service as a function of distance, or impose higher rates simply because traffic crosses national borders. Long usage times typical of Web surfing, compared to the short holding times for voice conversations, will force planners to retrofit and design networks that accommodate greater demand and new services. Moreover, voice telephony, with comparatively low bandwidth requirements, will no longer dominate and drive network design and engineering.

The telecommunications/information infrastructure must accommodate exploding new consumer demand for digital throughput. Accommodation must occur regardless of whether or when data will overtake voice in total traffic volume and bandwidth requirements. The development of networks that use TCP/IP is evidence that consumers and businesses already

zerland, available at http://www.andygrove.com/intel/people/asg/asg_davos.htm; see also http://www.intel.com/intel/paranoid/.

32. See Eli M. Noam, "Beyond Liberalization—From the Network of Networks to the System of Systems," *Telecommunications Policy*, Vol. 18, No. 4, 1994, pp. 286–294.

require flexible, broadband networks.[33] Today's voice-oriented PSTN does not always serve emerging consumer requirements because of the narrowband, analog local exchange loops that serve end users and the need for modems that take minutes to configure. Many forecasters expect major infrastructure upgrades to occur over the next 5 to 10 years. Cheaper bandwidth means that consumers can afford the broadband pipes needed to support multimedia applications made available via the Internet.[34]

Some carriers like Qwest and Level Three already have configured international networks to handle Internet traffic. If such new ventures have correctly predicted the future, an Internet-centric network will predominate. Carriers will convert conventional voice into packets and route them over diverse, space-available capacity using TCP to handle network management and IP to handle origination and destination addressing and numbering. While current quality-of-service problems exist in the Internet's best efforts at routing topology, business ventures are working to satisfy concerns about quality, reliability, security, and billing.

5.6 The Promise of E-Commerce

As the Internet matures and diversifies, it has become more commercial and private and less government sponsored and incubated. Commercialization of the Internet and the exit of government subsidization require Internet operators to find private revenue sources. Many of the Internet's favorite offerings finance attractive content primarily through advertisements, in lieu of subscriptions and direct payments from users. An increasingly important financial model involves electronic commerce, in which the ISP receives a share of revenues for serving as a medium for a commercial transaction or for funneling customers to another site where a transaction takes place.

33. See Organization for Economic Co-Operation and Development, Directorate for Science, Technology and Industry, Committee for Information, Computer and Communications Policy, Working Party on Telecommunications and Information Services Policies, *Internet Traffic Exchange: Developments and Policy*, DSTI/ICCP/TISP(98)1 (1998).

34. For a helpful annotated bibliography of materials addressing Internet economics issues, including interconnection and access pricing, see Bruce Klopfenstein, "Internet Economics: An Annotated Bibliography," *Journal of Media Economics*, Vol. 11, No. 1, 1998, pp. 33–48.

Like the Internet in general, e-commerce has created impressions of both too much change and not enough change. Because well over 50% of the consumer base, even in developed nations like the United States, do not have access to the Internet, one cannot say, at least for the time being, that e-commerce will change how we all do business. On the other hand, we can reasonably conclude that many businesses and consumers will seek to tap the benefits of cheap, fast, simple, and convenient access to a commercial network that defies time zones and operates all the time. "The growing popularity of this new media is driving development of new kinds of information-based products and services, and attracting business and consumers alike to the electronic marketplace" [1]. The investment banking firm of Morgan Stanley Dean Witter projects that by 2004 e-commerce will constitute 20% ($2 trillion) of the U.S. economy [2]. (Prudential Securities projects a slightly less steep global ramp-up: $55 billion in 1998, $330 billion in 2001–2002, and $1 trillion in 2003–2005 [3].)

E-commerce holds the promise for reducing cost, inconvenience, and other impediments to commercial transactions. Already, plenty of case studies provide empirical evidence that corroborates that point. By extrapolation, blue-sky visionaries predict Internet-mediation as promoting a friction-free economy, that is, one with few, if any, barriers and burdens to conducting business, presumably on a global basis. While the total elimination of transaction costs and economic drag may overstate the case, we can expect e-commerce to stimulate efficiency gains in terms of consumer access to information and markets, business access to consumers, and in general opportunities for interaction and open, unfettered commerce.

The positive consequences of e-commerce provide consumers and businesses with more and better resources. Those tools empower decision-makers by providing greater opportunities to acquire and analyze data that, in turn, should enhance the decision-making process. The Internet provides a multimedia forum for individual consumers to access resources that previously did not exist or that commanded prices only large businesses could afford. The Internet provides a global set of prospective consumers, thereby aggregating the demand necessary to support a service at the scale and scope necessary to maximize efficiency. That one-to-many function means that a service developer can make available an Internet-mediated feature at low cost per use in terms of both the cost to provide the service to a user and the user's cost of access. Consumers often have free access to many high-powered Internet functions that took millions of dollars to develop.

The one-to-many function also makes it possible to multicast the same content to a large number of simultaneous users.

5.7 Netheads and Nethead Companies: Major Forces

By itself, the Internet's ascendancy does nothing to diminish the importance of telecommunications links and the power of the incumbent carriers that operate them. However, the Internet success stories largely herald ventures that add value to the basic transport function provided by telecommunications. The creative and entrepreneurial drive for Internet innovation in large part did not originate in the telecommunications sector, but rather in the information technology and data-processing environment.

The term *Nethead* refers to an individual who works in the information technology world and who shares common values with others operating in that sector of the economy. Netheads make vastly different assumptions about the role of government, the nature of markets, and the role of businesses in society than do their telecommunications counterparts, referred to as *Phoneheads*. Netheads view any government involvement with suspicion and claim that much of the Internet's success results from a governmental, hands-off approach. Netheads thrive on a no-holds-barred, competitive environment and have great confidence in the marketplace's ability to reward innovation. Netheads rarely assume social responsibilities beyond the view that their innovations will make life better for everyone.

With the rise of Internet-mediation, Netheads have achieved substantial marketplace success and influence. Many leading Internet firms did not exist 10 years ago. Likewise, the highest stock appreciation and capitalization have accrued to information hardware, software, processing and e-commerce ventures, not telecommunications incumbents. Networking equipment manufacturers like Cisco and e-commerce players like Amazon.com, Priceline.com, and eBay have captured both the headlines and investors' fancy.

5.8 Basic Telecommunications: A Low-Margin, Commodity Business

Investors have rewarded ventures that add value to the telecommunications lines that link consumers with e-commerce and other Internet companies. With the onset of competition, margins for basic telecommunications transport have dropped, consumers have shown less brand loyalty, and the increasing use of fiber optic transmission makes one company's service technologically equivalent to that offered by others.

For the most part, telecommunications links have become fungible, that is, largely interchangeable with little perceivable difference among companies. Consumers will need more bandwidth to serve as the medium for delivering broadband applications like streaming video and wireless Internet access. The likely scenario for telecommunications players that do not embrace the Internet and offer new services still points to financial health achieved through increasing volume. Far greater opportunities, however, lie in adding services, features, and value to the telecommunication pipeline.

5.9 The Rush to Merge and Provide Global, One-Stop Shopping

Telecommunications companies, including incumbents, have to revise or devise new business plans. They must respond to convergence of technologies and markets, competition in a formerly cozy, insulated sector, and declining margins. At the risk of stretching across too many product and service lines, many telecommunications companies have recognized the need to diversify. Diversification attempts to accrue economies of scale and scope and reflects the ability of the Internet to mediate a variety of services. As such, we can consider diversification a defensive strategy that reflects the need to shore up revenue streams lost to competition. It also serves as a proactive strategy that aims to generate more revenues by becoming a one-stop shop for many services consumers previously secured from multiple vendors.

One-stop shopping may satisfy both consumer and service provider interests. Consumers may opt to rely exclusively on a single provider for a broader array of services if the package accrues financial savings and if the consumer has confidence that no greater risks result from such exclusivity. Service providers seek to expand their range of offerings, particularly when they perceive the need to become a more expansive presence on a consumer's radar screen, or lose visibility.

Companies often recognize that they lack the internal resources and skills to provide everything consumers may require. The spate of mergers and acquisitions in telecommunications reflects the decision to buy market share and expertise rather than build it internally. Mergers provide a faster way for ventures to pursue a market diversification and globalization strategy. For example, in the United States, AT&T has spent billions to acquire market share in wireless, Internet, and cable television markets. AT&T's one-stop shopping strategy seeks a triple-digit monthly bill for any and all of the following services: local and long-distance telephone service, cellular radio, Internet access, cable television, and premium programming like pay per view. In addition AT&T's investment in wireless and cable television ventures provides the company with an opportunity to reduce the access charge payments it must make to now unaffiliated local exchange carriers for originating and terminating wireless and long-distance traffic. Throughout the world, incumbent telecommunications carriers strive to find new revenue generators that will offset lost market share at home and maturing, core markets that no longer offer vigorous growth potential. (Chapter 10 explores the privatization opportunity in depth.)

5.10 E-Commerce: Shifting the Locus of Control to Consumers

As a ubiquitous global medium, the Internet provides consumers with new, powerful opportunities to deal more directly with product and service providers. *Disintermediation* is a complex-sounding word for a simple concept: the ability of consumers to eliminate intermediaries (middlemen) via the Internet. Closer, if not direct, links between end users and manufacturers largely eliminate a product chain in which distributors, wholesalers, and retailers add expense and possible delay. If an intermediary does not add value to a transaction, then a consumer increasingly has a bypass option via the Internet. For example, several Internet-mediated travel reservation search engines offer the capacity to find the lowest fares, sometime ones available only through the Internet. Air travelers might pursue the Internet option because it offers greater control over the transaction and all preliminary steps leading to the purchase of a ticket. Travelers also might pursue the Internet option to bypass the travel agency commercial arrangement, in which the amount of financial compensation flowing to the agent depends on the ticket cost, with the possibility of additional inducements for booking a particular carrier, hotel chain, and car rental company. After recouping the cost of establishing a World Wide Web presence and developing an on-line reservation system, airlines stand to save millions in reduced travel agent commissions and in the ability to process a reservation and ticket at lower cost than in rendering paper tickets and receiving payments through travel agents and ticket clearinghouses.

5.11 Revenge of the Phoneheads: What the Internet Does Not Change

With all the buzz and hype surrounding the Internet's ascendancy, one could easily conclude that a new world order will offer services that incumbent players somehow cannot or will not offer. Slogans like faster, better, smarter, cheaper, and more convenient describe the characteristics of what new, out-of-the-box thinking generates, a mind-set largely lacking in the regulation-numbed, competition-insulated world of incumbent telecommunications operators. The conventional wisdom largely relegates telephone companies to low-margin wireline and wireless transmission of bitstreams generated by others who add value and information processing features. Nethead information economy players apparently have the mental nimbleness and other skills to thrive, while their Phonehead counterparts have fallen from grace.

Pity the Nethead who underestimates the Phonehead's ability to game the system and slow the process of change sufficiently so that Phoneheads can adjust, adapt, and respond. Netheads offering the next best thing typically fail to appreciate fully the role of telecommunications in making or breaking their business plans. While the Internet changes a lot, it has yet to diminish powers held by enterprises providing the underlying bit transmission, particularly where traffic must traverse a different network, thereby requiring physical interconnection of facilities and often triggering a payment. While Phoneheads may bank too much on their continuing ability to maintain toll booths along the information superhighway, Netheads largely fail to appreciate that unless and until they vertically integrate throughout the service chain, they must rely on Phoneheads who are ready, willing, and able to exploit bottlenecks and choke points. Indeed, the Nethead managers of major Internet backbone providers recognize the power of gateway access control and have adopted some of the pricing and interconnection tactics of their telecommunications counterparts.

5.11.1 A Clash of Cultures

Phoneheads and Netheads do not understand each other, largely because they speak different languages, operate from fundamentally different assumptions about their businesses, and have generated substantially different cultures. Few in either camp realize that success depends on the acquisition of skills resident in both groups, for example, the Netheads' keen interest in competing and embracing change, coupled with the Phoneheads' ability to attend to details, manage complex networks, and effectively provide all the functions required by their business, including customer service, accurate billing, and maintaining high quality of service.

Because of the distrust between Phoneheads and Netheads, perhaps both camps should work harder to understand the other. Technological and marketplace convergence means that different types of people will work within a single company or have to reach consensus in the joint provisioning of a service.

5.11.2 The Phonehead Persona

For over 100 years, Phoneheads have dutifully worked within the box and inside the system to provide public utility–type services. By operating "businesses affecting the public interest," they executed a public compact: trading off pricing flexibility and accepting unprofitable service commitments in exchange for significant insulation from competition and regulatory process designed in part to ensure their ongoing financial health. No wonder Phonehead stocks have offered rock-solid investment opportunities for widows and orphans: these companies generate substantial, predicable cash flows and dividends. At the risk of resting on their laurels, Phoneheads could afford to move incrementally and manage change. Prior to the onset of technological innovations and new procompetitive regulatory policies, Phoneheads enjoyed the ability to manage change and to plan for the future at a leisurely pace.

As much as they might disparage regulation, Phoneheads actually benefited far more than they suffered. The government served as the guarantor of a stable revenue flow, even when insisting on long depreciation schedules and prescribing rates of return. Regulation pervades the Phonehead mind-set as a necessary evil and also as a mutually beneficial mechanism for both regulator and regulatee. Regulation offers a check against some of the most perverse marketplace forces, making it possible for Phoneheads to live in a safe, cautious, and unremarkable environment. Wall Street and investors historically have not rewarded such conduct with speculative valuations based on future performance and prospects. What Phoneheads offered in the past constitutes what they will offer in the future: continuity, predictability, and safety. In the process, Phoneheads have generated reliable and effective billing and customer service systems. Even without aggressive and creative marketing, incumbent telephone companies have managed to generate billions in revenues, with plenty of retained earnings to fund acquisitions required by the new world order and converging markets.

5.11.3 The Nethead Persona

Netheads operate "out of the box," given their nonconformist, libertarian bent. They deem anathema generally any form of government involvement, because Netheads operate on Internet time and have no patience for external forces that might cause delay. If Phoneheads thrive on continuity, Netheads find chaos stimulating and risk something to embrace, not strive to control. Netheads may risk reinventing the wheel or having to learn lessons the hard way. They could not tolerate the sleepy and measured life of most Phoneheads.

Netheads expect to grow rich quickly and dramatically through stock options. They have the energy to persevere and accept a high-risk, high-reward posture. Netheads take as a given the absence of barriers to market entry. They live the American dream by conceptualizing and building the better mousetrap, which in this age often attracts ample venture capital funding.

Bill Gates could serve as the archetypal Nethead: a flawed genius able to achieve incredible accomplishments and wealth while at the same time unwilling or unable to avoid legal and regulatory quagmires. A less doctrinal Nethead CEO might have found ways to avoid having to defend Microsoft from an antitrust suit brought by the U.S. Department of Justice and many state attorneys general. Yet the sense of purpose and relentless execution of a vision foreclosed any self-induced reticence or discretion. Netheads would rather shoot first and aim later. They are not afraid to fail, but they are surprised when failure occurs, because they truly believe they have created the next best thing.

5.11.4 Phoneheads and Their Control of First- and Last-Mile Access

The new world order heralds a time of robust competition in which consumers have sovereignty and the upper hand. While we are moving to that scenario, the fact of the matter remains that many businesses and just about all residential consumers still have but one primary telecommunications link to the rest of the world. A twisted pair of copper wires provides most residential subscribers with access to the Internet. In time, the incumbent telephone company will have upgraded the bandwidth of this plant at roughly the same time as cable television, electric utility, and possibly wireless options also become widespread. Until that time, incumbent telecommunications operators control a bottleneck through which most residential and business traffic originates and terminates.

Bottleneck control provides the last best shot for incumbent telecommunications carriers to thwart and delay the onset of robust competition. We should not underestimate the power of incumbents to slow the transition —as inevitable as it may be—from the old world order to the new world order.

5.11.5 Incumbency's Deep Pockets and Resources to Respond to Changed Circumstances

Even now, incumbent telecommunications players have the best chance for surviving, if not thriving, in the new world order. They simply have to establish the desire and the resolve to change throughout the corporation, not just in the office of a newly installed, Nethead-savvy CEO. Incumbent carriers typically have ample retained earnings to pursue strategic mergers, acquisitions, alliances, and new carrier privatization opportunities. Many will be content to limit their expansion to new telecommunications opportunities abroad. A few will more boldly embrace the Internet and the Nethead world, no matter how awkward it might feel. Set out in the next section is a tale of two giant incumbents that so far have achieved different degrees of success in this time of transition.

5.12 A Tale of Two Telcos

Two of Europe's incumbent telephone companies have responded to the imperatives of changed circumstances in similar ways, but with different results. Deutsche Telekom (DT), of Germany, and Telefonica (TEF), of Spain, face the challenges and opportunities of adapting to a vastly more complex and contested environment highlighted by the migration from government-conferred monopolies to a competitive marketplace. On the negative side, both companies must confront the reality of lost domestic market share as regulators pry open captive markets. They also must adjust to a marketplace that favors volume over margins and one in which size and one-stop shopping matters. On the positive side, those companies have the opportunity to diversify and take advantage of new market access opportunities abroad, as well as burgeoning growth in wireless and Internet markets.

5.12.1 Old Dogs, New Tricks

Although it may lack the size and staying power needed to operate in a global, convergent marketplace, Telefonica has executed an "affinity strategy," with investments throughout the Spanish-speaking world. DT has yet to fashion and execute a coherent strategy, despite having the resources to do so—a capitalization of $215 billion, which exceeds that of AT&T. However, in 2000 DT acquired two mid-sized U.S. wireless telephone companies, VoiceStream and Powertel.

Both incumbent carriers have the resources and the wherewithal to survive, if not thrive, under changed circumstances. But they have to make substantial changes in their corporate culture and strategy. Foremost, these companies have to get over the shock of changed circumstances. So many of the fundamental assumptions about their business have changed:

- Telephone companies have to make up in volume what they have lost in margin as telecommunications carriage becomes a commodity business.
- Customers have greater sovereignty than ever—they can vote with their feet and their currency.
- Borders and markets are porous—a unified Europe reduces barriers to market entry and forces greater symmetry of regulatory policies.
- Wireless and Internet markets offer the greatest growth opportunities, even at the risk of service cannibalization, for example, Internet-mediated telephone calls at a fraction of previous per-minute rates.
- Multinational alliances and investments require a global orientation and other skills heretofore not in great demand or in significant presence at incumbent telephone companies.

5.12.2 The New Game Plan

Before they lose substantial market share, a deep reservoir of retained earnings, and appreciated stock, incumbent telephone companies like DT and TEF have a number of strategic opportunities:

- Monetize noncore market holdings;
- Diversify globally;
- Embrace the Internet;
- Recognize the value in providing one-stop shopping.

5.12.2.1 Monetize Noncore Market Holdings

Like TeleCommunications, Inc. and AT&T, both DT and TEF have spun off portions of their wireless, Internet, cable television, and other noncore holdings. Those offerings generate ample funds for the carriers to make strategic investments in existing companies and to participate in the privatization and new franchise sweepstakes underway in many nations.

5.12.2.2 Diversify Globally

Lost market share at home requires companies like DT and TEF to acquire market share in developing markets, like wireless and e-commerce, and to make strategic investments in other companies. Neither company has successfully become part of a big globalization alliance. The Global One alliance of Sprint, France Telecom, and DT fell apart, as did a possible alliance between DT and Telecom Italia. Also, DT's bid for Qwest/US WEST came across as too little, too late and was rife with regulatory and litigation problems. Telefonica has chosen to work incrementally with significant, but not massive, investments throughout Latin America and South America, several wireless plays, and more than a minor stake in content creation. It failed to respond favorably to a more ambitious alliance proposal from MCI and could not reach closure with KPN of The Netherlands.

5.12.2.3 Embrace the Internet

Remarkably few incumbent telcos recognize the importance of the Internet to their future. As domestic wireline revenues stagnate or decline, the Internet becomes a key vehicle for new revenues. The Yahoo! portal strategy, mimicked by TEF's Terra Networks, provides a way to become the preferred cyberspace/e-commerce entry point. Additionally, Telefonica acquired a majority interest in the Lycos search engine and portal site.

5.12.2.4 Recognize the Value in Providing One-Stop Shopping

If telecommunications transport becomes a commodity business, then adding value to bit transfers will become an increasingly important strategy. The successful incumbent recognizes its limitations and has no reluctance to buy expertise and market share through acquisitions. DT serves some credit for recognizing the importance of Internet-mediated telephone services through an investment in VocalTec, a leader in Internet telephony software. Likewise, TEF has established e-commerce alliances with a variety of banks and travel industry players, including Banco Bilba Vizceya of Argentina.

5.12.3 Coming Down from the Summit

In the last few years, investors have rewarded both DT and TEF, making those companies look more like growth stocks and less like the utility value stocks they were in a monopoly environment. Both stocks have evidenced greater volatility than typical public utility companies as investors recognize the challenges posed by competition, globalization, and diversification. Perhaps investors wonder whether DT and TEF have what it takes to survive in a marketplace possibly dominated by a half-dozen multinational and multimedia ventures. One would think DT would have a strong advantage compared to TEF and other less capitalized companies. However, it appears that so far TEF has executed better, with its investment in telephone companies and ISPs operating in Spanish-speaking nations.

5.12.4 The Way Forward

The major question confronting both DT and TEF is whether they have what it takes to thrive in the future. No one enterprise by itself can grow, diversify, and exploit future opportunities. AT&T has spent billions acquiring market share and expertise in cable television and wireless services. Failing to execute an effective one-stop-shop strategy, AT&T will divest itself into four separate ventures—wireless, cable, business, and retail telephone services. NTT in Japan is struggling to find and execute an international strategy, even as KDD, the incumbent international carrier, is executing a domestic wireless strategy. DT must establish and execute a strategy that reaches a fairly large critical mass with a global or strong regional footprint like that achieved by TEF.

It would not come as a surprise to see DT spend billions to acquire a major U.S. local exchange carrier, one or more interexchange carriers, or several ISPs in addition to its global wireless acquisitions. Both DT and TEF have to allocate more attention and resources abroad, even as they have to contend with nimble competitors at home who have acquired significant local market share.

5.13 Greater Opportunities for Consumers to Save Money—and to Lose It

Technological innovations empower consumers. The Internet provides unprecedented opportunities for individuals to negotiate almost at parity with large corporations and to secure the terms and conditions traditionally available only to primary customers. Technological innovations also provide consumers with choices and opportunities to secure cheaper or more customized services.

But technological innovations have a down side as well. The wide reach of the Internet provides unprecedented opportunities for consumers to suffer real and immediate financial and personal harm. Internet-mediated commerce means that some percentage of transactions will involve deception and fraud. Consumers may not have all the same legal resources available to secure an adequate remedy, particularly since the perpetrator may operate outside the jurisdiction of a local court.

In the telecommunications world, consumers may have greater choices and opportunities to save money, but such choices can cost more than expected. The uninformed consumer may be lured to incredibly attractive per-minute long-distance calling rates only to learn later that hidden or unemphasized additional charges raise the effective per-minute rate. Even a straightforward pitch offering 20 minutes of long-distance calling for $1.00 requires close examination—the rate applies even if the caller reaches an answering machine and terminates the call well before 20 minutes have elapsed.

Consumers may need regulatory safeguards at the very time many nations embrace a deregulatory philosophy. In streamlining and removing regulatory underbrush, legislators and regulators should consider the greater need for consumer protection.

5.14 The Future

Marketing hype and blue-sky visions aside, a bright future does appear for new services and technologies that blend aspects of information processing and telecommunications.[35] The merger of technologies and markets results when creative engineers and entrepreneurs think of new ways to use equipment like the telephone and the personal computer and facilities like the twisted-wire pair (local loop) and cable television coaxial cable. The future promises to make it possible for television sets and computer terminals to provide a more diverse array of services and applications than what we now

35. See U.S. Congress, Office of Technology Assessment, *Electronic Enterprises: Looking to the Future*, OTA-TCT-600 (Washington, D.C.: U.S. Government Printing Office, May 1994).

expect from them. An information appliance may integrate the functions of both television sets and computer terminals. Likewise, service providers will identify and serve new markets with telephone companies entering entertainment markets like cable television and cable television companies entering telecommunications markets.

No-holds-barred competition contrasts with an international telecommunications industrial structure that until quite recently emphasized monopolies, discrete markets, and rules designed to maintain the status quo. Depending on one's perspective, the status quo supported worthwhile public policies like universal service at affordable (and often subsidized) rates or a sleepy, insular, and coddled world of industries administered by government bureaucrats who provided centralized mismanagement. Even in the United States, less than 20 years ago, FCC policy prevented AT&T from providing text services and the now-defunct international record carriers like ITT World Communications from providing voice services. Well after technological innovations made it feasible for the same line and the same carrier to provide both text and voice services, the FCC sought to render the markets mutually exclusive, ostensibly to prevent AT&T from dominating the telegram and telex business, as it had done in voice markets.

Governments throughout the world have attempted to insulate their national carriers from competition, but technological innovations make such protection less likely. The merger of telecommunications and information-processing technologies means that unregulated data-processing companies and systems integrators will become involved in the provisioning, or at least ordering for customers, of leased telecommunications lines. A volatile clash of market philosophies occurs between such relative newcomers to telecommunications as Global Crossing, 360 Networks, Viatel, and CarrierOne and government-predominated and monopoly-oriented incumbents.

The stakes have grown in telecommunications because the sector now constitutes an integral part of information-based economies and the perceived locomotive for jump-starting national economies.[36] Companies like EDS have entered the telecommunications arena to provide users with one-stop shopping access to efficient and versatile global networks maintained by professionals, enabling users to concentrate on their particular line of business that happens to require cutting-edge telecommunications capabilities. Global markets operating 24 hours a day have no tolerance for a 9-to-5

36. See S. Pitroda, "Development, Democracy, and the Village Telephone," *Harvard Business Review,* Nov.–Dec. 1993, pp. 66–79.

service mentality. Additionally, technological innovations make it possible for EDS to apply its systems-integration skills for a larger set of customers who rely on telecommunications. Little if any exclusivity remains among information, entertainment, data processing, financial services, message transport, catalog sales, movies, video games, massaging, and software delivery. Put another way, there remain few stand-alone markets and product or service distribution channels. A business executive can send a hard-copy facsimile via a machine connected to a telephone line, or a soft-copy resident in a computer can be transmitted via the same line. One can view a movie on a pay-per-view basis via satellite, cable television, wireless cable television, or telephone company delivery.

Versatility in communicating means that more options will exist and more players will have opportunities to seek a market niche. Barriers to market entry will fall as a function of declining facility costs and robust demand generated by multinational enterprises and by small businesses and individuals who, for the first time, can qualify as potential consumers at home or at work with relatively minor investments. The proliferation of cable television settop converters, personal computers, modems, widescreen televisions, and satellite dishes attests to consumer interest in new methods for engaging in commerce, communications, and social interaction. The technological innovations that foster merged or newly competitive industries also provide new opportunities for different media to serve different human senses. Multimedia telecommunications blend sight, sound, and human perception into a virtual reality that accentuates the enjoyment of video games, but also provides a more realistic simulation environment to test designs (e.g., in manufacturing) and also to test human competency (e.g., jet-piloting skills).

Increased interactivity allows people to control their computer-mediated environment. It remains to be seen whether governments can—or should try to—control the telecommunications environment. They surely cannot achieve a measured and centrally managed introduction of new technologies and services.

References

[1] Organization for Economic Co-Operation and Development, Directorate for Science, Technology and Industry, Committee on Consumer Policy, *Consumer Protection in the Electronic Marketplace*, DSTI/CP(98)13/Rev2, Sept. 21, 1998.

[2] Morgan Stanley Dean Witter, *The Internet Data Services Report*, Aug. 11, 1999, p. 2.

[3] Gabriel, C. A., Jr., and J. P. Lucier, Jr., "Net Assessment," Prudential Securities, April 26, 1999.

6

Players in International Telecommunications Policy Making

International telecommunications require coordination, consensus building, and compromise among stakeholders. Because resources like radio spectrum and satellite orbital parking places constitute shared global resources, national governments participate in multilateral decision making, some of which results in binding treaty-level commitments. Common rules of the road can promote efficiency, reduce interference, and support ease in interconnection of diverse networks.[1] Governments generally believe that uniformity, in terms of technical standards, spectrum allocations, and operational rules, reduces misunderstanding, streamlines equipment production lines, supports universal access to essential services, and fosters market growth. Nations achieve a shared view on international telecommunications issues based on the pragmatic assessment that relinquishing a degree of national sovereignty will accrue ample dividends in terms of enhanced consumer welfare and speedy deployment of equipment and services.

1. For background on the law, regulation, technology, and policies affecting international satellites, see Heather E. Hudson, *Communications Satellites: Their Development and Impact* (New York: Free Press, 1990); Elke A. Hoffmann, "The World Telecommunications Policy Forum: Globalization, Liberalization, and Privatization in the Provision of Satellite Services," *Law and Policy in International Business*, Vol. 28, Spring 1997, p. 929.

Despite the logistical and financial benefits accruing from consensus building, the process has never proceeded smoothly. Some nations already may have allocated spectrum and issued licenses for uses inconsistent with the consensus allocation subsequently made under the auspices of the ITU. For information-processing market segments operating on "Internet time," the slow process of consensus building at the ITU or other multilateral forums may be overtaken by unilateral national and corporate decisions regarding technical and operating standards. Because spectrum allocations and standard-setting decisions can make or break a business plan, stakeholders experience great ambivalence over a process both essential and frustrating.

The ITU faces extraordinary challenges to its viability as the key global policy maker, standard setter, and spectrum allocator. Some stakeholders would rather take their chances on marketplace forces and consumer choice to determine winners and losers rather than rely on the ITU. The ITU standard-setting process either results in an "all or nothing" outcome or generates a compromise that may blunt the "first mover," the first-to-market leadership a venture may have assumed it had. Under the ITU process, an organization wins with endorsement of a proposed spectrum allocation or technical standard, partially loses by having to compromise and reengineer, or completely loses when the ITU rejects a proposed standard or spectrum allocation. Given such high stakes, national industrial policies and politics may prevent the ITU from making rational decisions that optimize consumer welfare by adopting the best technologies and most efficient use of scarce resources.

A rapidly changing telecommunications environment and a convergence of markets and technologies also challenge the ITU's ability, as a multinational and bureaucratic organization, to operate swiftly and efficiently. As carriers and service providers globalize and become more competitive, individual companies and national governments perceive increasingly attractive incentives to enter markets first and establish rules, regulations, and standards later, if ever. The accelerating pace of technological innovations and the merger of telecommunications and information-processing technologies increase the types of constituencies the ITU must serve and accommodate. Participants in information-processing markets more frequently operate in a marketplace environment in which enterprises promote incompatible systems operating on different standards (e.g., Microsoft's Windows versus Linux and Apple's operating system) rather than the ITU model, which contemplates cooperation, even among competitors, to establish a single standard (e.g., a uniform telephone number dialing system, compatibility among the modems of different manufacturers, and standards for real-time delivery of voice and audio packets via the Internet).

6.1 Why Nations Cooperate on Telecommunications Policy Matters

Nations cooperate based on enlightened self-interest: a pragmatic assessment that common rules of the road reduce costs, stimulate efficiency, and ensure interference-free use of spectrum and satellite orbital slots. Simply put, nations cooperate based on a cost-benefit analysis: Nations will cooperate and participate in a multilateral forum when they conclude that they have more to gain by agreeing to a consensus than what may be lost in terms of preferential market share and earnings potential for national ventures. When nations act unilaterally, they expect the innovations and technical standards of national manufacturers and carriers to dominate, at least in-country and perhaps internationally as a de facto, market-driven standard. When nations go along with an international consensus, they either support widespread sharing of the financial and logistical benefits from single rules and standards, or they lack confidence that their domestic companies could dominate in a market-driven process.

Rather than unilaterally establish standards, frequency allocations, and satellite orbital slot assignments, nations typically collaborate bilaterally and multilaterally. They do so out of both enlightened self-interest and the desire to be good global citizens. Self-interest recognizes that a single or reduced number of standards may promote scale economies in production of equipment and reduced transaction costs, for example, less inconvenience, delay, and expense in linking national networks. The desire to be good global citizens stems from the recognition that many of the resources involved, like spectrum and the orbital arc for satellites, are shared, with no ownership rights flowing to any single nation.

Nations collaborate on matters as diverse as these:

- Reduction of actual or potential spectrum interference; frequencies;
- Standards that, if globally applied, would reduce the number of product lines and improve the potential for manufacturers to achieve production scale economies;
- Universal rules of the road that facilitate interconnection of networks and reduce the potential for "balkanization," that is, rendering separate networks inaccessible, because they operate on

incompatible formats, follow different rules, or comply with different standards;

- Integration of technological innovations and procedures that make telecommunications more efficient and user friendly.

Individual nations risk spoiling the prospects for universality when they act on the temptation to deviate from the consensus. The decision to opt out typically results when a single nation believes its national hero manufacturer or telephone company can capture a larger share of financial rewards by establishing a global standard, frequency allocation, or satellite orbital slot claim outside multilateral forums like the ITU. While intellectual property rights like patents and copyrights were designed to reward innovation, uniform rules of the road allow for multiple manufacturers and carriers while also reducing costs, inconvenience, and misunderstanding.

Other national decisions not to embrace an international standard may result from political or industrial strategies. For example, a nation might enhance the geographic reach and impact of its radio stations by operating at higher-than-agreed-to power levels, or on unauthorized frequencies. A similar outcome would result if a nation permitted launch of a satellite into an unregistered orbital slot. The nation may achieve bolstered coverage[2] or attempt to register phantom satellites, so it could resell the orbital slot or delay the onset of competition. It also may improve opportunities to proselytize or, defensively, to prevent duly registered operators from transmitting a signal receivable by its citizens.

A nation also may attempt to insulate domestic markets from foreign competition by mandating a standard inconsistent with the international consensus. That strategy may generate short-term benefits by closing procurements or ensuring that national heroes win "public" and "open" requests for proposals. In the long term, however, such insulation from competition and deviation from international standards may handicap the national hero in competitive tenders outside the home country. Also, such policies may comparatively disadvantage domestic companies, because they may not have access to the latest and most efficient technology and accordingly may not operate the most economically.

---

2. The FCC's International Bureau maintains a collection of many international agreements with Canada and Mexico at http://www.fcc.gov/ib/pnd/agree/welcome.html.

In other instances, nations want to support international consensus building, but enterprises chafe at the pace of standard setting and policy making. The ITU process may appear to reach an untimely "consensus by exhaustion," despite speedy technological evolution and marketplace changes. Other concerns include the potential for the process to result in a least common denominator or the involvement of governments and industrial policy, rather than the marketplace to set rules and standards.

Another factor supports consensus building in telecommunications policy: the composition and culture of delegations to the ITU. Until the convergence of telecommunications and information-processing technologies became more pronounced, ITU delegations had a relatively cohesive composition. Government representatives, versed in public diplomacy, joined a few international telecommunications carriers, led by the major incumbent carriers and equipment manufacturers. With convergence, both the range of issues requiring consideration and the composition of ITU delegations expanded.

6.2 Why Nations Do Not Readily Cooperate on Information-Processing Policy Matters

Global consensus building does not readily occur on information policy, including Internet governance and nontechnological matters. The Nethead culture favors marketplace solutions and rejects just about any government involvement, particularly involvement that would have a profound impact on the market prospects for individual companies and technologies. Netheads lack tolerance for the time it takes to reach consensus and a full appreciation of the benefits that can accrue.

The standard-setting process for 56,000-bps modems provides an instructive case study. Part of the ITU's standard-setting process includes the promulgation of recommendations regarding the basic technical parameters for modems. Because modems interconnect with the PSTN, the ITU provides a helpful forum to ensure modem compatibility and minimize the potential for harm to individuals and the telecommunications network when consumers use modems.

The ITU has performed a standard-setting role for every generation of modems, typically setting a standard after which many manufacturers introduced devices conforming to the next-generation standard. However, manufacturers of 56,000-bps modems displayed much greater reluctance to await an ITU standard. Worse yet, two major companies, Rockwell and U.S.

Robotics, later acquired by 3Com, had developed incompatible modems. Rather than work within the ITU standard-setting process to develop a standard compatible to both modem types, Rockwell and U.S. Robotics aggressively courted computer manufacturers, ISPs, and retailers with an eye toward dominating the modem marketplace and setting a de facto standard.

An unexpected problem beset both manufacturers and the other companies that secured licenses to build modems based on one or the other unrecognized standard. Consumers largely balked at having to pick one uncertified and unendorsed modem standard. At least some consumers remembered the fallout from having opted for the Betamax video tape recorder/player standard only to find the marketplace over time favoring the VHS format. Other than early adopters and other users keen on the better performance of next-generation modems regardless of the risk for premature obsolescence, consumers and at least some manufacturers typically await the development of a de facto, market-driven standard or one driven by ITU consensus. Once the ITU developed the V.90 modem standard, which ensured compatibility between the two major modem designs, sales picked up dramatically.

The 56,000-bps modem case study demonstrates the ongoing importance of standard setting, notwithstanding the increasing reluctance of stakeholders to accept the delay and potential diminished market prospects built into the process. While a Nethead may thrive in the chaotic, no-holds-barred world of unfettered marketplace competition, some restraint and coordination, like those occurring in telecommunications, accrue ample dividends.

6.3 A Brief History of the ITU

Despite growing disincentives and the proliferation of regional policy and standard-setting bodies, nations still rely primarily on the ITU to establish common rules, regulations, standards, and policies in telecommunications. Founded in 1865, the ITU serves as the world's oldest continually operating international policy making forums.[3] It continues to function because

3. For a general description of the ITU prior to its major reorganization in 1993, see G. Codding, Jr., and D. Gallegos, "The ITU's 'Federal' Structure," *Telecommunications Policy,* Vol. 15, No. 5, Aug. 1991, p. 353. For coverage of recent issues affecting the ITU's ability to shape policy, see Captain Roscoe M. Moore III, "Business-Driven Negotiations for Satellite System Coordination: Reforming the International Telecommunication Union to Increase Commercially Oriented Negotiations over Scarce Frequency Spectrum,"

nations recognize the need for conflict management and resolution in telecommunications.[4]

Uniform rules of the road make it possible, for example, to reach any telephone throughout the world, for facsimile machines and other devices operating over different networks to work, and for transmitting facilities usually to operate without harmful interference.

The ITU evolved when the Austro-German and Western European Telegraph Unions merged in the late 1800s to supervise and establish standards for an interconnected, regional network. Operators recognized the need to agree on frequencies and operating rules. Agreement on frequencies meant that international telegrams could be transmitted over radio spectrum in addition to closed-circuit cables. Consensus on operational rules meant, for example, that telegraph companies could designate a particular emergency frequency for 24-hour monitoring, and use a single code (i.e., Morse code) and common shorthand letter sequences to stand for frequently used sets of words (e.g., "QRZ" for the query "Are you receiving my transmission?"). Such shared rules and recommendations provided measurable benefits, for example, a reduction in the number of fatalities resulting from delays in mobilizing for emergency and disaster assistance.

The motivation for collaboration expanded when radio-based services proliferated in the early 1900s. In 1903, a group of nations with maritime interests met to coordinate implementation of radio systems. Twenty-nine nations agreed in 1906 to form the International Radiotelegraph Union (IRU) to coordinate usage, agree on common frequency bands, register station operations, and work to avoid or resolve cases of radio interference. National governments perceived the need to participate, because privately operated systems had sought to create equipment and service monopolies by refusing to allow communications with users of other equipment, even in an emergency. It has been alleged that a ship was in the vicinity of the *Titanic* as it struck an iceberg in 1912, but that the radio operator had gone off duty.

Journal of Air Law and Commerce, Vol. 65, Winter 1999, p. 51; Henry Wong, "The Paper 'Satellite' Chase: The ITU Prepares for Its Final Exam in Resolution," *Journal of Air Law and Commerce*, Vol. 63, May–June 1998, p. 849; Krishna Jayakar, "Globalization and the Legitimacy of International Telecommunications Standard-Setting Organizations," *Indiana Journal of Global Legal Studies*, Vol. 5, Spring 1998, p. 711; Jannat C. Thompson, "Space for Rent: The International Telecommunications Union, Space Law, and Orbit/Spectrum Leasing," *Journal of Air Law and Commerce*, Vol. 62, Aug.–Sept. 1996, p. 279.

4. For an overview of the ITU, see http://www.itu.int.

The IRU created a treaty-level document and began to establish a formal collection of radio regulations, such as requiring radio operators on ships exceeding a certain weight to monitor a designated emergency frequency all the time.

6.3.1 The ITU's Mission

The ITU conference held at Atlantic City in 1947 substantially expanded the Union's mission as nations sought to rebuild and as post–World War II alliances formed. The ITU voted to become a specialized agency in the United Nations system and to adopt its structures and procedures. The ITU's mission grew to encompass several activities:

- Support cooperation among nations for the improvements and rational allocation and use of spectrum;
- Promote development of technical facilities;
- Coordinate efforts to eliminate harmful interference;
- Facilitate worldwide standards;
- Foster international cooperation in the delivery of technical assistance;
- Promote adoption of measures for ensuring safety of life;
- Undertake studies, make regulations, adopt resolutions, formulate recommendations and options, and collect and publish information concerning telecommunications matters.

The ITU emphasizes the right of each member nation to participate in studying issues and expressing views at various world or regional conferences.[5] Nations ratify the final acts of conferences, and most subsequently codify in domestic laws and regulations the spectrum allocations, rules, and regulations established at ITU conferences.[6] The future viability of new

5. For an outline of United States preparation for the *2000 World Radiocommunication Conference*, see, for example, "The FCC's Advisory Committee for the 2000 World Radiocommunication Conference Offers Additional Draft Proposals on WRC-2000 Issues," Public Notice, DA 99-1364, 1999 W.L. 496688 (F.C.C.) (rel. July 14, 1999).

6. For an example of how the United States implements international spectrum allocation decisions, see "Redesignation of the 17.7–19.7 GHz Frequency Band, Blanket Licensing of Satellite Earth Stations in the 17.7–20.2 GHz and 27.5–30.0 GHz Frequency Bands, and the Allocation of Additional Spectrum in the 17.3–17.8 GHz and 24.75–25.25 GHz Frequency Bands for Broadcast Satellite-Service Use," IB Docket No. 98-172, Re-

services and spectrum allocations depends in large part on the willingness of the ITU's community of nations to reach consensus decisions. Domestic regulatory agencies may await a global decision or take unilateral action to expedite availability of new services and technologies and perhaps also to affect future global decision-making.[7]

6.3.2 Structure and Function of the ITU

The ITU structure can be divided into permanent, plenary, and ad hoc elements. Supreme authority lies with the more than 185 member nations, who agree to comply with promulgated rules and regulations and who financially underwrite ITU operations.[8] The infrequently convened Plenipotentiary Conference (Plenipot)[9] reviews the ITU's basic documents, the convention, which is subject to revision at each Plenipot, and the constitution, a permanent document infrequently revised.[10]

At the Plenipot, the nations, holding one vote each, establish budgets for future conferences and for the ITU's permanent staff, based primarily in

---

port and Order, FCC 00-212, 2000 WL 796768 (F.C.C.) (rel. June 22, 2000) (allocating 400 MHz of spectrum at 17.3–17.7 GHz for primary broadcast satellite service uses, effective in 2007, and also allocating additional spectrum for fixed satellite feeder links in the 24.75–25.05 GHz band).

7. See, for example, "Amendment of the Commission's Rules with Regard to the Establishment and Regulation of New Digital Audio Radio Services," GEN Docket No. 90-357, Report and Order, 10 FCC Rcd. 2310 (1995); see also "Digital Audio Broadcasting Systems and Their Impact on the Terrestrial Radio Broadcast Service," MM Docket No. 99-325, Notice of Proposed Rule Making, FCC 99-327, 15 FCC Rcd. 1722 (1999).

8. See Constitution of the International Telecommunication Union, CS/Art. 6-34, "Execution of the Instruments of the Union," in *Final Acts of the Plenipotentiary Conference*, Minneapolis, 1999 (Geneva: ITU, 2000), p. 7: "The Members are bound to abide by the provisions of this Constitution, the Convention and the Administrative Regulations in all telecommunication offices and stations established or operated by them."

9. See J. Savage, "The High-Level Committee and the ITU in the 21st Century," *Telecommunications Policy*, Vol. 15, Aug. 1991, p. 365.

10. See *Final Acts of the Plenipotentiary Conference, Constitution and Convention of the International Telecommunications Union*, Optional Protocol, Decisions, Resolutions, Recommendations and Opinions, Minneapolis, 1999 (Geneva: ITU, 2000). The constitution contains the basic provisions and purposes of the ITU. The convention complements the constitution and addresses more functional provisions relating to the operation of the ITU and its conferences.

Geneva. The Plenipot also elects the ITU's secretary general and other officials. It selects representatives from 46 geographically diverse nations to participate in the ITU Council, which performs executive board functions. The Plenipot also schedules the many specific conferences that modify rules, regulations, standards, protocols, and recommendations. Conferences can address one of the ITU's three major geographic regions or have global application.[11] The meetings have addressed such diverse issues as mobile radio, use of the orbital arc, telecommunications development, high- and middle-frequency radio, satellite frequencies, and the terms and conditions for provision of information services. Reforms adopted in 1992 created a more routine and frequent meeting schedule so that conferences will not have either a single issue on which to focus or too many matters to address.

The Council implements ITU policies and regulations, oversees senior management, and establishes questions and issues to be considered at regularly scheduled conferences by the ITU's three sectors:

- *Development:* "[T]o encourage international cooperation with a view toward harmonizing and enhancing the development of telecommunication services and facilities" [1, p. 3];
- *Telecommunications standardization:* "[T]o study technical, operating and tariff questions and to issue recommendations on them with a view to standardizing telecommunications on a world-wide basis" [1, p. 4];
- *Radiocommunication:* To achieve "efficient management of the radio-frequency spectrum in terrestrial and space radiocommunications ... [including] examining and registering all notices for frequency assignments liable to cause interference outside the territory of the country in which the station in located ... [and] all notices for orbital positions of ... satellites" [1, p. 5].

As part of reforms adopted by the ITU in 1992 to reflect changes in the telecommunications environment, the three sectors have established advisory boards to promote participation by nongovernmental players. The ITU already had authorized participation by recognized private operating

11. For an analysis and criticism of U.S. participation in ITU conferences, see U.S. Congress, Office of Technology Assessment, "The 1992 World Administrative Radio Conference Issues for U.S. International Spectrum Policy—Background Paper," OTA-BP-TCT-76 (Washington, D.C., 1991).

agencies[12]: nongovernmental carriers and service providers that provide international telecommunications capable of causing harmful interference (e.g., AT&T). The ITU additionally has authorized participation by scientific and industrial organizations[13] and nongovernmental organizations that study telecommunications problems or design and manufacture equipment (e.g., IBM). But the creation of advisory boards marked a commitment to solicit participation by former outsiders representing a wide cross-section of interests and expertise.

6.4 Spectrum Management by the ITU

A key function that the ITU performs is serving as a global traffic cop of the airwaves. This role involves defining services and functions that use spectrum, allocating spectrum for particular services and functions on an exclusive or shared basis, and providing a forum to coordinate and register spectrum use. World and regional radio assemblies meet on an increasingly frequent and regularly scheduled basis primarily to address issues involving spectrum definitions, allocations, and procedures for coordinating multiple users and services. The Radiocommunication Bureau (BR) performs most ongoing spectrum registration, coordination, and conflict prevention and resolution functions. It maintains a master list that theoretically contains all nations' actual and proposed frequency and orbital arc uses, exclusive of military and national security applications.

Before recording a new use, the BR reviews it for compliance with the ITU constitution, convention, and applicable rules, regulations, and frequency allocations on a regional or worldwide basis. It also assesses the potential for interference with other registered uses (in operation or planned)

---

12. A recognized private operating agency operates a telecommunications or broadcasting service and complies with the obligations imposed by Article 6 of the ITU's constitution, that is, the treaty-level agreement to be bound by the ITU's constitution, convention, and administrative regulations. See ITU Constitution Annex, "Definition of Certain Terms Used in This Constitution, the Convention and the Administrative Regulations of the International Telecommunication Union," CS/An 1008 in *Final Acts of the Plenipotentiary Conference* (Minneapolis, 1999).

13. Scientific or industrial organizations are nongovernmental agencies "engaged in the study of telecommunication problems or in the design or manufacture of equipment intended for telecommunication services," CS/An 1009 in *Final Acts of the Plenipotentiary Conference* (Minneapolis, 1999).

and issues an advance publication of the proposed new use. The issuance of the advance publication document triggers a time period within which nations may report potential interference and express their desire to participate in future meetings convened to resolve such problems. Upon successful conclusion of this coordination process, the BR officially registers the use and notifies all member nations. A reconstituted, part-time Radio Regulations Board provides conflict resolution services, which used to be provided on a full-time basis by its predecessor, the International Frequency Registration Board (IFRB).[14]

Reforms adopted by an additional Plenipot (APP) in 1992 merged all spectrum planning and management functions into a single radiocommunication sector. Those functions include the spectrum management activities of the IFRB and the radio regulation drafting and study functions of the International Radio Consultative Committee and the various world and regional administrative radio conferences. World radiocommunication conferences convene every two years, along with a radiocommunication assembly that performs a plenary function, including the provision of technical support for the conferences and determination of priorities for study groups, small issue-specific assemblies of government and industry representatives.

APP consideration of IFRB reforms proved to be the most controversial, because the board was reformulated to become a part-time body comprising non-ITU employees. In view of increasing congestion in some frequency bands, the launch of more satellites vying for geostationary orbital arcs and the need to oversee coordination between geostationary and LEO satellites, some nations expressed concern that a part-time body could not provide adequate and timely consideration of pressing matters.

The Radio Regulations Board "will approve the Rules of procedure which are used in the application of the Radio Regulations to register frequency assignments, will consider any matter which cannot be resolved through the application of the Rules of procedure and will perform any duties related to the assignment and utilization of frequencies and to the equitable utilization of the geostationary-satellite orbit" [1, p. 6]. The board holds up to four meetings a year, meaning that much of the day-to-day operations are performed by the director of the Radiocommunication Bureau, who investigates claims of harmful interference and formulates recommendations for their resolution.

---

14. As part of the ITU's streamlining efforts, the IFRB was reconstituted to become a part-time forum comprising nine members instead of a full-time operating component of the ITU with five full-time members.

Domestic spectrum management and policy making cannot operate outside the context of a parallel and interdependent international process:

> Domestic allocations ... generally conform to the international Table of Allocations and the Radio Regulations maintained by the ITU and revised at the W[orld] A[dministrative] R[adio] C[onference]s. Those international allocations and regulations, in turn, are the product of negotiation among many countries, each pursuing national goals. [2, p. 9]

Beginning with the onset of wireless telegraph service, spectrum management has been a key function of the ITU.[15] Its intermittent World Radio Conferences (WRC; sometimes still referred to by the older acronym WARC, for World Administrative Radio Conference) define services, allocate spectrum on a global or regional basis, and promulgate binding regulations on frequency use. Nations typically implement the consensus decision reached at a WRC,[16] notwithstanding the sovereign right of domestic regulatory authorities to formulate their own rules and policies on spectrum issues.

Even if domestic regulatory action precedes the ITU process,[17] nations expect the ITU to establish an international or at least a regional frequency allocation based on the consensus decision that particular frequencies best meet the requirements of specific services. "Radio regulations that result from ITU conferences have treaty status and provide the principal guidelines

15. For a comprehensive history of the ITU, see G. Codding, Jr., *The International Telecommunication Union: An Experiment in International Cooperation* (New York: Arno Press, 1972), and G. Codding, Jr., and A. Rutkowski, *The International Telecommunication Union in a Changing World* (Dedham, MA: Artech House, 1982). A more concise summary is available in "The ITU and the Radio Frequency Spectrum: Use and Management of a Shared Universal Resource," in J. Savage, *The Politics of International Telecommunication Regulation* (Boulder, CO: Westview Press, 1989), Chap. 2, pp. 61–129.

16. Decisions taken at WARCs "will determine how and when new radio services will be implemented and will influence the development of new technologies and applications." U.S. Congress, Office of Technology Assessment, "The 1992 World Administrative Radio Conference: Issues for U.S. International Spectrum Policy—Background Paper," OTA-BP-TCT-76, Nov. 1991, p. 1.

17. The prospect of an upcoming WARC can prompt the FCC to expedite its consideration of a spectrum allocation matter to ensure that the United States will have reached a position it can then advocate for other nations to endorse. "[I]t is necessary to make such authorizations now to clarify future United States domestic satellite interests at the upcoming Space WARC." "Licensing Space Stations in the Domestic-Fixed Satellite Service," 101 FCC 2d 223, 224 (1985).

for world telecommunications operations" [3, p. 655]. In that way, other nations can make similar spectrum allocations and assignments, thereby coordinating spectrum usage and reducing the potential for harmful interference.

Conflict results when usage in particular nations exhausts the amount of spectrum allocated for a particular service or when users find the operational rules unduly restrictive. For example, the ITU establishes a hierarchy of access rights and entitlements to protection from interference. "Secondary" service users are subordinate to both existing and future users of the "primary" service designated for a particular frequency band. Under that system, existing and prospective users of primary service have a superior right of access to the allocated spectrum and the privilege to operate free from interfering users. Existing or future secondary service users operate on a subordinate basis and must not cause interference to primary service users.

In many instances, the spectrum allocation process involves a win-lose, zero-sum situation in that one or more authorized users will be designated to the exclusion of other uses. Spectrum is earmarked in blocks for specific services with an eye toward matching its technical characteristics with particular user requirements. For example, point-to-point microwave applications need extremely high frequencies (in the gigahertz range), in which radio waves propagate in a line-of-sight pattern. Long-distance broadcast applications need middle or high frequencies that enable signals to bounce off the ionosphere. More numerous, geographically separated, local broadcasters can operate on the same frequencies at lower power or at higher frequencies (e.g., in the FM radio band) that typically do not skip off the ionosphere, instead traveling close to the ground.

Matching technical characteristics of the spectrum with user requirements requires answers to several questions.

What Application(s) Should Occupy This Frequency Band?

For any particular frequency band, certain uses appear best suited, but other commercial, social, or national security factors may prevail, particularly when more than one application can use a frequency band. In the United States, commercial interests supported allocation of ultra high frequency (UHF) channels 14 to 82 to augment the number of television channels previously available in the very high frequency (VHF) band. But more compelling public safety and mobile communications requirements combined with incomplete use of the band by broadcasters have eroded the UHF band so that the upper portion of the band (channels 70 to 82) was reallocated in 1970, the lower portion (channels 14 to 20) is shared in some regions, and channels 59 to 69 will migrate to mobile telecommunications use over the next few years.

Who Should Use This Allocation? How Should Those Users Be Selected?

The technical characteristics of the frequency band will target certain types of users for particular allocations. For example, telephone companies and large, geographically dispersed corporations such as railroads have requirements best served by microwave allocations. International broadcasters need allocations that support long-distance transmission via ionospheric skip, while local broadcasters can operate with lower power or higher frequencies. National regulatory authorities are responsible for matching spectrum requirements with suitable users and subsequently licensing qualified users. Increasingly, nations have opted to allow marketplace forces to dictate who can use spectrum. Few have decided to allow market forces to direct both who can use spectrum and for what purposes the spectrum can be used.

How Will the Spectrum Be Used?

The ITU radio regulations establish spectrum rules of the road that include specifications on what types of transmissions are permitted and what operators must do to avoid causing harmful interference.

Where Can the Spectrum Be Used?

ITU conferences attempt to allocate spectrum on a global basis, but allocations frequently are limited to one of three regions: Region 1—Europe and Russia west of the Urals; Region 2—North, Central, and South America; and Region 3—Asia-Pacific. Nations may take a reservation to any allocation, in effect refusing to go along with the consensus. Alternatively, they can request the ITU to attach a footnote to the allocation, contained in the ITU's radio regulations, specifying limitations to the consensus allocation or a different allocation altogether. Such departures from the consensus typically are rare, because individual nations realize that deviations run the risk of causing or incurring interference from the predominant spectrum use and because of the sense that national manufacturers will have less to gain in manufacturing equipment for frequencies usable in only one nation or region.

The location of spectrum use is also important, because the first duly authorized, registered, and operating user is entitled to continue operating interference-free when other operators subsequently seek to operate.

Alternative uses for the frequency typically are permissible only on a restrictive, subordinate basis whereby the newcomer must not interfere with incumbents or future operators providing the service for which the band was allocated.

Incumbent users of spectrum are reluctant to relinquish spectrum; to agree that the spectrum can be shared with new services, despite technological innovations that make it possible to make do with less spectrum; or to migrate to higher frequencies. For some, particularly users in developing nations, such technological innovations are too expensive, particularly where less efficient but cheaper equipment is operational and in short supply. Users in developed countries are disinclined to change frequencies or share, based on the view that they should not have to incur additional expense merely to accommodate users in nearby developed nations.

The international spectrum allocation process primarily involves decisions on how spectrum will be used, not who will use it. The process assumes that nations for the most part will register spectrum requirements as they do for satellites on an a posteriori, first-come, first served basis. Nations should register uses with the ITU when and if they exist or will soon develop. Developed countries typically have a head start when it comes to actual use of newly allocated or reallocated spectrum, because they have the earliest requirements and the resources necessary to convert an allocation into a registered use.

Developing countries have voiced objections to the a posteriori system and have proposed that the ITU allocate some spectrum and satellite orbital slots on an a priori basis, that is, nations agree to a plan for reserving shared global resources for use by one or more specific countries in advance of actual requirements. While such a system might foster distributional equity, it could result in a glut of unused spectrum for developing countries and a scarcity for developed countries. The matter of equity versus efficiency in the spectrum allocation process, like that for satellite orbits, has on occasion led to sharp disagreements between developed and developing nations.

Spectrum management has become more controversial at both the international and the domestic level, because of increasing demand by both incumbent users and proponents of new services for what is a finite resource. WRCs and the domestic regulatory process must balance the duty to ensure that existing users do not experience harmful interference, with the obligation to accommodate new spectrum requirements that would serve the goals of national interests and international comity.

WRC decisions on spectrum allocation can either validate previous domestic actions or isolate nations whose previous or subsequent spectrum allocations do not jibe with the international consensus.[18] "The failure to aggressively link long-term international policy efforts with domestic needs could threaten U.S. technological and policymaking leadership and could undermine future success in U.S. international spectrum policymaking" [2, p. 9]. However, with limited exceptions, the United States concurs with the consensus decisions reached on spectrum allocation and service definitions reached at WRCs.

6.4.1 National Spectrum Management, Licensing, and Regulation

The efficient management of spectrum constitutes an essential element for effective use of a nation's telecommunications infrastructure.[19] Some countries have opted to treat spectrum like real estate and create a market for its sale, subject to restrictions on use analogous to zoning regulations. Even when auctions are not used, spectrum has substantial, if unrealized, value,[20] particularly when demand far exceeds the amount of bandwidth allocated.[21]

18. "The more advanced our technology becomes, and the more complicated our frequency utilization, the more apparent it is that there must be complete correlation of the national and international aspects of frequency use." H. Felloes, in testimony at hearings before a subcommittee of the Committee on Interstate and Foreign Commerce on allocations of radio spectrum between federal government users and nonfederal government users, 86th Cong., 1st Sess., June 8–9, 1959, p. 36; quoted in U.S. Congress, Office of Technology Assessment, "The 1992 World Administrative Radio Conference: Issues for U.S. International Spectrum Policy—Background Paper," OTA-BP-TCT-76, Nov. 1991, p. 9.

19. The U.S. government has begun to realize the need for effective and market-oriented spectrum management. See U.S. Dept. of Commerce, National Telecommunications and Information Administration, *U.S. Spectrum Management Policy: Agenda for the Future*, NTIA Spec. Pub. 91-23 (Washington, D.C.: Government Printing Office, Feb. 1991), p. 13.

20. NTIA estimated the spectrum value of cellular radio services, which consumes 50 MHz, to be over $79 billion. See U.S. Dept. of Commerce, National Telecommunications and Information Administration, "Estimating the Value of Cellular Licenses," Appendix D in *U.S. Spectrum Management Policy: Agenda for the Future*, NTIA Spec. Pub. 91-23 (Washington, D.C.: Government Printing Office, Feb. 1991). Just the license to operate in a major metropolitan or national market may cost several hundred million dollars. Spectrum auctions for mobile radiotelephone licenses have generated billions of dollars in the United States, the United Kingdom, and Germany. See Rob Frieden, "Get Yer Spectrum Here," *The Industry Standard*, Vol. 3, No. 17, May 8, 2000, p. 136, available on-line at http://www.thestandard.com/article/display/0,1151.14525.00.html; see also Brian C. Fritts, "Private Property, Economic Efficiency, and Spectrum Policy in the Wake of the C

Some spectrum uses have the characteristic of a public good in that one person's consumption of, for example, an educational program on broadcast television, does not exhaust or reduce what can be received by others. Spectrum also can constitute a common-pool economic resource, like offshore drilling sites owned by the government in that it is exhaustible, is subject to congestion, can be allocated for specific uses, and can be sold or leased to particular users. Technological innovations have enabled productive use of progressively higher frequencies and the ability to derive usable channels with less bandwidth. But along with innovations that conserve spectrum and provide more throughput are new ideas and services that generate additional spectrum requirements.

Because of increasing demand for spectrum and the costs incurred by incumbents or newcomers to conserve it, international and national agencies must conserve and manage spectrum. Such an endeavor involves allocating spectrum among competing uses and serving as a traffic cop of the airwaves to avoid interference and to resolve conflicts. Spectrum managers need to fashion compromises based on a number of factors, including the following:

- *Technology:* The duty to prevent harmful interference and to achieve efficient activation of channels. For example, in allocating spectrum for broadcast television in the VHF band, the FCC had to create large geographic spacing between stations to prevent interference. That limited the number of available stations in any locality, thereby generating demand for an additional allocation in the UHF band.
- *Regulatory policy:* Regulation may direct spectrum allocations in ways designed to serve public policies. For example, the FCC sought to promote the doctrine of localism by reserving broadcast channels for as many different localities as technologically possible. That

---

Block Auction," *Federal Communications Law Journal*, Vol. 51, May 1999, p. 849; William Kummel, "Spectrum Bids, Bets and Budgets: Seeking an Optimal Allocation and Assignment Process for Domestic Commercial Electromagnetic Spectrum Products, Services, and Technology," *Federal Communications Law Journal*, Vol. 48, June 1996, p. 511.

21. Sales of VHF television stations in major markets can exceed $500 million, far in excess of the physical assets involved. See Gregory L. Rosston and Jeffrey S. Steinberg, "Using Market-Based Spectrum Policy to Promote the Public Interest," *Federal Communications Law Journal*, Vol. 50, Dec. 1997, p. 87.

policy reduced the number of stations available in urban localities that otherwise could have served nearby towns.

- *Commerce:* The need to conduct a comparison of spectrum requirements by services with an eye toward allocating spectrum to uses that will maximize social welfare primarily and individual profitability of firms secondarily. For example, the FCC reallocated portions of the UHF television band for mobile radio services when it determined that most localities could not support the full inventory of UHF television stations the Commission had reserved. The pressing need for public safety and private wireless spectrum triggered a reallocation.
- *Social welfare:* The public interest merit in allocating spectrum for a particular service in the face of other requirements that accordingly have to make do with less, different, or possibly no spectrum. For example, in allocating spectrum for new wireless mobile services like personal communications networks, the FCC forced existing microwave users, like railroads and public utilities, first to share the spectrum and subsequently to move to higher, less congested frequencies.
- *National security:* Compelling requirements for safety, public welfare, national defense, and emergency applications. For example, the ITU has allocated particular emergency calling frequencies that are always monitored.

Spectrum allocation decision-making blends engineering with social sciences to provide the basis for making cost-benefit analyses. Policy makers need to determine who deserves spectrum allocations and what entitlements accrue to incumbent users. Managers must devise procedures for assessing new spectrum requirements and determining how to accommodate incumbent users and operators.

Currently, most nations generally allocate spectrum on the basis of consensus decisions reached at global or regional conferences convened under the auspices of the ITU. In the United States, the FCC uses the public interest as the basis for determining whether to implement ITU decisions by incorporating changes in the national table of spectrum allocations.[22] The FCC also typically uses the public interest standard to assign spectrum, but

22. See 47 C.F.R. § 2.106 (1999).

in 1993 it received Congressional authority[23] to auction portions of the spectrum. The Commission also imposes user and licensing fees to compensate it for processing applications, granting licenses, participating in international conferences, and the cost of regulation. Demand typically will exceed supply without a market mechanism[24] for clearing out spectrum inventory to the highest bidder.[25]

New Zealand leads all nations in the creation of a spectrum marketplace with passage of the Radiocommunications Act of 1989, which privatized frequencies between 44 MHz and 3.6 GHz. The law created a management right equivalent to an exclusive, transferable 20-year ownership interest. By auctioning unoccupied frequencies, New Zealand established a property-based alternative[26] to public interest–based spectrum allocation and assignment. Most

---

23. See Omnibus Budget Reconciliation Act of 1993, Title VI, Communications Licensing and Spectrum Allocation Provisions, PL 103-66 (HR 2264), 107 Stat. 312, 379 *et seq.* (1993). The FCC has authorized cash payments from newly licensed applicants to incumbent users for relocating to another frequency band before spectrum reallocation subordinated their status and required such relocation. See "Redevelopment of Spectrum to Encourage Innovation in the Use of New Telecommunications Technologies," ET Docket No. 92-9, Notice of Proposed Rulemaking, 7 FCC Rcd. 1542 (1992) (proposing to reserve 220 MHz in the 2-GHz band for new services like personal communications networks while accommodating incumbent microwave licensees). Further Notice of Proposed Rulemaking, 7 FCC Rcd. 6100 (1992), 1st Report and Order, 7 FCC Rcd. 6886 (1992); 2d Report and Order, 8 FCC Rcd. 6495 (1993) (reallocating spectrum above 3 GHz to ease migration of incumbents in 2-GHz bands to accommodate emerging technology mobile services); 3d Report and Order and Memorandum Opinion and Order, 8 FCC Rcd. 6589 (1993) (adopting a plan for fair sharing in the 2-GHz band between incumbent microwave users and emerging technology services and for accelerated relocation by incumbents). The Commission also can collect fees to recoup the cost of regulation. See Implementation of Sec. 9 of the Communications Act, MD Docket No. 94-19, Report and Order, 9 FCC Rcd. 5333 (1994).

24. Until passage of enabling legislation, the FCC could consider marketplace forces when making spectrum allocations but could not auction off spectrum to the highest bidder. "Nothing on Sections 303(A)-(C)[of the Communications Act] suggests that the Commission is not permitted to take into account marketplace forces when exercising its spectrum allocation responsibilities under the public interest standards." Amendment of Parts 2, 15, and 90 of the Commission's Rules and Regulations to Allocate Frequencies in the 900 MHz Reserve Band for Private Land Mobile Use, GEN Docket Nos. 84-1231, 1233, 1234, Report and Order 2 FCC Rcd. 1825, 1939 (1986).

25. In 1959, R. H. Coase argued for definitive spectrum ownership rights as a market-driven way to foreclose chaos and interference. R. H. Coase, "The Federal Communications Commission," *Journal of Law and Economics,* Vol. 2, 1959, p. 1.

nations that embrace market-based forces endorse spectrum assignment by competitive bidding but balk at abandoning the spectrum allocation function and allowing the unlimited use that New Zealand permits.

6.4.2 The Role of Scarcity in Spectrum Management, Licensing, and Regulation

Policy makers and courts have justified regulation of spectrum usage because of spectrum scarcity and the inability of all parties desiring access to secure a license. The term *scarcity* has different meanings, and some critics of regulation dispute that it exists at all in an age of spectrum-conserving technological innovations and closed circuit options.[27] One can consider spectrum scarcity from a number of perspectives: technological, financial, marketplace, allocational, and resource.

6.4.2.1 Static and Dynamic Technologies

From the technological perspectives, we can look at innovation as forestalling scarcity with ways to achieve more throughput in a digital format at faster transmission speeds carried via closed-circuit fiber optic lines that do not use spectrum. This dynamic view of technology considers spectrum shortages unlikely, just as innovations in agriculture prevent widespread famine despite massive increases in population. For example, compression technology provides the basis for deriving four channels or more from the bandwidth of spectrum that previously could generate only one channel. Compression provides the basis for 500-channel cable television systems and for 100-channel direct-to-home satellite broadcasting. The technology conserves spectrum by

---

26. Douglas Webbink identifies the following rights accruing from property-based ownership of spectrum: (1) exclusive access to use of the property; (2) the right to generate income from the property; (3) freedom to transfer the property right to someone else; and (4) freedom from confiscation of the property absent some extreme situation (e.g., war, nonpayment of tax, natural catastrophe). See D. Webbink, "Radio Licenses and Frequency Spectrum Use Property Rights," *Communications and the Law,* Vol. 9, June 1987, pp. 3–29; see also Pablo T. Spiller and Carlo Cardilli, "Towards a Property Rights Approach to Communications Spectrum," *Yale Journal on Regulation,* Vol. 16, Winter 1999, p. 53; Thomas W. Hazlett, "The Rationality of U.S. Regulation of the Broadcast Spectrum," *Journal of Law and Economics,* Vol. 33, Apr. 1990, pp. 133–175.

27. See, for example, N. Negroponte, "Products and Services for Computer Networks," *Scientific American,* Vol. 265, Sept. 1991, pp. 106–113; G. Guilder "Into the Telecosm," *Harvard Business Review,* Vol. 60, No. 2, Mar./Apr. 1991, pp. 150–161 (supporting the

allocating processing power only on changes detected from the preceding frame of information, sound, or video.

A static view of innovation considers spectrum scarcity a very real threat as the population grows and increasing requirements for frequencies exceed spectrum conserving innovations. For example, in many urban corridors, the demand for microwave frequencies has reached a point where additional licenses cannot be issued due to the likelihood of interference. New or additional requirements typically can be served only if the applicant agrees to operate on higher frequencies. The migration to another frequency band can impose higher costs, because existing operators must procure new equipment, and higher frequencies often require more expensive equipment and closer spacing of relay towers due to poorer transmission characteristics.

The technological template of scarcity considers spectrum from either a positive or a pessimistic viewpoint. For example, increases in radio modem processing speeds mean that equipment using the same amount of bandwidth can process more data and achieve greater throughput. It also means that less spectrum would be needed to operate at the slower, formerly cutting-edge speeds. The dynamic view believes some users will do more with less, while the static view believes that most users will occupy the same or more bandwidth, despite the increase in throughput and productivity.

6.4.2.2 Financial Scarcity

The financial approach looks at the individual's ability to pay for desired spectrum. Spectrum cost varies as a function of the cost of the technology needed to activate the spectrum, whether the user must pay for the right to use the spectrum, and the services that can be provided.[28] Significant transaction costs create disincentives for individuals to seek spectrum, but the extent of those costs varies with the type and amount of spectrum sought. Relatively few individuals can secure the funds necessary to buy a broadcast station or to pursue an application through the FCC regulatory process for a new, major market broadcast license. On the other hand, virtually anyone can secure spectrum for such applications as private microwave radio networks

---

view that no real spectrum scarcity exists, because many users of radio spectrum can migrate to closed-circuit media such as fiber optic cables).

28. While broadcasters continue to secure spectrum without having to bid for it, operators of personal communications network services competed in a high-stakes auction that raised over $7.7 billion for the general treasury. See D. Rohde, "Big Winners Thrown a Curve as the PCS Auction Closes," *Network World*, Vol. 12, No. 12, Mar. 20, 1995, pp. 8, 85.

and unlicensed services like citizen's band radio, cordless telephones, baby monitors, model airplane controllers, and garage openers.

For high-cost, contested spectrum, the FCC assigns frequencies on an exclusive, noninterference basis only to qualified individuals. For low-cost, largely unused spectrum, the FCC makes spectrum available without a screening process and on a nonexclusive basis. The FCC does not even require a license for operation of low-powered devices in specific frequency bands.

6.4.2.3 Marketplace-Determined Scarcity

The marketplace approach to spectrum scarcity applies economic analysis and determines where an equilibrium exists that accounts for overall supply and service-specific demand. That approach is academic, because policy makers have yet to allocate and assign most spectrum on the basis of marketplace resource allocation. It appears likely that the FCC and other national regulatory agencies will use competitive bidding to assign some unused or underused spectrum allocated for private or government use. However, major portions of the spectrum, for example, broadcasting, public safety, and police bands, are not likely to be auctioned, because of countervailing public interest concerns.

Currently, spectrum availability is tied to perceived demand and the comparative public interest merits of the services to be provided. The process of calibrating spectrum bandwidth allocations with need and merit is far from exact. Even as some observers claim that spectrum scarcity has grown acute, U.S. UHF television channel allotments for specific localities—even for major metropolitan areas—remain unclaimed. Arguably, the FCC has allocated too much spectrum for UHF television relative to what the broadcast television marketplace can support in any particular region and relative to available alternative, nonspectrum-using services like cable television. In 1970, the FCC reallocated UHF television channels 70 to 83 (806 to 890 MHz) to mobile radio services. UHF television overcapacity has begun to reach equilibrium with the FCC's decision to allow broadcasters the opportunity to acquire a second 6-MHz channel to simulcast a high-definition television signal[29] and the Commission's decision to auction off spectrum in the 700-MHz range.[30]

---

29. See "Advanced Television Systems and Their Impact Upon the Existing Television Broadcast Service," MM Docket No. 87-268, Fifth Report and Order, 12 FCC Rcd. 12809 (1997); *on reconsideration,* 14 FCC Rcd. 1348 (1998); *further reconsideration dismissed,* FCC 00-59 (rel. Feb. 23, 2000); see also "Review of the Commission's Rules and

For other services, the claim can be credibly made that the FCC and other national regulatory agencies have not allocated sufficient spectrum relative to the bandwidth needed and the public and private benefits that could accrue. Spectrum advocates for essential requirements like public health, protection, and safety demand spectrum reallocations to accommodate rising demand. To satisfy those demands, agencies must balance the interests of incumbents, who may have failed to activate all the spectrum allotted, for example, UHF television broadcasters versus newcomers with compelling requirements.

6.4.2.4 Allocational Scarcity

Allocational scarcity results when the decision-making process fails to balance marketplace imperatives and other possibly countervailing, public interest considerations. A regulatory agency may reserve spectrum for particular applications on the basis of a public interest determination that does not necessarily gauge the extent of the beneficiary's actual or prospective use relative to the needs of others. For example, the FCC has reserved spectrum in the FM radio band from 88 to 92 MHz for noncommercial and educational radio, irrespective of the requirements in any particular region.

In many major U.S. broadcast markets, additional commercial broadcasters cannot operate, because the FCC has assigned all usable channels. A disenfranchised broadcaster might be willing to pay the federal government for the privilege to activate an unused noncommercial channel. The prospective broadcaster could get on the air for less than the marked-up price an operating commercial broadcaster would charge, and the number of allotted channels in the noncommercial FM band appears ample enough so that no existing or prospective noncommercial broadcaster would be foreclosed from operating. Despite what appears to be a win-win situation, public interest mandates prevent the FCC from permitting such an arrangement. Spectrum allocational decisions impose real scarcity in terms of access when the regulatory agency reserves otherwise usable and sought-

---

Policies Affecting the Conversion to Digital Television," Notice of Proposed Rule Making, FCC 00-83, 15 FCC Rcd. 5257 (2000).

30. See Service Rules for the 746-764 and 776-794 MHz Bands, and Revisions to Part 27 of the Commission's Rules, WT Docket No. 99-168, First Report and Order, FCC 00-5 (rel. Jan. 7, 2000); Second Report and Order, 15 FCC Rcd. 5299 (2000).

after spectrum for a particular service or application that may not have takers for all of the available spectrum.

6.4.2.5 Resource Scarcity

Spectrum constitutes an exhaustible resource, like air, space on a highway, or offshore oil drilling sites. As a finite resource, spectrum must be managed, but scarcity need not be a foregone conclusion. Licensing can ration spectrum allocated on a service-specific basis. On the other hand, spectrum cannot be replenished or expanded, as would be the case by planting more trees for more newspaper or widening the highway to accommodate more cars.

Resource scarcity is real and becomes acute when spectrum users have no duty or incentive to conserve and maximize efficiency. Without a financial incentive, existing spectrum users may delay replacing outmoded and inefficient technologies, and proponents of new services may face a lengthy administrative process for demonstrating that the public interest favors requiring incumbents to share spectrum, use narrower channels, or even relocate to another frequency band.[31]

The extent and the severity of spectrum scarcity depend on a number of variables, including the number of incumbent users, the scope of sunk investment, the suitability of relocating to another frequency or migrating to a closed-circuit wireline option, and the costs that would be incurred.

6.4.3 Block Allocations

WRCs and, in turn, the FCC allocate spectrum in blocks of frequency bandwidth earmarked for a particular service. Politics and nontechnological factors may become factors in the spectrum allocation decision-making process, particularly when advocates of two or more uses vie for the same bandwidth. The process places a premium on incumbency. Once a spectrum allocation occurs, the licensees operating on that spectrum expect to have unlimited rights of use, without the prospect of having to share the allocated spectrum or move to another frequency. New services, technological innovations, or user constituencies with expanded spectrum requirements must vie with

---

31. See, for example, "Redevelopment of Spectrum to Encourage Innovation in the Use of New Telecommunications Technologies," ET Docket No. 92-9, Third Report and Order and Memorandum Opinion and Order (rel. Aug. 13, 1993) (adopting a plan for sharing of 2-GHz spectrum between incumbent microwave users and new personal communications service operators, including a mechanism by which newcomers could compensate incumbents for expedited movement to a different frequency band).

incumbents for spectrum. In the best circumstances, newcomers receive a spectrum allocation on a "co-primary" basis, meaning they must share the spectrum with incumbent, primary users but enjoy interference protection from subsequent users even with primary service requirements. A secondary allocation would subordinate the newcomer not only to existing primary service users but also to subsequent ones.

The block allocation process awards bandwidth on the basis of the technologies, services, and user requirements effectively advocated at the time of decision. For example, satellite services have been divided as a function of orbital location (e.g., geostationary or low Earth) and transmitter and receiver locations (e.g., fixed or mobile). There are maritime, land mobile, fixed, and aeronautical services with separate allocations. Discrete service definitions and spectrum allocations made sense when users could not easily move terminals, but now they can easily operate the same portable transceiver in a variety of fixed and mobile environments.

The FCC acknowledged the flexibility afforded by miniaturization and portability of new satellite transceivers. At a 1987 and a 1992 WARC, the United States proposed a generic mobile satellite service that would eliminate separate spectrum allocations for maritime, land, and aeronautic mobile uses. When the international consensus persisted in maintaining separate, geographically specific allocations, the United States "took a reservation ... with respect to these allocations, indicating its continuing desire to implement MSS in an appropriate manner to satisfy U.S. requirements" [4].

Frequency spectrum users and commentators have begun to recognize the financial stakes in the spectrum planning and allocation process:

> The United States is at a crucial turning point in the history of spectrum use and management. Technological, economic and political forces are converging to radically alter the context within which domestic and international spectrum decisions and policies are made. [2, p. 2]

6.5 The Spectrum Auction Alternative

In March 2000, the U.K. government organized the latest in a series of multibillion-dollar auctions for radio spectrum over the last few years.[32]

32. See http://www.spectrumauctions.gov.uk/.

Governments throughout the world have found a new, big-money revenue source: awarding the exclusive right to use portions of the radio spectrum, in a manner that confers just about everything actual ownership would.

These auctions have triggered a gold rush mentality for three primary reasons. First, a venture lacking spectrum cannot provide a wireless service. Second, certain kinds of spectrum have increasing value, because wireless applications, like mobile telephones and Internet access, will continue to grow in importance. Third, the auction mechanism guarantees interference-free access in an efficient manner without the delay and uncertainty of previous licensing regimes.

6.5.1 Spectrum as a Franchise

Governments have only recently considered spectrum as something assignable to the highest bidder, like oil drilling and grazing rights. A sense that the public owned the spectrum previously prompted governments to assign spectrum on public-interest grounds. Prospective commercial broadcasters and cellular radio operators alike had to participate in a comparative hearing, not unlike a beauty contest, to determine who would best enhance the public welfare. A blend of extremely high stakes, a regulatory agency staff, and a cast of lawyers, economists, engineers, and other consultants resulted in a multibillion-dollar process. Even worse, the process took a long time to complete, often rewarded less qualified but craftier players, and subsequently may result in private auctioning as licensees "flip" the franchise to another operator without cutting the government in on the windfall.

6.5.2 Bound by the Rules of Physics

Governments used to think the rules of physics foreclosed the option of granting bidders free reign on the use of spectrum. Spectrum continues to have propagational limits to use, and most auctions limit winners to specific types of services. Nevertheless, technological innovations make certain types of spectrum usable for a broader range of services. With frequency-agile radios, digital signal processors, and a wireless infrastructure, new ventures offer a range of services previously available only via a wire or fiber optic cable. The proliferation of cellular radio and other wireless substitutes for fixed, wireline services has accrued growing market share, and stock markets have amply rewarded holders of shares in wireless companies like Qualcomm, Orange, and the subsidiaries of major telephone companies.

This wonderful story about economic efficiency, painless government revenue enhancement, and expedited time to market for wireless entrepreneurs does not necessarily end well. Spectrum auctions can work only when a keen demand exists and any anticipated use will remain within one nation. Even as billions of dollars chase wireless telephone and Internet access spectrum, other types of spectrum have little or no market value or should remain in government hands. Given their ability generally to operate without causing interference, cordless telephone users should not have to bid for the use of spectrum that they can share with other users. Federal, state, and municipal governments should not have to compete with private users for spectrum. Likewise, try demanding a fee from commercial broadcasters for their spectrum franchise. They would respond that they pay for spectrum through the programming they offer. The U.S. Congress continues to endorse that view, having awarded television broadcasters an additional channel for digital television services.

6.5.3 Serving Global Interests

National auctions also may run afoul of the ITU's spectrum management process, particularly where, as in the New Zealand auctions, the winning bidder has complete freedom to use the spectrum for whatever purpose the bidder considers most profitable. The ITU process serves the perceived need for clarity, certainty, and conflict resolution. It identifies specific spectrum uses and allocates spectrum in service-specific slivers. What users lose in flexibility they gain in knowing what services use which frequencies and having a registrar. The registration function works to preempt and resolve disputes over who has priority rights of access to a shared global resource.

While nations can enhance flexibility of use for spectrum that does not cross borders, many services by design and by the laws of physics do. The ITU attempts to provide structure and order, because the alternative would lead to chaos. The blockbuster movie *Titanic* underscores the need for uniform rules of the road regarding at least some spectrum use, like that allocated for maritime distress and safety.

Increasingly, the ITU has encountered difficulty in making its voluntary system work. Stakeholders, like prospective mobile radio service providers, recognize the financial advantages in being first to market. In markets operating on Internet time, opportunities can appear and disappear during the time it takes the ITU process to run its course. One of the reasons the $5-billion Iridium LEO satellite venture failed was the time it took the ITU to reallocate spectrum and satellite orbit registration policies to accommodate the new technology. By the time Iridium began commercial service, cheaper terrestrial options had so proliferated that insufficient numbers of road warriors remained willing to pay $3 a minute for access to the rest of the world. What started as an incredible concept of providing "anytime-anywhere telephone service" became passé as cellular radio service became available in most places where people wanted service and could pay for it.

6.5.4 Satellite Spectrum Auctions

The laws of physics make it possible for satellites to provide service to as much as one-third of the Earth. What if each and every nation covered by a satellite decided to capture the financial benefits that spectrum auctions generate? They might have some legal rationale to do so based on the concept of sovereignty, which one nation to ask permission from another nation to land an airplane and even to fly over the country.

Auctioning satellite signal landing rights may logically follow the competitive bidding for spectrum and other operator franchises. But it might price many satellite services out of the market: no Internet access, no HBO, no MTV to remote locations. If the cellular radio spectrum sweepstakes extends to satellite service, then the ITU will become even less effective in providing uniform rules of the road. Because of their point-to-multipoint, multicasting function, satellites provide a competitive alternative to fiber optic cables. Think of transborder satellite coverage as applying Metcalfe's law to telecommunications: The overall benefit of a satellite network to users grows as a function of the number of users able to access the satellite. Auctioning satellite access in a manner similar to auctioning spectrum jeopardizes the public benefits satellites generate. But some buy into the rationale that orbital slots for satellites have the same characteristics as real estate and spectrum.

6.5.5 Going to the Well Too Often

The primary problem with spectrum auction may be the mind-set it creates. Auctioning spectrum becomes a painless revenue enhancer that governments presumably can tap endlessly. Already, the FCC has found that too many auctions yield less than impressive returns, while the rules for auctions of most desired spectrum led to bids that some ventures could not finance. Some bidders have come away with spectrum grants for pennies per prospective user, while others have bid hundreds of dollars. Throughout the process, other nations have watched with interest, waiting their turn to cash in.

6.5.6 Auctions: Only a Partial Solution

Auctions clearly provide opportunities for governments to capture value previously secured by private parties. That money can go to good uses, and auctions by nature allow winning bidders to put spectrum to good use. However, the proper and effective use of auctions does not mean that all spectrum lends itself to this type of allocation mechanism, or that spectrum has become the functional equivalent of real estate.

Spectrum matters in the business plans of wireless ventures. Operators need certainty, a timely administrative process, the right to operate without interference from others, and the opportunity to provide a marketable service. However, in freeing operators and markets, governments need to safeguard spectrum uses and users who should not have to bid for the privilege.

6.6 Managing the Satellite Orbital Arc

Governments justify their management, regulation, and operation of satellites, because such facilities use both scarce frequency spectrum and orbital parking places.[33] To maximize the usefulness of those resources, nations must establish policies and rules for interference avoidance in the same manner they must coordinate radio frequency use.

The geostationary orbital arc nears saturation in some regions, like North America, because of the increasing demand for telecommunications satellite service, growing numbers of nations opting for regional or domestic systems instead of global or regional cooperatives, and physical limitations on the number of satellites that can be positioned without causing harmful frequency interference. The danger of satellite collision is remote (the closest spacing of 2 degrees still maintains approximately a 900-mile separation),

33. The ITU Constitution directs member nations to "bear in mind that radio frequencies and the geostationary-satellite orbit are limited natural resources and that they must be used rationally, efficiently and economically, in conformity with the provisions of the Radio Regulations, so that countries or groups of countries may have equitable access to both, taking into account the special needs of the developing countries and the geographical situation of particular countries." ITU Constitution, Art. 33, Para. 175, "Use of the Radio-Frequency Spectrum and of the Geostationary-Satellite Orbit," (Geneva: ITU, 1990).

but the potential for interference exists, because adjacent satellites typically use the same spectrum allocated for fixed, mobile, maritime, aeronautic, and other satellite services.

Earth stations using expensive, state-of-the-art technology can communicate interference free with satellites located as close as 2 degrees apart. That means that for any particular satellite frequency band the theoretical maximum number of operational satellites in the geostationary orbit is 180 (360 degrees divided by 2). However, the actual number is significantly less because of several factors:

- Not all nations will mandate the use of sensitive equipment to enable 2-degree satellite spacing.
- Operators seeking maximum transoceanic, international coverage will cluster satellites in middle ocean regions.
- Satellites can be equipped with antennas that concentrate the transmission beam, thereby increasing the probability of interference even from satellites many degrees away.

6.6.1 ITU Conflict Resolution in Satellite Orbital Slots

Notwithstanding shared interests in consensus and conflict prevention, the ITU regularly bears the duty to broker compromise and lends its "good offices" to resolve conflicts. Particularly for shared resources like satellite orbital slots, nations vie for a finite number of available positions in the manner of a zero-sum game: one nation's orbital slot use often can occur only at the expense of another nation's current or future use.

With only recent, limited qualification, the ITU rules favor incumbents and prior registered uses. Typically, developed nations like the United States and global cooperatives like INTELSAT[34] and Inmarsat[35] have established satellite requirements. That means incumbents will have completed the registration process well before later market entrants in developed nations

34. The International Telecommunications Satellite Organization (INTELSAT) was organized as a global cooperative by intergovernmental agreement with a mission of providing ubiquitous satellite communications service via GSO satellites. See *Multinational Communication Satellite System,* opened for signature Aug. 20, 1964, 15 U.S.T. 1705, 514 U.N.T.S. 26 (19-nation agreement establishing interim arrangements for a global satellite cooperative); *Agreement Relating to the International Telecommunications Satellite*

and before lesser developed nations have accumulated the finances to begin procuring a first satellite.

Nations notified of a future satellite deployment have an affirmative duty to avoid causing harmful interference should they seek to use the same orbital location. Developing nations may have to settle for less than optimal orbital slots, because their later-launched systems must not interfere with already operating networks and preoperational, but registered, satellites. The congested orbital arc and limited finances have prompted such nations to lease capacity from cooperatives' satellites, even for domestic applications. Alternatively, they have formed regional coalitions like Palapa, which serves the nations of southeast Asia.

6.6.2 Inequity in Access to Satellites and Information Resources

The matter of access to orbital slots takes on even greater significance when we consider its impact on nations' collective opportunities to tap information resources and to use telecommunications as a catalyst for economic development. Inequitable and inadequate access to orbital slots can exacerbate the gap between nations with rich and diverse telecommunications infrastructures and ones with second-rate or leased facilities. If we subscribe to the view that the wealth of information resources has a direct and substantial effect on national financial wealth, then access to the orbital arc may significantly affect a nation's overall social and economic welfare. Accordingly, the stakes in the orbital slot access sweepstakes involve more than how many television channels a nation can access. It affects the broader issue of whether and how nations share information resources and the technologies for access and distribution. Likewise, it raises questions of whether developed nations should promote parity of access to resources, both public (orbital slots) and private (programming and databases sent via satellite).[36]

Organization ("INTELSAT"), opened for signature Aug. 20, 1971, 23 U.S.T. 3813, 1220 U.N.T.S. 21 (INTELSAT Agreement); *Operating Agreement Relating to the International Telecommunications Satellite Organization ("INTELSAT")*, opened for signature Aug. 20, 1971, 23 U.S.T. 4091, 1220 U.N.T.S. 149.

35. The International Maritime Satellite Organization (Inmarsat) was organized as a global cooperative by intergovernmental agreement with a mission of providing ubiquitous maritime telecommunications to ships in the high seas, with aeronautic and land mobile services available on an ancillary basis. See *Convention of the International Maritime Satellite Organization*, opened for signature, July 16, 1979, 31 U.S.T. 1, T.I.A.S. No. 9605.

The ITU regularly becomes the forum for fact finding, arbitration, and conflict resolution in the face of growing demands for orbital slots. Developed nations require additional satellites to satisfy demand for more video program options, while developing nations may need a first orbital slot. Congestion grows when nations cannot or will not commit to reduced spacing between satellites,[37] which would expand the number of satellites.

The ITU fashions remedies at the macrolevel by convening conferences to revise spectrum allocations and to consider changes to the method for reserving and registering orbital arc slots. At the microlevel, the BR publishes prospective uses, coordinates the necessary technical and operational assessment of a nation's interference claims, resolves real interference problems, and formally notifies the ITU membership of newly registered orbital arc uses.

Satellite orbital arc policy raises political questions and juxtaposes equity and efficiency concerns. For its part, the ITU must fashion a compromise that balances financial concerns regarding satellite spacing and efficient frequency use with equity concerns about parity of orbital slot access by developing nations. Under the customary first-come, first-registered process, the later-filed registrations of developing nations receive subordinate status. That process creates an incentive for both incumbent and prospective satellite operators to overestimate their orbital slot requirements, despite a good-faith standard for pursuing slots. To prevent warehousing of orbital slots, the ITU has established a timetable within which a registrant must actually launch a satellite to fill a registered slot.

Representatives from developing countries have advocated the need for equitable access to the orbital arc through a system that guarantees slots, even at the risk of leaving fallow a resource that other nations, singularly or collectively, could put to use sooner.[38] An a priori "allotment" plan for satellite

---

36. See, for example, B. Harris, "The New Telecommunications Development: Bureau of the International Telecommunication Union," *American University Journal of International Law and Policy*, Vol. 7, Fall 1991, p. 83; R. Saunders, J. Warford, and B. Wellenius, *Telecommunications and Economic Development* (Washington, D.C.: World Bank, 1983); R. Crandall and K. Flamm, *Changing the Rules: Technological Change, International Competition and Regulation in Communications* (Washington, D.C.: Brookings Institution, 1989).

37. In "Licensing of Space Stations in the Domestic Fixed-Satellite Service," 54 Rad. Reg. 2d (P&F) 577 (1983), the FCC ordered domestic satellite operators to position satellites within 2 degrees of each other. The reduced orbital spacing accommodates more satellites over the United States and enables the neighboring nations of Canada and Mexico to op-

orbital arc deployment[39] reserves slots for developing nations that typically have generated requirements for and ability to finance satellite telecommunications after developed nations.

Professor Harvey J. Levin attempted to quantify the cost handicap incurred by latecomers, who may have less attractive orbital slots available and who may have to incur added costs to operate on higher frequencies[40] to avoid causing harmful interference to existing satellites.[41] When developing nations have to migrate to higher frequencies to accommodate incumbent users and adhere to the rules of the road articulated by the ITU, they lose

---

erate their domestic satellites with wider separation. However, it required higher investment in more sensitive Earth stations.

38. See Harvey Levin, "Orbit and Spectrum Resource Strategies: Third World Demand," *Telecommunications Policy*, Vol. 5, No. 2, June 1981, p. 105.

39. "An a priori system of frequency and orbital position regulation uses administrative conferences to subdivide and allot radio frequencies and orbital positions to countries in advance of need or use. On the other hand, an a posteriori ["first-come, first-served"] system requires subsequent satellite operators to coordinate with pre-existing satellites to avoid harmful interference." Straubel, "Telecommunication Satellites and Market Forces: How Should the Geostationary Orbit Be Regulated by the F.C.C.?" *North Carolina Journal of International Law and Commercial Regulation*, Vol. 17, No. 30, 1992, pp. 205, 211.

40. The first commercial satellites operated exclusively in the C-band with Earth-to-satellite transmissions occurring in the 6-GHz band and satellite-to-Earth transmission occurring in the 4-GHz band. In the 1980s, satellite carriers also operated in the Ku-band with Earth-to-satellite transmissions primarily occurring in the 14-GHz band and satellite-to-Earth transmission occurring in the 11-GHz band. Recently, limited commercial and experimental use has occurred in the Ka-band with Earth-to-satellite transmissions occurring in the 30-GHz band and satellite-to-Earth transmission occurring in the 20-GHz band.

41. "More generally Third World resentment of the practice of awarding rights to build space satellite systems on a first-come, first-served basis, is seemingly based on what those nations perceive as the dwindling availability of slots or orbit spectrum assignments. The developing countries (LDCs) also fear the handicaps they suffer due to the higher R&D and engineering costs incurred to open up new bands at higher frequencies." Harvey Levin, "Latecomer Cost Handicap: Importance in a Changing Regulatory Landscape," in Donna Demac, ed., *Tracing New Orbits—Cooperation and Competition in Global Satellite Development* (New York: Columbia University Press, 1986), pp. 251–252. A more extensive and quantitative analysis is available in Harvey Levin, "Global Claim-Staking and Latecomer Cost in the Orbit Spectrum Resource," *Telecommunications Policy*, Vol. 13, June 1990, pp. 233–248. See also "Regulation of Transnational Communications," *Michigan Yearbook of International Legal Studies*: Part 1, Regulation of the Geostationary Orbit, pp. 3–70, and Part 2, Regulation of Satellite Communications, pp. 73–82 (New York: Clark Boardman, 1984).

much of the financial and operational benefits that would have accrued if they could use older, proven technology. Instead, the orbital slot registration and satellite coordination process may make it difficult for latecomers to find a suitable orbital slot, unless they agree to operate on higher frequencies. A developing nation with heavy rainfall would find that the ITU registration process would burden it with higher costs, because that nation would have to procure more powerful and expensive satellites and more sensitive and expensive Earth stations to overcome the signal degradation that occurs at higher frequencies during rainfall.

Access to the orbital arc raises questions of equity. While developed nations should not have to handicap or postpone their orbital arc development plans, decision makers will need to consider what, if anything, should be done to support developing nations' access to satellite technology and orbital slots. In view of efforts by INTELSAT and Inmarsat to privatize, developing nations have no certainty that the existing cooperative model will continue to provide access to satellite capacity at averaged rates for any nation, including those unable or unwilling to make a sizable investment in facilities ownership.[42]

6.6.3 Space WRCs

The ITU has found itself in the middle of a geopolitical battle of philosophies, particularly at radio conferences with a heavy agenda of spectrum allocation and orbital slot management issues. Such WRCs establish rules, regulations, and policies for various types of satellite services, for example, fixed (for telecommunications from and to many fixed locations on Earth), broadcast (for satellite broadcasting of video and audio programming directly to dispersed receiver locations), and mobile (for telecommunications between fixed locations and mobile stations or between mobile stations). They also determine at what frequencies such services should operate, ensuring that the operations do not interfere with other existing satellite networks or other operators whose terrestrial uses are authorized for the same frequencies. Most important, WRCs must anticipate and resolve future bilateral conflicts possibly involving nations whose representatives might lack the expertise or inclination to help shape a speedy and fair compromise. Accordingly, most member nations of the ITU invest significant resources for

42. See R. Frieden, "Should Intelsat and Inmarsat Privatize?" *Telecommunications Policy*, Vol. 18, No. 9, Dec. 1994, pp. 679–686.

extensive preconference preparations and send representatives to marathon international meetings that can run for six weeks or more. Recent WRCs have confronted such difficult issues as equitable access to the orbital arc, allocating spectrum for new and expanding mobile satellite services, developing procedures for accommodating satellites in LEO, and accommodating both the growing need for "feeder links" to satellites and new terrestrial microwave services, including local multipoint distribution services.[43]

While grandfathering existing registrations, the ITU has established a framework for reserving at least one orbital slot for the fixed and broadcast satellite service requirements of all nations and for resolving conflicts on a bilateral, multilateral, and regional basis. The ITU also has expanded the amount of spectrum allocated for mobile satellite services and has begun to establish procedures by which operators of LEO satellites can coordinate their use with operators of geostationary orbiting satellites and terrestrial applications. The last few WRCs have emphasized mobile telecommunications and have addressed a number of satellite coordination and spectrum allocation issues.[44]

Software can compute optimal slotting plans based on the fact that satellites can collocate if they operate in different geographic regions, use different frequencies, and have diverse service parameters, coverage plans, transmission power, and interference levels. But there will always be a human component that requires an honest and impartial broker like that performed at ITU conferences.

The Space WRCs have also considered the role of global cooperatives like INTELSAT and their status within the ITU. While cooperative satellites situated at mid-ocean can serve the telecommunications needs of many nations, the ITU must acknowledge the interest of individual nations in retaining options for national systems. A proliferation of satellite systems can result in excess capacity and inefficiency, much like what arguably has occurred in international commercial aviation, given the number of national flag carriers. On the other hand, nations with multiple satellite systems cannot seize the diplomatic high ground in orbital slot negotiations, given the number of slots their own satellites already occupy.

43. See, for example, "Establishment of Policies and Service Rules for the Mobile Satellite Service in the 2 GHz Band," IB Docket No. 99-81, Report and Order, FCC 00-302, 2000 WL 1209424 (F.C.C.)(rel. August 25, 2000).

44. See Federal Communications Commission, WRC 2000 Homepage, available at http://www.fcc.gov/ib/WRC00/.

Cooperatives like INTELSAT have official observer capacity at the ITU and use the services of the nation where the cooperative headquarters is located (the United States in the case of INTELSAT) for satellite registration and official advocacy. Conferring an opportunity for direct participation might bolster such cooperatives' negotiating leverage, but some nations believe that would confer "supersovereign" status. However, privatization efforts by both INTELSAT and Inmarsat will reduce the potential for conflict, because the remaining cooperative will have a small satellite constellation compared to the number spun off to private ventures. In any event, the ITU and its constituent nations must regularly address how to share the satellite orbital arc to avoid having to consider the tougher task of having to ration slots.

6.6.4 Orbital Slot Reservation Alternatives

Given the prospect for more satellite systems separate from international cooperatives, one can question whether the GSO registration system can remain intact and whether nations like the United States can continue to enjoy the luxury of supporting an open-skies, liberal market entry policy that currently fills in excess of 25 geostationary orbital slots. Noting the reservation of orbital slots by developing countries that might never use them, some policy makers[45] and academics[46] have proposed the use of auctions and other types of market valuation mechanisms that would enable a developing country to transfer a slot for financial compensation.

---

45. "A market for spectrum licenses or rights, if properly structured, can maximize both 'allocative efficiency' (i.e., prices bid for spectrum reflect the costs to society of spectrum use) and 'distributive efficiency' (i.e., those who value the spectrum most will use it)." U.S. Dept. of Commerce, National Telecommunications and Information Administration, *U.S. Spectrum Management Policy: Agenda for the Future*, NTIA Special Pub. 91-23, 1991, p. 98.

46. See, for example, H. Levin, "The Political Economy of Orbit Spectrum Leasing," *Michigan Yearbook of International Legal Studies* (New York: Clark Boardman, 1984), p. 41; H. Levin, "Emergent Markets for Orbit-Spectrum Assignments: An Idea Whose Time Has Come," *Telecommunications Policy*, Vol. 12, No. 1, Mar. 1988, p. 68; T. Schroepfer, "Fee-Based Incentives and the Efficient Use of Spectrum," *Federal Communications Law Journal*, Vol. 44, No. 3, May 1992, p. 411; F. G. Hart, "Orbit Spectrum Policy-Evaluating Proposals and Regimes for Outer Space," *Telecommunications Policy*, Vol. 15, No. 1, Feb. 1991, pp. 63–74; see also De Vany, et al., "A Property System for Market Allocation of the Electromagnetic Spectrum: A Legal-Economic-Engineering Study," *Stanford Law Review*, Vol. 21, 1969, p. 1499.

The Communications Act precludes United States' licensees of spectrum to view their authorization as property, but the FCC has received legislative authority to auction some spectrum [5]. In the satellite arena, the FCC has approved prelaunch sale of transponder capacity for the lifetime of the satellite [6] and so far has auctioned off two direct broadcast satellite orbital slots. By extension, should a developing nation decide against operating its own satellite system, its reserved slot might have significant resale value if nearby developed nations need additional space.[47] While the GSO has the characteristic of a *res communes* (shared global resource), consensus-reached decisions designating specified slots to specific nations could create a secondary market for conversion of a slot into a more desirable resource: hard currency.

6.7 Standard Setting

International telecommunications standard setting typically combines the hard science of technology with softer social science factors such as politics and industrial policy. The process does not always result in optimized technological performance, because economists, lawyers, business executives, diplomats, government officials, and bureaucrats join with engineers in national delegations participating at ITU-sponsored forums. Standards are developed internationally at the ITU, but increasingly national and regional standard-setting bodies undertake a parallel and sometimes independent process.

Standard-setting bodies provide common specifications that run the gamut from the type of plug used for telephones to the format for routing, signaling, and transmission to the rate elements in a tariff and the dialing procedure necessary to complete an international telephone call.[48] They provide the basis for uniform, or at least interconnected, operating systems.

---

47. The Kingdom of Tonga advanced published with the ITU a total of 31 "Tongasat" satellites to be located in 26 separate orbital locations.

48. "Standardization is a common toll of rational, control-oriented societies: it simplifies and brings order, releasing energies for other tasks. To this extent standardization is more than the dull agent of conformity: it is also essential for flexibility and diversity. . . . They become the means for organizing change as technologies change continuously in bursts of innovation and slower periods of modification." G. J. Mulgan, *Communication and Control: Networks and New Economics of Communication* (New York: Gilford Press, 1991), pp. 184–185.

There are two primary classifications for standards: (1) end-to-end standards that cover network systems and the facilities necessary to achieve a complete linkage and (2) interface standards that enable end users to attach equipment on the customer premises onto a network.

Standards provide timely solutions to problems by assigning precise meanings, definitions, and rules. Advocates for standard setting emphasize the need for and the benefits derived from a consistent, systematic, transparent, and comprehensive explanation of how to erect an international telecommunications and information-processing network. Common rules of the road make it possible to achieve shared goals such as efficiency, interoperability, ubiquity, portability, ease of use, a common interface so different equipment can access networks, expanded choice, and scale economies.

Successful implementation of a uniform standard can achieve the following benefits for businesses and society:

- Reduced equipment costs by making network elements interchangeable and available from a larger number of competing manufacturers and vendors;
- Enhanced vendor independence, making it less likely that a manufacturer of a particular network element can hold vendors captive to unfair terms and conditions;
- Improved opportunities for manufacturers and vendors to achieve lower per-unit costs by having fewer incompatible product lines to support;
- Network transparency, meaning that services can be provided seamlessly across borders even where different equipment and carriers are involved;
- A wealth of information on network functionality in the public domain;
- The ability of individuals and multinational enterprises to achieve reliable and efficient networking capabilities.

The failure to reach a single standard or the reaching of a consensus-driven standard leached of specificity and beaten down to a least common denominator may result in reduced efficiency and consumer welfare. When they believe their innovations and proprietary technology may establish a de facto, market-based standard instead of a mediocre, delayed, or unavailable ITU-generated standard, vendors and users are tempted to use self-help actions that bypass the ITU process. Vendors appear to support open systems and compatibility, but they may worry about having to disclose too much information to competitors and ceding too much power to users, who, with ample choice of options, will try to play one vendor against the others in search of a better deal.

Users articulate a need for interoperability and compatibility, but they may be tempted to use products and services with special features. If a core group of trend-setting manufacturers, service providers, and users, including governments, cannot agree on a single standard, then second-tier players will not or cannot follow. Breakaway forums and ad hoc user groups step into the void that is the result of incumbent bodies failing to reach a timely solution.

The near ubiquity and seamlessness of international direct dialing and facsimile connectivity attest to the potential for success in the standard-setting process. On the other hand, the failure to achieve a single broadcast color television standard, the vast array of incompatible computer languages, and competing formats (e.g., Beta versus VHS videotapes) point to the adverse consequences of failure.

6.7.1 Standard Setting in Perspective

Standard setting generates public goods in the sense that the product is designed for general application and one's use does not reduce what others have available. Because international telecommunications involve interconnection of networks, agreement on standards promotes the use of diverse types of equipment and the ability to communicate across national borders. Universal standards promote rationality, conformity, economies of scale, and lower risk in facility and service deployment.

On the other hand, the standard-setting process takes network and equipment design away from individuals, who might be in the best position to customize private applications. When forums like the ITU secure a consensus, they may reduce a standard to a least common denominator designed to serve the so-called typical user. The interest in uniformity may result in standards with less complexity but also less utility for sophisticated users requiring customized applications.

Once predominated by engineers seeking optimized technological solutions, the standard-setting process has increasingly become politicized and subject to national industrial policies as governments[49] and private

49. For a recent Congressionally commissioned assessment of U.S. interests in the standard-setting process, see U.S. Congress, Office of Technology Assessment, *Global Stan-*

enterprises recognize the stakes involved.[50] When individual government or corporate agendas drive or influence standard setting, the process and the output reflect a more diverse constituency. Some players may seek to capture the process for monetary or nationalist gain by advocating a homegrown standard developed by domestic manufacturers and service providers. Such advocates infer that regional or global endorsement of a domestic standard will accrue a marketplace advantage. Accordingly, when the stakes are high, as is the case with standard setting for global, integrated telecommunications and information processing, a greater risk exists that ulterior and self-serving motives drive much of the advocacy. While common rules of the road promote network connectivity and seamlessness, participants in the standard-setting process may advocate the solution developed by their domestic manufacturers and service providers.

6.7.2 Increasing Complexity in Standard Setting

A number of recent factors have increased the stakes and the complexity of the standard-setting process:

- Increasing numbers of competitors, including newcomers from the computer manufacturing and services industries, entering the marketplace for telecommunications equipment and services;
- The convergence of telecommunications and information-processing technologies requiring standard-setting participation by individuals and enterprises having less experience with and confidence in the process;

---

dards—Building Blocks for the Future, TCT-512 (Washington, D.C.: U.S. Government Printing Office, Mar. 1992).

50. "Standards have never been innocent or purely technical. They have been used a form of discreet (and indeed indiscreet) trade protection. . . . They have also served as tools of corporate strategy and governmental industrial policy. Because of this political dimension orthodox economics has proved ill-suited to describing the dynamics of standards processes; the economists' assumption of optimality offers little insight." G. J. Mulgan, *Communication and Control: Networks and New Economics of Communication* (New York: Gilford Press, 1991), p. 186. See also G. Wallenstein, *Setting Global Telecommunication Standards: The Stakes, The Players & The Process* (Norwood, MA: Artech House, 1990).

- Heightened recognition by governments of the economic rewards in setting standards with an eye toward making output from national heroes the de facto or ITU-sanctioned global standard;[51]
- Bolstered interest in serving global markets to compensate for lower margins resulting from heightened competition and the need to defray higher development costs.

The desire for universal interfaces, which permit interconnection of different equipment types, but with the requirement that such interfaces not prevent the use of proprietary equipment designs and novel services interconnected through the interface (e.g., setting standard audio tones for touch tone telephones but allowing answering machine manufacturers to design features activated by a sequence of tones). Standard setting is key to achieving global connectivity and the efficient, worldwide distribution of information, goods, and services at reasonable rates. Pressure has increased to accelerate the process and to specify standards in tandem with preliminary research and development of the underlying technology. Entrepreneurs want to rush products and services to market perhaps even before sufficient time has passed to debug systems and to resolve conflicts over how specific the standard should be. Increasingly, it appears that only a narrow window exists when sufficient operating experience meshes with an adequate timetable for international deliberation. Otherwise, operators may concentrate on marketplace concerns irrespective of whether it supports or bypasses the process for consensus building in the standard-setting process.

The failure to reach a global standard on a timely basis can have several results:

51. The Office of Technology Assessment reported that critics of the U.S. standard development process have concerns that "other countries are better organized and better able to influence the international standard setting process, to the detriment of U.S. trade. In particular, they fear that the harmonization of European trade law … [and the formation of the European Telecommunication Standards Institute] … will not only make it harder for U.S. companies to trade in Europe, but will also allow the Europeans to take the lead in setting international standards." *OTA Global Standards Study*, pp. 3–4, citing U.S. International Trade Commission, "Standards, Testing and Certification," Chap. 6 in *The Effects of Greater Economic Integration Within the European Community on the United States: First Follow-Up Report*, USITC Pub. 2288 (Washington, D.C.: USITC, Mar. 1990).

- Development of multiple, market-driven standards that may be incompatible with each other and, accordingly, that may prevent or raise the cost for interoperability, resulting in stifled innovation and decreased options for consumers;
- Establishment of regional standard-setting bodies, perhaps with an implicit mandate to favor indigenous manufacturers and service providers;
- Imposition of higher costs, greater risk, instability, and delay as various enterprises and standard-setting organizations attempt to work out differences that might not have evolved if a centralized standard-setting process had worked;
- Bolstered incentives for enterprises and users to concentrate on the short-term need for specialized networks that take advantage of liberalized regulatory policies rather than address long-term need for standards that help achieve major changes in infrastructure and the services available (e.g., conversion from analog to digital networking).

The kinds of negative consequences in the failure to reach a timely consensus may create incentives to create multiple standards or provisional standards. In most instances, participants in the process seem to believe that multiple or imperfect standards are better than none at all, particularly if official activity can preempt de facto standard-setting activities elsewhere. If users or manufacturers take matters into their own hands, the outcome may be immediate privately beneficial results, with only the possibility that larger public dividends will follow.

International standard setting, whether under the auspices of the ITU or another global forum, succeeds on consensus building and voluntary compliance. A mandatory standard, particularly one created within a single nation or region, may violate national sovereignty and may constitute a non-tariff barrier to trade. On the other hand, laissez-faire reliance on market-determined, de facto standards also can fail to maximize consumer welfare. Reliance on the marketplace, no matter how consistent with the prevailing philosophy in some nations, can result in stranded investment in equipment by consumers who bet on the wrong standard even though, on objective technical criteria, it might be superior (e.g., the Beta video standard).

6.7.3 Predicting When Consensus Will Occur

Stanley Besen and Garth Saloner provide an analytical framework[52] for assessing the standard-setting process through four quadrants representing high or low interest in: (1) promoting the universal adoption of any standard and (2) whether individual parties have a vested interest in one particular standard. Pure coordination occurs when the per capita rewards to participate in standard setting are large enough to induce participation, even though the parties will compete (e.g., determining where to place the steering wheel in automobiles). Parties have a high interest in a universal standard and no vested interest in any particular company's proposed standard.

When the per capita gain from standardization is too small for individuals to find participation worthwhile, government may have to intervene to promulgate one. In such a scenario, government involvement creates a public good and confers a societal benefit, as was the case when government promulgated standards for weights and measures, time, and language. Governmental intervention generates maximum public benefits when the marketplace or prospective manufacturers and service providers are so fragmented and less apt to have a vested interest in a particular standard.

The standard-setting process generates "pure private goods" when individual participants have a high interest in promoting a particular standard (e.g., a software platform for personal computers) but the industry lacks a single dominant firm that can establish a de facto standard. While society would benefit from a single standard, in terms of reduced costs and lower likelihood of stranded investment and incompatible technologies, interested parties may stall proceedings if their particular standard appears unlikely to receive official endorsement.

Conflict reflects dueling standards (e.g., Beta versus VHS videotape), when individual parties have a keen interest in promoting a particular standard, and society in general has an interest in and would benefit from a universal standard. A dominant firm may try to coerce others into accepting its standard, but it cannot organize a new standard-setting forum. Absent a formalized standard-setting process, manufacturers using divergent standards will vie for consumers in the marketplace.

52. The matter of incentives to promote consensus-driven standards is explored by S. M. Besen and G. Saloner, "The Economics of Telecommunications Standards," in R. W. Crandall and K. Flamm, eds., *Changing the Rules: Technological Change, International Competition and Regulation in Communications* (Washington, D.C.: The Brookings Institute, 1989), pp. 147–220.

Single standards enable manufacturers and service providers to reduce costs and more easily achieve scale economies. Standards facilitate access to new markets through product and service compatibility, in the absence of artificially created barriers to entry like closed procurements. Single standards promote lower product development costs, speed product rollout, and guard against balkanization (i.e., fractionalizing product and service markets into relatively impenetrable national or regional markets).

6.7.4 Standard-Setting Models

6.7.4.1 The Traditional Model

Traditionally, standard setting has occurred as part of the close collaboration and consultation undertaken by incumbent carriers, manufacturers, and government representatives. While national carriers and manufacturers might promote a standard that would favor national heroes, nations collectively reach consensus at the ITU on operational, international standards. Nations either fall in line and adopt a single complete standard or agree to a degree of standardization sufficient to make it possible for interconnection between different networks. Generally, nations install international gateways that comply with global standards to serve as the point of interconnection with other, possibly incompatible foreign networks. A nation may use incompatible or obsolete domestic facilities, but its installation of a satellite or submarine cable gateway that conforms to contemporary standards ensures access to and from the rest of the world.

The standard-setting process in the traditional model proceeds at a slow, measured pace and establishes a least common denominator that harmonizes existing national standards into an acceptable compromise. The process emphasizes voluntary consensus among like-minded players, all of which share a stake in managing change. The parties also share an interest in the incremental evolution of a standard over the long term so the cost of any innovation can be fully amortized.

The traditional model views standard setting as capping off a period of experimentation leading up to widespread diffusion of an innovation. It does not represent a forward-looking process in which innovators propose standards. Instead, it emphasizes a process in which manufacturers and service providers collaborate and then make their offerings comply with the agreed-on standard.

6.7.4.2 The New Model

The new model emphasizes speedy and market-influenced standard setting. In this model, some operators and manufacturers resort to regional bodies or none at all should the ITU fail to act on a timely basis. The new model reflects the more extensive integration of business enterprises and users with governments while also evidencing increasing tension over shortcomings in centralized standard setting. The process must closely track technological innovation rather than await a consensus to evolve on which innovation to use as the foundation for a standard.

In this model, standard-setting participants and outsiders who never achieved access to the process are less inclined to accept the ITU's centralized standard-setting process as establishing compulsory international rules.[53] More parties want to become proactive rather than reactive standard takers, particularly because they now recognize the marketplace advantages accruing from aggressive efforts to convert proprietary technology into the industry standard.[54] Manufacturers and service providers want to participate in the process at the outset to ensure that, even if their proposals do not become standards, their equipment can interconnect with other equipment using the established standard.

With increasing frequency, several enterprises involved in the standard-setting process join in strategic alliances even though individually they may have developed different products or services. Alliances form on the assumption that there is strength in numbers both to achieve economies of scale and scope and to improve the odds for expediting and dominating the standard-setting process. They anticipate improved odds for success by collectively achieving a global marketing presence. Absent such collaboration, multiple and incompatible standards fragment markets, making individual countries less likely to be served by multiple foreign vendors but at the same rendering each market segment less lucrative than a single, global market. For example, electricity-using appliances in the United States and the United Kingdom vary in terms of voltage, plugs, and power. For a manufacturer in

---

53. The old model elevated the ITU standard setting to virtual absoluteness. "They were the anchor of a regime that facilitated bilateral monopolistic bargains, reinforced national monopolies, and limited the rights of private firms in the global market. In short, the conventional view of the telecommunications regime as primarily a technocratic exercise in technical collaboration is wrong." P. F. Cowhey, "The International Telecommunications Regime: The Political Roots of Regimes for High Technology," *International Organization*, Vol. 44, No. 2, Spring 1990, pp. 169, 176.

54. In telecommunications and information processing, it appears that in increasing numbers manufacturers are willing to adopt a market posture geared to make their technology the de facto standard. Companies offering "the next best thing" or "a killer application" hope to establish a de facto industry standard that they can license to others.

either country to target and serve the other country, it would have to develop an entirely new product line or retrofit existing appliances with new plugs and adapters to convert the appliances for operation using a different level and type of electrical power.

6.7.5 Products of the Standard-Setting Process

Standardization typically results from a bottom-up development process, beginning with an innovation, leading to experimentation and testing, distilling into building block frames of reference, and finally leading to the formation of uniform standards. Alternatively, a de facto standard may be established when official bodies endorse an existing technology. Standards can evolve and become more comprehensive, particularly when they are structured in modules that establish a least common denominator but allow additional developments.

6.7.6 International Standard Setting

The Telecommunication Standardization Bureau of the ITU has the responsibility to issue recommendations on technical, operating, and tariff questions with a view to promoting telecommunications connectivity on a worldwide basis.[55] The ITU can issue only recommendations,[56] which nations and national standard-setting bodies can use in promulgating compulsory or voluntary domestic or regional standards.

The ITU's standard-setting process works in parallel but not necessarily in full coordination with other international standard-setting organizations. These bodies include the International Standards Organization (ISO), whose membership consists of about 90 national standards bodies. The ISO promulgates standards for worldwide use in numerous fields, not including electrical and electronics matters. The ISO has had a significant impact on information processing, including the creation of the Open System Reference Model, which presents a seven-tiered hierarchy of standards to

55. The radiocommunications sector has a narrow, nonduplicative standard-setting role relating to spectrum use. It does not include the issuance of recommendations on the interconnection of radio systems to public telecommunications networks and performance level of such interconnections. Those functions are specifically assigned to the telecommunications standardization sector.

56. The ITU's recommendations are collected in a series of "books," with each book's color reflecting the year of issuance.

foster connectivity between different information processing systems.[57] Other bodies include the International Electrotechnical Commission, the International Maritime Organization, the World Intellectual Property Organization, the Universal Postal Union, and the International Civil Aviation Organization.

An ITU-recommended standard typically constitutes the final product of a process that involves the generation of questions from the ITU's council and plenary assemblies addressing telecommunications standardization and radiocommunications. Those questions are assigned to one or more study groups, with a particular and sometimes narrowly drawn area of responsibility and expertise. Study group participants include government officials and representatives from individual companies and trade associations. The findings and recommendations of the study groups are considered at world telecommunications standardization assemblies and radiocommunications conferences, which are convened every four years. The ITU has arrangements for ratification by correspondence if the involved study group unanimously endorses an expedited process.

The ITU standard-setting process[58] primarily involves government and industry participants. The ITU governance structure permits participation by recognized private operating agencies (e.g., AT&T) and scientific or industrial organizations (e.g., IBM). Other intergovernmental or standard-setting organizations participate as observers, and the ITU recently encouraged ad hoc participation by interested parties to be deemed associates. Both the radiocommunications and telecommunications standardization bureaus of the ITU now have advisory groups, whose membership includes trade associations, manufacturers, and other nongovernmental entities. This new component reflects recognition by the ITU's governmental members that the forum must make the standard-setting process more representative of the

57. "The Open Systems Interconnect model is the computer industry's attempt to come to terms with the convergence of computing and telecommunications. A general law of information technologies states that in design, assembly, testing and networking costs rise with complexity and complexity rises with the number of objects to be managed. . . . The OSI project, launched by the ISO in 1978, had the aim of achieving the interworking of any computer system irrespective of its manufacturer, operating system or location." G. J. Mulgan, *Communication and Control: Networks and New Economics of Communication* (New York: Gilford Press, 1991), p. 199.

58. See Krishna Jayakar, "Globalization and the Legitimacy of International Telecommunications Standard-Setting Organizations," *Indiana Journal of Global Legal Studies*, Vol. 5, Spring 1998, p. 711.

increasingly diverse and nongovernmental set of players. The ITU advisory groups review priorities, strategies, and work progress. Additionally, they provide guidelines and recommendations on how to coordinate with other standard-setting bodies.

Nations and individual companies commit the resources required by the ITU standard-setting process primarily because it confers a recognized seal of approval. Adopting an ITU-recommended standard generates consumer confidence, reduces confusion, and promotes uniformity across borders. Nevertheless, in many high-stakes standardization contests, like broadcast color television and cellular radio, nations have failed to reach closure on a single standard.

Disincentives to an international consensus standard exist for several reasons.

- Two or more equipment manufacturers or service providers have developed—and heavily invested in—incompatible means of achieving the same technological goal. If a national government or regional alliance has invested heavily in research and development of a particular standard, for example, color television, they may attempt to foreclose foreign ventures operating on a different standard from acquiring the certification needed for lawful sales in-country.

- A technology is not perceived as international in nature and hence can be protected against foreign competition through using incompatible standards as nontariff barriers (NTBs). For example, nations initially viewed cellular radio as a local, noninterconnected facility. Because there was no perceived benefit in coordinating with other nations on a regional basis to support wide-area roaming by visiting business executives, a nation might decide to promulgate a standard incompatible with carrier equipment operating in nearby nations to prevent manufacturers in those nations from exporting equipment.

- A national telecommunications equipment manufacturing industry may be threatened by foreign competition, and incompatibility may serve as an effective NTB. For example, U.S. manufacturers of digital multiplexing equipment must set up a separate product line or software support for an incompatible system in Europe and elsewhere. U.S. T-1 circuits have 1.544 Mbps of throughput, while European E-1 circuits represent 2.048 Mbps.

- Basic philosophical differences concerning the need for standardization cannot be reconciled. For example, the FCC could not decide whether to sponsor a standard for AM radio stereo transmissions or to defer to a market-driven solution. The Commission's failure to promulgate a standard may have constituted a key reason why no single standard and no market developed.

The ITU standard-setting process is viewed by members as inequitable. At various times, developing nations have considered the process a means for developed nations to maintain market domination and to preclude developing countries from establishing indigenous manufacturing capabilities [7]. Nevertheless, in most cases, the need for uniformity and global connectivity constitutes strong incentives for differing factions to reach closure.

6.8 Development Issues

The 1989 Plenipot of the ITU, recognizing the need for greater emphasis in development assistance, created a new, permanent organ: the Telecommunications Development Bureau (BDT).[59] BDT operates in an environment in which the developing world still lacks access to basic telecommunications capabilities and the disparity grows between developed and developing nations in terms of overall access to telecommunications equipment and services. On the other hand, as the dichotomy of technology access worsens between developed and developing nations,[60] the ITU's one-nation, one-vote structure has motivated developing nations to give more attention to the problem and a commitment of more funding by developed nations.

---

59. For a summary of the proceeding that led to the formation of the Telecommunications Development Bureau, see B. Harris, "The New Telecommunications Development Bureau of the International Telecommunication Union," *American University Journal of International Law and Policy*, Vol. 7, 1991, p. 83.

60. See William P. Cassidy, Jr., "Universal Service in a Competitive Telecommunications Environment: The Current State of Universal Service in the European Union and the United States," *North Carolina Journal of International Law and Commercial Regulation*, Vol. 10, Fall 1999, p. 107; Edward R. Leahy and Michael O'Brien, "Telecommunications Law and Technology in the Developing World," *Boston College International and Comparative Law Journal*, Vol. 22, Winter 1999, p. 1; Christopher J. Sozzi, "Project Finance and Facilitating Telecommunications Infrastructure Development in Newly-

International development issues tend to arise in a crisis-like atmosphere, replete with acrimony and recriminations. The 1982 Plenipot nearly became captive to political issues related to development and equitable access to the satellite orbital slots. At the behest of developing nations, the conference amended the ITU Convention to obligate the organization to initiate efforts aimed at serving the special needs of developing countries. The 1982 Plenipot became a politicized forum for various expressions of what constitutes special needs and the scope of obligations developed nations should assume to provide favorable treatment to their less developed counterparts.

In a less politicized environment, the BDT serves as a legitimate forum to address development issues, including access to capital, technical and managerial training, educating policy makers on the importance of telecommunications, and seeking private industry's participation in telecommunications financing. The ITU Convention authorizes the BDT to convene world and regional conferences to address such issues.

6.8.1 Telecommunications Development Assistance

Rapid advances in telecommunications have not accrued the same financial and social dividends throughout the world. While the world economy has grown increasingly integrated and interconnected, developing nations continue to lag behind in terms both of statistical indices of technology diffusion,[61] like the number of telephone lines per 100 inhabitants, and of the economic stimulation that an efficient and modern telecommunications infrastructure generates.[62]

Industrializing Countries," *Santa Clara Computer and High Technology Law Journal*, Vol. 12, Aug. 1996, p. 435.

61. The number of telephone lines per 100 inhabitants is a frequently used statistical index of market penetration in a nation. In 1993, Sweden had 68.3 telephone lines per 100 inhabitants, Canada 59.7, the United States 57.8, Australia 47.8, the United Kingdom 46.1, Thailand 3.7, China 1.5, Indonesia 1.0, and India 0.9. G. Staple, ed., *TeleGeography 1994, National Telecommunications Indicators* (Washington, D.C.: International Institute of Communications, 1992), pp. 164–167. More extensive statistics are available in ITU, *Yearbook of Common Carrier Telecommunication Statistics* (Geneva: ITU, 1999).

62. See, for example, R. Saunders, J. Warford, and B. Wellenius, *Telecommunications and Economic Development* (Baltimore: Johns Hopkins University Press, 1983); B. Wellenius, P. Stern, and T. Nulty, eds., *Restructuring and Managing the Telecommunications Sector* (Washington, D.C.: World Bank, 1989).

The telecommunications infrastructure in many developing countries suffers from a combination of underdevelopment, underfinancing, and overuse where available. Accordingly, many of the nations that are most apt to benefit from telecommunications development have inferior opportunities to improve national cohesion, achieve equitable access to information and social discourse, reduce disparity between urban and rural telecommunications options, and enhance the quality of life. Nations least able to finance telecommunications development are the ones most in need of it. Those nations may not easily afford such development, but they can ill afford not to commit scarce hard currency to make them more attractive targets for foreign investment.

Ithiel de Sola Pool considers development in three different contexts:

1. Economic measures like gross national product, productivity, and living standard indices;
2. Multiplication of centers of initiative, that is, diversification and increased complexity in organization and division of labor in terms of where and how initiatives may occur;
3. Modernization through the transfer of technology, that is, the acquisition of technical skills to adapt and use innovations, thereby enhancing social welfare.[63]

Professor de Sola Pool identifies a number of characteristics that a telecommunications system should have for effective diffusion.

> [It should be] an adjunct to expression by those persons who have credibility in the culture. It should use the language and the symbols of the culture. Its contents should be capable of local adaptation. It must be cheap. It should require as little foreign exchange as possible. It must also be reliable and relatively rugged, and not require highly sophisticated maintenance and operating personnel. It must operate even in the absence of an elaborate infrastructure … [and be capable of] link[ing] the underdeveloped region. [8]

---

63. See I. de Sola Pool and E. Noam, eds., "Communications for Less Developed Countries," in *Technologies Without Boundaries—On Telecommunications in a Global Age* (Cambridge, MA: Harvard University Press, 1990), pp. 167–204.

The lag in telecommunications development persists notwithstanding the commonly held view that an expanded infrastructure and more accessible telecommunications networks will stimulate demand, increase revenues, and help jump start other sectors of a nation's economy. On the other hand, we should recognize that telecommunications must compete for scarce financial resources, including foreign currencies, with life-maintaining infrastructures like sanitation, health care, water, housing, transportation, and electricity. Even with heightened attention to the benefits from a viable telecommunications infrastructure, only about 3% of the World Bank's loan portfolio finances telecommunications projects [9].

While intuitively pleasing, the favorable contribution made to a nation's gross domestic product from telecommunications investment is not easily measured.[64] Studies show a statistically significant positive correlation between gross national product (GNP) and the level of telephone service penetration, that is, nations with higher GNPs have more telephone lines per 100 inhabitants.[65] Most studies, however, have not determined definitively the cause and effect resulting in such a correlation. Do improved telecommunications stimulate economic performance, or do improved economic conditions, measured by increases in GNP, make it easier for a nation to finance development, thereby making the telecommunications infrastructure more accessible and reliable?

Once governments decide to invest in telecommunications, they must consider how best to allocate funds. That kind of decision making blends economic concerns for maximizing utility and value with equity and political concerns (e.g., achieving a balance between serving the needs of urban residents and the needs of rural residents and between the complex requirements of corporate users and the need for POTS for residential users).

Often, policy makers must decide whether to allocate funds to maximize the number of lines and population served, or to upgrade the array of features available in locales already having access to POTS. Service to rural and residential users typically costs substantially more on a per-user basis than service on dense routes and in populated areas. As such,

---

64. See, for example, S. Pitroda, "Development, Democracy and the Village Telephone," *Harvard Business Review*, Nov.–Dec. 1993, pp. 66–79.

65. See, for example, H. Hudson, *When Telephones Reach the Village* (Norwood, NJ: Ablex, 1984); H. Hudson, *A Bibliography of Telecommunications and Socio-Economic Development* (Dedham, MA: Artech House, 1988); ITU, *World Telecommunications Development Report* (Geneva: ITU, 1999).

government must decide the public welfare merits of extending for the first time a basic lifeline service to remote areas versus upgrading facilities and replacing deteriorating equipment to improve unreliable or overtaxed service to urban users.[66]

Governments also must respond to the requirements of corporate users, who typically need more than POTS. Corporate users have the financial wherewithal and possibly the political clout to erect their own internal telecommunications facilities should the incumbent operator fail to respond adequately and quickly. If a large corporate or government user bypasses public network facilities, the incumbent carrier may have substantially less revenue available to underwrite improvements, because the largest users contribute disproportionately to the revenues generated. Once a large-volume user opts for a private facility, it may seek to share or resell its excess capacity. While such facilities may provide sophisticated options not available from the incumbent operator, they also can duplicate available options, possibly resulting in underuse of incumbent carrier facilities, known as stranded investment.

In 1984, the Independent Commission for Worldwide Telecommunications Development reported to the ITU[67] a substantial disparity in access to telecommunications between nations:

---

66. Telecommunications entrepreneurs eager to tap developing markets like to think that half of a nation is waiting for a phone line, while the other half is waiting for dial tone. "The shortage of sets and lines creates two situations that encourage over-use [of existing facilities]. First, because individuals and businesses cannot readily add telephones, each installed station serves a much larger number of people than is the case in the industrialized world. Thus, telephones are, to exaggerate a bit, seldom at rest during the business day. Second, over-utilization, in turn, breeds ... sequential dialing in an effort to seize the [often busy] local line once it becomes disengaged. These repetitive calling attempts, but not completions, also generate busy signals that frustrate others attempting to reach the original callers. In an attempt to get through to the party called, some subscribers resort to automatic dialing equipment which further exacerbates the number of busy signals in the system." Robert Bruce, et al., "The Telecom Mosaic—Assembling the New International Structure," Chap. VI in *Telecommunications Structures in the Developing World: An Essay on Telecommunications and Development* (Kent, England: Butterworths, 1988).

67. *The Missing Link*, Report of the Independent Commission for Worldwide Telecommunications Development (Geneva: ITU, 1984). Sir Donald Maitland chaired the commission created by the administrative council of the ITU as directed by the 1982 Plenipot. For a review of the commission's findings, see Rowan and Waite, "International Communications Law, Part I: Maitland Commission, Economic Development and the United States," *International Law*, Vol. 19, 1985, p. 1339.

> While telecommunications is taken for granted as a key factor in economic, commercial, social and cultural activity in industrialized countries and as an engine of growth, in most developing countries the telecommunication system is not adequate even to sustain essential services. [10, p. 3]

The commission noted that nine nations accounted for 75% of the then-available 600 million telephones in the world: the United States, Japan, Germany, France, the United Kingdom, the Soviet Union, Italy, Canada, and Spain. It recommended that "by the early part of the next century virtually the whole of mankind should be brought within easy reach of a telephone, and in due course, the other services telecommunications can provide" [10, p. 4]. The commission estimated that an annual investment of $12 billion is necessary to achieve even incremental progress in reaching that goal [10, p.13]. It asserted that profitability alone constituted an "inappropriate criterion for assessing the merits of telecommunication facility loans and investments, because indirect benefits have to be taken into account" [10, p. 7].

6.8.2 Failed Strategies

Individual nations and international organizations have pursued different strategies to achieve improved telecommunications in lesser developed nations. Immediately following the Second World War, telecommunications constituted a small part of the technology transfer rewards for aligning on either side of the East-West political axis. By the late 1960s, "many Third World nations began to see Western dominance of the international economic and communication system as a source of their 'underdevelopment'" [11].

To bring about greater parity to the international environment, Third World leaders have called for a new world information order (NWIO),[68] an environment in which nations of all types have relative parity of access to telecommunications and information resources. While the concept has varied in both definition and application among advocates, many strategies for achieving parity of access were perceived by officials in developed nations as threatening commercial transactions involving intellectual property and even freedom of the press at the hands of authoritarian political regimes.

Scholars have identified several new approaches that avoid the stridency of calls for absolutely free access in a short period of time. New viewpoints

68. See Merehoo Jussawalla and C. W. Cheah, *The Calculus of International Communications* (Littleton, CO: Libraries Unlimited, 1987); Meheroo Jussawalla, ed., *Global Telecommunications Policies: The Challenge of Change* (Westport, CT: Greenwood Press, 1993).

embrace "grassroots or participatory communication [and political] policies for achieving economic growth" [12] "shaped by … the widespread proliferation of new information/communication technologies" [11, p. 488]. In application, that means that because development funding will remain inadequate, countries in need must achieve greater self-reliance based on available funding and training while striving for more liberal technology transfer, new sources for loans, and greater decision-making roles in lending and development organizations, as at the ITU. A World Bank technical paper notes a rise in entrepreneurship, a "trend deserv[ing] all the domestic and international support it can possibly get, because its strengthening promises to greatly contribute toward the 'normalization' of industrial and business conditions … [where] moderate incentives can be very effective" [13].

6.8.3 The Payoff from Telecommunications Investments

Telecommunications joins with energy and transportation as an essential component in social and economic development:

> The need … [for a] telecommunication infrastructure to guarantee the success of any capital investment is very obvious. The relocation of developed countries industrial estates to the developing countries will require a high quality and reliable telecommunication [network] for control and transfer of information between the new industrial location and the head office. [14]

These projects usually turn a profit[69] and enhance social welfare as measured by statistical methods[70] and economic analysis:

69. "The reason for inadequate investment in the telecommunications sector in developing countries is also not that telecommunication entities lose money or require government subsidies. In general, reasonably well managed telecommunications entities can generate large financial surpluses in local currency." R. Saunders, J. Warford, and B. Wellenius, *Telecommunications and Economic Development* (Baltimore: Johns Hopkins University Press, 1983), p. 12.

70. Even anecdotal evidence confirms the view that properly targeted and managed investments in telecommunications will enhance efficiency and other investments throughout a national economy. "The introduction of long distance telephones in the Amazon region of Peru in the late 1970s resulted in substantial cost savings and increased revenue in river transportation; in another region in 1983, the use of telex for reservations increased average hotel occupancy in a tourist town from less than 50 percent to about 70 percent. In Sri Lanka in the 1970s, tele-

> In a macroeconomic sense, the large unsatisfied demand for telecommunications services in developing countries and the high returns on new investment is evidence that the perceived communication needs of both producers and consumers are not being met. Not meeting this demand may worsen the extent to which information is distributed unequally between parties to transactions and may diminish the opportunity for information transfer. [15]

Common sense and anecdotal information confirm the view that a poor telecommunications infrastructure can retard national economic development:

> Inadequate telecommunications reduces efficiency throughout the economy, diminishes the effectiveness of investments in priority sectors and development programs, causes a comparative disadvantage in trade and in attracting investment, and lowers the quality of life in terms of personal access to emergency services and communication with kin and friends. In Uganda in 1983, for example, because of inadequate rural and provincial telecommunications and postal services, trucks collecting coffee and cotton from a large union of cooperatives made trips estimated to be 20 percent ineffective; also about 200 workyears were wasted in otherwise unnecessary administrative travel by senior government officials. [16]

In response to allegations that we cannot quantify with any precision the economic stimulation accruing from telecommunications development, Sir Donald Maitland restated the question:

> [H]as economic and social progress occurred in any country without accompanying investment in the communications infrastructure? The evidence available to the Commission left us in no doubt that there was indeed a correlation between economic development and investment in telecommunications. Which was the chicken and which was the egg seemed an esoteric question. [17]

---

phone access to market information allowed farmers to place their product at 80–90 percent of Colombo prices, as compared with 50–60 percent before; in the 1980s, Colombo food merchants relied on the telephone and telex for operations in international markets." Bjorn Wellenius, "Beginnings of Sector Reform in the Developing World," in B. Wellenius, P. Stern, and T. Nulty, eds., *Restructuring and Managing the Telecommunications Sector* (Washington, D.C.: World Bank, 1989), p. 91.

Telecommunications development means more than expanding the number of telephone lines per 100 inhabitants. To become attractive to foreign investment and to operate in an integrated global economy, nations also must deploy enhancements to POTS, such as high-speed data networks, cellular radio, satellite services, e-mail, and Internet access. Developing nations face the same pressure to upgrade and diversify the telecommunications sector as developed nations, but they are far less able to do so when even the basic infrastructure is incomplete, obsolete, or in disrepair.

To take advantage of the development and economic stimulation opportunities presented by telecommunications, developing countries must find solutions to several problems:

- Limits on access to borrowed or self-financed capital;
- The goal of providing universal, basic service while at the same time installing state-of-the-art overlay networks, including cellular radio, digital microwave, and very small aperture terminal satellite systems;
- Regulatory limitations on the pricing of services that may translate into mandatory underpricing of certain services;
- Compulsory contributions of telecommunications revenues into the general treasury;
- Institutional limitations that generate inefficiency, lower productivity per employee, and higher costs per unit of output;
- Limits on access to equipment, resulting from domestic trade and procurement policy or "tied-aid," the linking of loans and other financial assistance with the obligation to buy equipment only from suppliers in the nation providing the aid;
- The frequent requirement to pay for at least a portion of any procurement with hard currency;
- The diversification of constituencies (e.g., residential versus corporate) and interests (e.g., universal service versus business networks) mandating a new regulatory and management structure that promotes fairness.

Chapter 10 examines whether and how developing nations might benefit when government partially or completely exits from owning and centrally managing the telecommunications sector.

6.9 Reforming the ITU

Some nations perceive increasingly significant disincentives to participate in international forums like the ITU. The pace of technological innovation has taxed the ability of the ITU to respond to market developments and user needs with timely spectrum allocations, recommendations on operating procedures using new technologies, and ways to coordinate services. The heightened competitiveness in telecommunications industries means that individual carriers and equipment manufacturers may attempt to introduce products and services in advance of an ITU-promulgated standard with an eye toward establishing a de facto standard and dominant market share.

Much of the efficiency gains from global rules of the road could be lost without efforts by the ITU to address challenges to its efficiency and relevance. Many of the newfound incentives not to participate stem from fundamental changes in the composition of the telecommunications marketplace. A small number of monopoly carriers and national hero manufacturers dominated the old world order, a model that emphasized stability, incremental change, and collaboration among a few players with end-to-end network responsibility and control:

> Like the networks themselves, the old telecommunications world had a relatively simple, hierarchical, star architecture. There were only two tiers: at the national level there were monopoly providers; and for international coordination there was the ITU around which everyone clustered. It was a relatively small club of constituents, and no one else really mattered. [18, pp. 286, 291]

The ITU's role in the old environment was one of bureaucratically lending "good offices" for the expert participants to share information, develop standards, and adopt rules of the road geared for public facilities and services provided by single national carriers. Most participants in the process shared a common bond as technocratic elites operating a nation's natural monopoly PTT administration and pursuing similar objectives:

- Maximizing revenues;
- Using revenues from overpriced international services to cross-subsidize socially desired services like universal and inexpensive local telephone service;
- Maintaining legislatively conferred or de facto barriers to market entry.

The ITU well served those participants by emphasizing careful deliberation and continuity over speed and flexibility [18, p. 292].

The increasingly volatile, complex, and competitive telecommunications environment closely linked with information service markets supports a new world telecommunications order. The model supports telecommunications as a key vehicle to stimulate national economies and to promote commerce between nations. High-volume users need—and increasingly demand—an open, flexible, and diversified telecommunications infrastructure priced on the basis of cost and demand elasticity, rather than social policy.[71] Deregulation, privatization, liberalization, globalization, and conditional market entry have become the key elements of change.

6.9.1 The ITU and the New World Order

The new world order emphasizes responsiveness to consumer requirements rather than obedience to the incumbent carrier constituency. Speediness, accessibility, and flexibility are key factors, because new carriers, resellers, and equipment manufacturers have secured no reserved market or monopoly franchise. Likewise, they may have little faith in or loyalty to the ITU membership and staff that did not always welcome or accommodate them.[72]

---

71. Beginning in the late 1970s and early 1980s, large users began to consider the consequences of high telecommunications costs and the absence of competition. While few enjoyed a legally possible option to erect a private, bypass network, users began to advocate change and recognize technological options in traffic routing, the ability to bargain over terms and conditions, the benefits from reduced regulation and increased competition, and the possibility that resale of valued-added telecommunications services could constitute a new profit center. See P. Cowhey and J. Aronson, "The ITU in Transition," *Telecommunications Policy*, Vol. 15, Aug. 1991, pp. 298, 300.

72. "It may be argued that the ITU is a specialized agency, and therefore may be immune from changes created by geopolitical shifts. This is a comforting illusion. The ITU may well be narrow in focus but telecommunications is now becoming much too central for competitive global business to be left exclusively to the ITU. Indeed, in many respects the ITU is in danger of suffering sclerosis from its age. In an increasingly borderless global village the ITU is still border bound, reflecting the era of nation states and administering regimes (e.g., accounting rates) which are based on strict respect for national sovereignty with an arm's length type of carrier relationship. At a time when innovation and global networking are essential to maintain competitive edge, intergovernmental multilateral negotiations which cannot keep pace with business practices are in danger of becoming increasingly irrelevant." Jonathan Solomon, "The ITU in a Time of Change," *Telecommunications Policy*, Vol. 15, Aug. 1991, pp. 372, 373.

Outsiders to the ITU are more inclined to test their innovations in the marketplace, without resolution of spectrum allocation, standards, service coordination, and interference-avoidance issues. As an alternative to a marketplace-driven de facto solution, new players may support bilateral negotiations by their governments or policy making in other forums geared primarily to set standards on a domestic or regional basis.

In an age of divergent yet interconnected networks, a multiplicity of operators, and a more activist user community, the ITU "is no longer on top of the pyramid, but part of a geodesy of organizations all representing different communities that comprise today's complex telecommunications universe. In addition, the organization is compelled to be considerably more effective at what it has traditionally done" [19]. The ITU now must work in a subject area that involves numerous regional standard-setting organizations, such as the European Telecommunications Standards Institute, international and regional satellite cooperatives, commercial satellite enterprises (e.g., PanAmSat), submarine cable consortia, numerous trade organizations (including the WTO), international and regional development bodies, user advocacy groups, and regulatory agencies.

In 1989, the Nice Plenipot passed a resolution calling for the creation of a high-level committee (HLC) to carry out an in-depth review of the structure and functioning of the ITU, with an eye toward recommending ways to make it more efficient and effective in a fast changing environment [20]. In particular, the ITU needed to respond to an overtaxed standard-setting process, which could not issue timely recommendations, privatization, liberalization, deregulation, and market entry by private enterprises.

The HLC made 96 recommendations for reform, most of which were considered and adopted, including a more routinized scheduling of conferences.[73] The ITU has operated under a federalist, nonhierarchical structure that confers substantial power and independence to standard setting and radio consultative committees.[74] The diffusion of power will remain in effect, but the Secretary General has acquired greater resources, including a small

---

73. The ITU will convene one major conference a year, with a Plenipot for long-term policy planning, followed by standards, development, and radiocommunications conferences in the subsequent three years. The ITU contemplates convening a Plenipot and a standardization conference once every four years, with world or regional radiocommunications conferences and assemblies occurring once every two years.

74. See G. A. Codding, Jr., and D. Gallegos, "The ITU's 'Federal' Structure," *Telecommunications Policy,* Vol. 15, Aug. 1991, pp. 351–363.

strategic planning and research group, and somewhat more power to lead the ITU into a future where its legitimacy and authority are not a given.

In 1999, the ITU once again reassessed its role, particularly in light of the Internet's ascendancy and the view that the ITU cannot keep up with convergence issues or matters more effectively considered in a trade policy multilateral forum like the WTO. A high-level reform advisory panel recognized the need for the ITU to take affirmative steps to "preserve and strengthen its international credibility" [21]. Amid the general language, the panel did identify specific and acute problems, including the need for greater private sector participation, management of the ITU, and the need to eliminate registration of "paper satellites," that is, satellites unlikely ever to be launched but nevertheless filed with the ITU to secure a place in the queue and possibly to provide an opportunity to secure compensation. The reform advisory panel also noted the importance in streamlining and expediting the ITU's standard-setting process and the need for the ITU to serve as a proactive knowledge center for dissemination of information about best-practice regulatory policies and for benchmarking statistics.

6.9.2 Achieving True Reform

The ITU has available procedures for achieving greater effectiveness. It can modify rules and recommendations to promote the integration of new technologies, and it can reallocate spectrum to accommodate new requirements, while still deferring to individual nations on the decision whether to permit such alternatives. However, that "consenting adults policy means that there is no true [single] global framework for commercial competition" [21]. Instead, three elements have evolved:

1. The traditional model of jointly provided service by foreign correspondents, comprising primarily incumbent telephone companies;
2. Niche market service providers, primarily value-added networks, that lease lines from incumbent carriers and whose acceptability arises on a bilateral basis when two nations negotiate, in ITU lexicon, "special arrangements";
3. New standalone facilities-based carriers (e.g., satellite systems separate from INTELSAT and operators of private cables, including new ventures like Pangea), not a consortium of existing operators.

The ITU has belatedly "adapt[ed] the Union's structure, management practices and working methods to the changes in the world of telecommunications and to the increasing demands placed upon it to keep pace with the ever-accelerating progress in telecommunications." The HLC recommendations as endorsed by the additional Plenipot in 1992 created a better foundation for the ITU to become more responsive and effective. Likewise, it placed the Union in a better position to respond to

- Other international and regional trade and standard-setting forums;
- The need to respond more speedily to technological change;
- The obligation to find ways for involving the increasingly vocal and active nongovernmental players in telecommunications.

However, the "larger obstacle to the ITU's future may be its 'international civil service' mentality. As the international telecommunications world becomes increasingly commercialized and deregulated, how will the inter-governmental ITU cope?" [22]. The answer lies in the ITU responding to changed circumstances in a timely and professional fashion. Likewise, it requires the ITU to welcome more eagerly the participation of nongovernmental service providers and users.

6.10 Internet Standard Setting and Governance

The Internet and the rising importance of data communications and e-commerce raises the stakes for ITU reform. The ITU must do a better job in its core areas of competency or risk losing legitimacy to other multilateral forums or to a hybrid processing that combines marketplace results with voluntary regional and industry-specific forums. Additionally, the ITU must consider whether and how it might play a role in areas blending telecommunications with information processing and the Internet.

So far, the ITU has not made much of an impact on Internet governance, but its standard-setting process has incorporated issues pertaining to the technical infrastructure. In the former, the ITU tried to situate itself as a central conflict-resolution forum and a locus for the registration of Internet domain names,[75] among various national registrars. At the insistence of the U.S. government and, presumably, other stakeholders that favor a private

75. See ITU, Generic Top-Level Domain Memorandum of Understanding, available on-line at http://www.gtld-mou.org/.

alternative to an intergovernmental option, the Internet Corporation for Assigned Names and Numbers (ICANN) has assumed responsibility for administering the domain numbering system.

Theoretically, the ITU could serve as a fair-minded, international forum for Internet governance. Many of the issues involving the Internet have a telecommunications analog. For example, the ITU organized a global telephone numbering system whereby any telephone caller can reach any other telephone in the world by dialing the proper sequence of digits. Telephone company switches of various manufacturers have the capacity to recognize country and city codes and to route traffic to the intended destination. The domain numbering system can be analogized to telephone numbers, particularly because the letters of a World Wide Web site name are translated by IP into a specific and distinct sequence of digits. Nevertheless, the ITU's generic top-level domain policy oversight committee noted that

> little progress has been made because of both an extremely complex political environment and the economic interests of the parties involved in the multibillion dollar business of selling domain names. The Policy Oversight Committee also notes that despite a multiyear effort involving the United States Government and ICANN, little progress has been made in the core issue—the necessary economic model for [coordination of] registry/registrar relationships [where multiple competing registrars operate]. [23]

It remains to be seen just how viable the ITU will remain in both telecommunications and information policy.

References

[1] ITU, Press and Public Relations Service, "The New ITU: Round-Up," Geneva: ITU, Jan. 1993.

[2] U.S. Congress, Office of Technology Assessment, "The 1992 World Administrative Radio Conference: Issues for U.S. International Spectrum Policy—Background Paper," OTA-BP-TCT-76, Nov. 1991, p. 9.

[3] U.S. Dept. of Commerce, "National Telecommunications and Information Administration," Appendix A in *Telecom 2000—Charting the Course for a New Century*, Washington, D.C.: Government Printing Office, 1988, p. 655.

[4] "An Inquiry Relating to Preparation for the International Telecommunication Union World Administrative Radio Conference for Dealing with Frequency Allocations in Certain Parts of the Spectrum," GEN Docket No. 89-554, Second Notice of Inquiry at

26, para 56, citing Notice of Proposed Rulemaking, GEN Docket No. 89-103, 4 FCC Rcd. 4173, 4178 (1989); see also Report and Order, FCC 91-188 (rel. June 20, 1991).

[5] 103d Congress, 1st Sess., Omnibus Budget Reconciliation Act of 1993, Title VI, Communications Licensing and Spectrum Allocation Provisions, PL 103-66 (HR 2264), 107 Stat. 312, 379 *et seq.* (1993).

[6] Domestic Fixed-Satellite Transponder Sales, 90 FCC 2d 1238 (1982), *aff'd sub nom.*, Wold Comms., Inc. v. FCC, 735 F.2d 1465 (DC Cir. 1984).

[7] Savage, J., *The Politics of International Telecommunications Regulation*, Boulder, CO: Westview Press, 1989, pp. 215–216.

[8] de Sola Pool, I., and E. Noam, eds., "Communications for Less Developed Countries," in *Technologies Without Boundaries—On Telecommunications in a Global Age*, Cambridge, MA: Harvard University Press, 1990, p. 177.

[9] Saunders, R., J. Warford, and B. Wellenius, *Telecommunications and Economic Development*, Baltimore: Johns Hopkins University Press, 1983, p. 67.

[10] Independent Commission for World Wide Telecommunications Development, *The Missing Link*, Executive Summary, Geneva: ITU, 1984, p. 3.

[11] Ayish, M., "International Communication in the 1990s: Implications for the Third World," *International Affairs*, Vol. 68, No. 3, 1992, p. 487.

[12] Jussawalla, M., "Is the Communications Link Still Missing?" *Telecommunications Policy*, Vol. 16, Aug. 1992, p. 485.

[13] Ivanek, F., T. Nulty, and N. Holcer, "Manufacturing Telecommunications Equipment in Newly Industrialized Countries, The Effect of Technological Progress," World Bank Technical Paper No. 145, Washington, D.C.: World Bank, 1991, p. 31.

[14] Soerjodibroto, T., "ASEAN and World Telecommunication Development," paper presented to *PTC '92*, Honolulu, p. 1.

[15] Mendis, V. C. B., "Phased Privatization with Proposed Foreign Participation: The Sri Lanka Experience," in B. Wellenius, P. Stern, and T. Nulty, eds., *Restructuring and Managing the Telecommunications Sector*, Washington, D.C.: World Bank, 1989, p. 103.

[16] Wellenius, B., "Beginnings of Sector Reform in the Developing World," in B. Wellenius, P. Stern, and T. Nulty, eds., *Restructuring and Managing the Telecommunications Sector*, Washington, D.C.: World Bank, 1989, p. 90.

[17] Maitland, D., "'The Missing Link': Still Missing 8 Years Later?" *Proceedings of Seminar on Telecommunications and Its Role in Socio-Economic Development*, Copenhagen: Eigtveds Pakhus, Nov. 3–5, 1992, p. 5.

[18] Rutkowski, A., "The ITU at the Cusp of Change," *Telecommunications Policy*, Vol. 15, Aug. 1991, pp. 286, 291.

[19] Solomon, J., "The ITU in a Time of Change," *Telecommunications Policy*, Vol. 15, Aug. 1991, pp. 372, 373.

[20] ITU, *Final Acts of the Plenipotentiary Conference (Nice, 1989)*, RES/80, Geneva: ITU, 1989.

[21] ITU Reform Advisory Panel, "Observations and Recommendations for Reform," available on-line at http://www.itu.int./newsroom/reform/02-RapRecs.html.

[22] Savage, J., "The High-Level Committee and the ITU in the 21st Century," *Telecommunications Policy*, Vol. 15, Aug. 1991, pp. 365, 370.

[23] ITU, Generic Top Level Domain Memorandum of Understanding, available on-line at http://www.gtld-mou.org/#readthisfirst.

7

Players in International Telecommunications Trade Policy Making

The international telecommunications marketplace represents an amalgam of four basic models: government monopoly, private monopoly, private (domestic or multinational) enterprise, and international strategic alliances. The first two models flourish with restricted market access and various barriers to trade, while the latter two thrive with the abandonment of restrictions and expanded opportunities to serve external markets.

Historically, the telecommunications marketplace has emphasized the first model with "national monopolies tied together by an international cartel that legally sanctioned administered prices, equal splits of international revenues, and rules that forbid competition for international traffic" [1]. Even now some basic principles of free trade do not apply to international telecommunications, because of these factors:

- National security concerns about relying on foreign suppliers;
- Industrial policy grounds on which nations strive to insulate indigenous manufacturers and service providers from competition while exploiting market access opportunities in other nations;
- The view that international telecommunications, like aviation, requires diplomatic agreements (e.g., the conferral of landing rights

for aircraft and satellite signals) and accordingly cannot rely solely on marketplace forces;

- Concerns about cultural imperialism, such as excessive foreign language on culture and sovereignty.

In the services arena, other factors work against open markets. The cost of providing telecommunications service favored collaborative ventures to achieve economies of scale. While such joint procurements in transmission facilities mean that carriers will have roughly the same per-unit costs, it does not automatically lead to a noncompetitive market. But in the case of international telecommunications, until just a few years ago association in satellite cooperatives and submarine cable consortia made it possible for carriers to avoid having to compete on price. An industrial structure dominated by government PTTs managed the rollout of technological innovation, controlled access to markets and facilities, established service definitions, institutionalized cross-subsidies, and largely captured or rendered ineffectual the regulatory process.

Recently accelerating privatization initiatives have changed the trade climate and prospects for market access by replacing government managers with entrepreneurs. Even in many nations where a private monopoly substitutes for one operated by the government, the climate has begun to change from centralized, government command and control to some degree of reliance on marketplace forces. At an accelerating pace, nations have subjected at least part of the telecommunications industry to competition.

The following conditions are behind the driving force for change:

- Consumers are less willing to accept excuses for poor, expensive, or unavailable service and are more likely to vote with their feet and their currency, that is, to pursue new options.
- Technological innovations make it easier for incumbents, newcomers, and users to customize network solutions and to provide alternatives to inflexible, expensive, or incompetent service.
- Globalization initiatives of governments encourage incumbents to find new markets to compensate for lost domestic market share.

- Increasingly diverse and global service requirements of multinational users and market entry by "systems integrators" aim to provide one-stop shopping and turnkey services.
- Trade initiatives have encouraged most nations to make commitments to liberalized market access opportunities.

Those forces cannot trigger government abandonment of regulatory oversight or convert telecommunications into a fully competitive marketplace. But they do create the basis for market entry and expanded options.

Improved trading conditions evolve not because governments have opted for smaller roles, but because governments, carriers, and manufacturers have created new market opportunities and have responded to consumer requirements. Government representatives may be instrumental in securing better and more transparent market access opportunities, but the drive for such advocacy comes primarily from users and entrepreneurs. Users no longer are resigned to tolerating service that is expensive, unreliable, and lackluster. Likewise, telecommunications companies, which have achieved limited market niche opportunities, will lobby for further liberalization. They argue that all markets can support competition and that trade, economic, and regulatory policies designed to insulate incumbents from competition actually serve as a crutch that renders incumbents less responsive to the need for streamlining and efficiency.

Multinational enterprises, such as banks, airlines, oil exploration firms, and financial service companies, need global, integrated, and reliable telecommunications networks. Incumbent carriers, which have lost domestic market share, and new, nimble service providers strive to become full-service operators providing one-stop-shopping convenience to users, both domestically and on a regional or global basis. Many companies with global requirements recognize that logistical and cultural challenges support reliance on experts, particularly alliances of several carriers who can provide worldwide service.

Nations that permit corporate alliances and foreign investment in telecommunications foster economic interdependence, which can become a force for trade policy adjustments. As a middle ground between free trade/competition on the one hand and monopoly/cartel on the other hand, better service by multinational enterprises and the proliferation of international strategic alliances "implicitly recognizes that it is no longer possible for any country to unilaterally run its domestic market" [1, p. 229] or to keep it impervious to technological and financial penetration from outsiders.

7.1 Trade in Telecommunications Network Services[1]

Until the mid-1980s, nations did not use trade negotiation machinery to address services like telecommunications. In 1986, the signatory nations to the General Agreement on Tariffs and Trade (GATT) [2] agreed to consider trade in services in the context of the then-upcoming Uruguay Round of Multilateral Trade Negotiations.[2] At the end of 1993, the GATT formalized the General Agreement on Trade in Services (GATS),[3] with particular attention to a number of sectors, including telecommunications, construction-engineering, financial services, transportation, tourism, and professional services, including lawyering.[4]

Increasing interest in applying trade negotiation machinery to services in general and telecommunications in particular resulted from the growing appreciation that national economies have become integrated and that the

---

1. For more comprehensive examination of the WTO's work on trade in telecommunications services, see Taunya L. McLart, "Liberalized Telecommunications Trade in the WTO, Implications for Universal Service Policy," *Federal Communications Law Journal*, Vol. 51, Dec. 1998, p. 1; Laura B. Sherman, "'Wildly Enthusiastic' About the First Multilateral Agreement on Trade in Telecommunications Services," *Federal Communications Law Journal*, Vol. 61, Dec. 1998, p. 61; Niels Quist, "The WTO Agreement on Basic Telecommunications: Will European Union Telephone Service Finally Become Competitive?" *Journal of International Legal Studies*, Vol. 4, Winter 1998, p. 133; John J. Alissi, "Revolutionizing the Telephone Industry: The World Trade Organization Agreement on Basic Telecommunications and the Federal Communications Commission Order," *Connecticut Journal of International Law*, Vol. 13, Spring 1999, p. 485; Phillip L. Spector, "The World Trade Organization Agreement on Telecommunications," *International Lawyer*, Vol. 32, Summer 1998, p. 217.

2. GATT established a Group of Negotiations on Services "to establish a multilateral framework of principles and rules for trade in services, including elaboration of possible disciplines for individual sectors [such as telecommunications], with a view to expansion of such trade under conditions of transparency and progressive liberalization and as a means of promoting economic growth of all trading partners and the development of developing countries." Ministerial Declaration on the Uruguay Round, Punta del Este, Uruguay, Sept. 1986.

3. General Agreement on Trade in Services, Dec. 15, 1993, 33 I.L.M. 44 (GATS). In the acronym-rich trade world, the Group of Negotiations on Services (GNS), conducted in parallel with the Uruguay Round of GATT negotiations, addressed issues on how to extend the GATT approach to services sectors.

4. See Trade Negotiations Committee Meeting at Ministerial Level, Montreal, Dec. 1988, MTN.TNC/7 (MIN); Organization for Economic Co-Operation and Development, *Elements of a Conceptual Framework for Trade in Services*, Paris, 1987.

ease with which nations can transact commerce in services can significantly affect national welfare:

> Today's consensus points to the necessity of information and telecommunications in the more complete establishment of a market economy to replace barter and subsistence agriculture in rural areas of the developing world. There is increasing recognition that information infrastructure is a vital prerequisite for economic development rather than merely a desirable side effect. [3]

The GATT was negotiated in 1947 with an eye toward creating four permanent organizations: (1) the United Nations, to address diplomatic and political issues and to resolve conflicts before they result in wars; (2) an International Trade Organization, to establish policies designed to promote commerce between nations and to resolve trade disputes; (3) a vehicle to support trade and development, particularly in developing nations, now generally referred to as the World Bank; and (4) a basis for stabilizing currencies through the International Monetary Fund.

A permanent International Trade Organization came into existence in 1995, preceded by years of interim arrangements initially stylized as a protocol of provisional application (PPA) [4]. The PPA became the de facto permanent basis of the GATT, and over time the GATT became both an agreement between nations and an organization with a permanent staff and the legitimacy to schedule periodic rounds of tariff reduction negotiations.

7.1.1 The WTO

The WTO[5] was established on January 1, 1995, to implement the provisions of the GATT, including those in the area of intellectual property protection.[6] Membership in the WTO is available only to those countries that (1) are signatories to the GATT; (2) agree to adhere to all the provisions of the Uruguay Round of negotiations; and (3) submit schedules of market access commitments for industrial goods, agricultural goods, and services. The WTO agreement sets forth rules that implement the Uruguay Round

5. The WTO Web site provides helpful background on the organization. See http://www.wto.org/wto/about/about.htm.

6. For an introduction to the WTO and its functions, see President Bill Clinton's submission to Congress of documents concerning Uruguay Round Agreement, Dec. 15, 1993, Chap. 17 (World Trade Organization).

accords, including a formal basis for settling disputes and the option for imposing trade sanctions if members violate their obligations.[7]

The WTO provides a legal and institutional foundation for establishing, on a multilateral basis, national commitments for framing and implementing domestic trade legislation and regulations that collectively reduce barriers to trade.[8] A key objective lies in expanding trade under conditions of transparency and progressive liberalization, with an eye toward promoting the economic growth of all trading partners and, in particular, developing countries. The WTO also provides a forum for the resolution of trade disputes through collective debate, negotiation, and adjudication. Negotiations under the aegis of the WTO help establish a multilateral framework of principles and rules for trade in goods and services, including specific sectors such as telecommunications.

7.1.2 Trade-in-Services Issues

Telecommunications has generated special interest by the WTO,[9] which followed through on GATS examination of trade-in-services issues [5]. The GATS established a framework of general rules and disciplines, with annexes addressing special conditions relating to individual sectors and schedules that provide a timetable of market access commitments by specific nations. The GATS sought to extend fundamental open-trade principles like nondiscriminatory national treatment [6] (which requires countries to avoid discriminating in favor of domestic service providers) and most-favored nation (MFN) [7] treatment to service providers operating in nations that have acceded to the agreement. The GATS required countries to promote open-market

7. See Memorandum of Dec. 15, 1993, for the U.S. Trade Representative: "Trade Agreements Resulting from the Uruguay Round of Multilateral Trade Negotiations," 58 Fed. Reg. 67,263, 67,289 (1993); "Understanding on Rules and Procedures Governing the Settlement of Disputes," GATT Doc. MTN/FA II-A2 (Dec. 15, 1993).

8. The WTO is based in Geneva, Switzerland, and provides the following services: administering and implementing multilateral trade agreements; acting as a forum for additional multilateral trade negotiations leading to progressive liberalization of trade; seeking to resolve trade disputes; overseeing national trade policies; and cooperating with other international institutions involved in global economic policy making. For additional background on the WTO, see http://www.wto.org/wto/2_1_0_wpf.html.

9. The WTO was created by the Marrakesh Agreement establishing the World Trade Organization (*International Legal Materials*, Vol. 33, Sept. 1994, p. 1125).

access by refraining from imposing quotas and other quantitative restrictions, setting local incorporation requirements, or otherwise discriminating in favor of domestic providers in specific service sectors listed by nations in a schedule of covered services.

The WTO pursued liberalization of telecommunications trade through a Group on Basic Telecommunications, established in the Decision on Commitments on Basic Telecommunications adopted by the WTO's Council for Trade in Services on April 30, 1996 [8]. Basic telecommunications constitutes all public and private telecommunications services that involve the end-to-end transmission of customer-supplied information, such as the transmission of voice, data telex, telegraph, and facsimiles linking sender and receiver. National commitments apply to both facilities-based and resale services using leased private lines, as well as fixed and mobile satellite systems and services, cellular telephony, mobile data services, paging, and personal communications systems. Value-added services such as on-line data processing, credit card verification, electronic data interchange, e-mail, and voice mail were not formally part of the negotiations, but were already liberalized in 44 schedules (covering 55 countries) at the end of the Uruguay Round. Market access commitments made by nations apply to both international telecommunications services and the provision of services by foreign companies operating in-country, including the installation of a network infrastructure in addition to that operated by incumbent carriers.

Until a breakthrough in early 1997, progress had been slow[10] in the delivery of nations' telecommunications market access commitments. Few countries produced a schedule of market-liberalizing commitments, despite several meetings in 1996 and a recommended February 1997 deadline. Nations that previously had liberalized telecommunications markets were reticent to commit to even greater market access opportunities unless and

---

10. "The fulfillment of the objectives agreed at Marrakesh for negotiations on the improvement of market access in services—in financial services, movement of natural persons, maritime transport services and basic telecommunications—has proved to be difficult. The results have been below expectations. In three areas, it has been necessary to prolong negotiations beyond the original deadlines. We are determined to obtain a progressively higher level of liberalization in services on a mutually advantageous basis with appropriate flexibility for individual developing country Members, as envisaged in the Agreement, in the continuing negotiations and those scheduled to begin no later than 1 January 2000. In this context, we look forward to full MFN agreements based on improved market access commitments and national treatment." World Trade Organization: Singapore Ministerial Declaration, WTO Doc. WT/MIN(96)/DEC, Dec. 18, 1996 (reprinted in *International Legal Materials*, Vol. 36, Jan. 1997, pp. 218, 224).

until other nations at least made their first commitments toward eliminating foreign investment restrictions, establishing a regulatory system for resolving market access disputes, and imposing interconnection and fair competition rules. Likewise, the Group on Basic Telecommunications had to consider particularly vexing issues like the transborder delivery of video programming in view of many nations' reluctance to permit "television without frontiers," that is, unfettered opportunities to export programming despite cultural and financial concerns about the viability of indigenous programming. Other challenging issues included the role of intergovernmental satellite cooperatives, like INTELSAT, that have no single national identity, satellite orbital slot/spectrum management, and carriers' use of different international accounting rates for traffic between nations.[11]

By the February 15, 1997, deadline, 70 nations had made commitments in three major areas:

- *Market access*—the opportunity for domestic and foreign ventures to provide local, long-distance, and value-added services by reselling the facilities and lines of incumbents and by installing new facilities;
- *Foreign investment*—the option for foreign companies to invest in a telecommunications service provider;
- *Procompetitive regulatory principles*—the creation of a regulatory authority able to resolve disputes among competitors on such matters as interconnection, fair competition, transparency of rules and tariffs, and national treatment.

Nations differed on the range of services included in their commitments and the timing of liberalization. Likewise, countries differed on the extent of permitted foreign investment and how they would establish and implement a common set of regulatory principles. Nevertheless, the nations making commitments represent over 95% of world telecommunications revenues, leading the acting U.S. Trade Representative to project a doubling or tripling of revenues from the current level of $600 billion and the development of a global information highway.[12] At the very least, new opportunities

11. See World Trade Organization, Group on Basic Telecommunications, *Report of the Group on Basic Telecommunications*, S/GBT/4, Feb. 15, 1997.

12. See U.S. Trade Representative, Statement of Ambassador Charlene Barshefsky, Basic Telecom Negotiations, Feb. 15, 1997.

will exist for companies to establish an operating presence in foreign countries on either a facilities-based or resale basis. Such new ventures will stimulate some degree of competition with incumbents, but a "transparent," nondiscriminatory dispute resolution system must exist to ensure that market entrants have fair opportunities to interconnect with the facilities and services of incumbents.

7.2 Basic Trade Principles

The trade principles articulated in the GATT and now implemented by the WTO, whether applied to goods or services, fall into two major categories:

- Market access;
- The equivalency and reciprocity of such access as well as classifications used by nations to identify the degree to which they will support open markets.

The former typically involves bilateral negotiations among nations, with an eye toward reducing market access barriers for particular products or services exchanged between two nations. The latter involves a multilateral concept like MFN treatment, the commitment to apply the most open and liberal trade policies to WTO signatories.

7.2.1 Comparative Advantage

The WTO supports the fundamental view that nations should engage in open and free trade to exploit their comparative advantages, that is, their strengths relative to the strengths of other countries. With an unfettered ability to trade goods and services, nations theoretically can enhance overall welfare and wealth by trading the products and services that a nation has a comparative advantage in producing for other goods and services for which a nation lacks a comparative advantage.[13] For example, the United States can

13. The GATT preamble provides the following statement: "Recognizing that their relations in the field of trade and economic endeavor should be conducted with a view to raising standards of living, ensuring full employment and a large and steadily growing volume of real income and effective demand, developing the full use of the resources of the world and expanding the production and exchange of goods."

trade aircraft, satellites, and services in management consulting, engineering, and finance, goods, and services in which it has a comparative advantage, for labor-intensive products, like clothing, from nations with cheaper labor costs.

The comparative-advantage view considers the exchange mutually advantageous rather than unfair exploitation of cheaper labor and a nation's lower standard of living. On the contrary, a nation that does not trade for items and services for which it lacks a comparative advantage risks collectively lowering its national wealth and welfare. For example, in the 1970s, Brazil severely restricted imports of information-processing technology, with an eye toward jump-starting an indigenous "infomatics" industry. Its refusal to trade coffee for software may have retarded growth in its GNP. Likewise, Japan's insistence on maintaining a domestic rice production capability on land perhaps bettered suited for reducing a national housing shortage arguably elevated concerns for national security and market insulation over the welfare of its citizens. In economic terms, the concept of comparative advantage stimulates economies of scale and efficiency. Trade can stimulate growth in the demand for products and services, leading to their generation at the lowest per-unit costs. Likewise, a nation might depart from pursuing markets in which it lacks a comparative advantage and in which it could never reach an efficient level of output. Presumably, the workers and inputs committed to inefficient production could be redeployed to more efficient outputs. However, the transaction costs, dislocation, and human suffering in such transitions cannot be discounted. For example, the U.S. work force undertook a painful transition from an economy predominated by agriculture and heavy industrial production (e.g., steel manufacturing) to one dominated by information services and light industry (e.g., semiconductors).

7.2.2 Most Favored Nation Treatment

The MFN clause of the GATT provides that

> Any advantage, favour, privilege or immunity granted by any contracting party to any product originating in or destined for any other country shall be accorded immediately and unconditionally to the like product originating in or destined for the territories of all other contracting parties.

In application, that means that once a nation decides to confer a trade benefit, privilege, or concession to one nation, it must do so unconditionally to all signatories to the GATT and as well by and for signatories to the GATS.

MFN treatment provides a strong incentive for nations to become members of the WTO, because they can accrue market access benefits secured by any other member nation. That "free rider" opportunity can level the trade negotiation playing field in the sense that regardless of negotiating strength a nation can acquire new market access opportunities. On the other hand, the very fact that a market access initiative will accrue to every nation tends to reduce incentives for proposing new access opportunities. With MFN treatment, it becomes difficult for a nation to use a market access opportunity as leverage for negotiating a reciprocal opportunity.

For example, if the United States agrees to increase the percentage investment for British individuals or companies in U.S. broadcasters, MFN treatment theoretically would require the United States to confer such expanded ownership opportunity to all WTO member nations. U.S. trade negotiators lose a major bargaining chip if they cannot offer a market access opportunity on a nation-specific basis. Additionally, they risk causing U.S. firms to lose revenues and investment opportunities without offering a reciprocal opportunity for investment abroad if other nations do not similarly expand broadcast investment opportunities for U.S. citizens. The MFN concept may create incentives for trade negotiators to settle on a least common denominator in ambiguous general language with specific exceptions for particular industries and market segments (e.g., procurement activities of government agencies including PTTs).

7.2.3 National Treatment

Membership in the WTO also establishes a requirement of national treatment, that foreign suppliers of goods and services shall be treated no less favorably than what national laws, regulations, and administrative practices apply to domestic enterprises. The national treatment concept attempts to prevent nations from nullifying MFN status through the use of discriminatory internal measures, such as local content legislation and other nontariff barriers (NTBs), as well as discriminatory taxes. However, national treatment as applied by the WTO recognizes the need, in some situations, to maintain nonparity as might exist in national regulations, for example, a cap on foreign investment in domestic broadcast facilities. Whether for legitimate concerns or simply to maintain trade barriers, regulatory exemptions to national treatment mitigate the effectiveness of the national treatment concept.

A nation that conscientiously embraces the concept of national treatment risks creating unilateral market access opportunities that go unreciprocated, because nations can invoke national security or other regulatory requirements as sufficient reasons not to match the initiative. For example, the United States might perceive its nation treatment obligation to include elimination of barriers to foreign investment in telecommunications, including opportunities to market service and accept overseas traffic from locations within the United States. Other nations would not have to provide a reciprocal opportunity for U.S. businesses to operate in-country if domestic regulations prohibited such a presence, and membership in the WTO does not absolutely require repeal of such regulations. Arguably, a nation that confers a telecommunications equipment or services monopoly does not violate the national treatment concept, because it discriminates equally between domestic and foreign businesses by denying a market access opportunity to both types. Similarly, a country may appear to support national treatment, but the empirical evidence shows a "buy domestic" cultural heritage. Likewise, technical and regulatory requirements may favor the domestic incumbent or prevent foreign enterprises from meeting all certification requirements needed prior to market entry.

7.2.4 Transparency and Other Trade Principles

Other fundamental trade concepts include an expectation of parity in legal and regulatory treatment between domestic and foreign entities. The concept of transparency means that nations must establish fair, unbiased, and understandable laws and regulations that affect trade. Additionally, practices, procedures, and restrictions should serve some rational purpose and not restrict or impede trade, including the transit of goods and services that merely cross through national territory en route to a third nation.

When trade disputes arise, the WTO provides a forum for dispute resolution and treaty-based powers of enforcement. Still the efficacy of such a process depends on the willingness of signatory nations to relinquish a degree of sovereignty in recognition that the WTO forum will lend its "good offices" in a fair manner aimed at enhancing trading opportunities. A member state triggers the dispute resolution mechanism by identifying how benefits due it are being "nullified or impaired" [9]. The mechanism involves a number of procedures designed to secure a remedy through consultation and coordination as opposed to retaliation and trade sanctions. The complaining nation first attempts to secure informal support for a proposal designed to remedy the problem. If that route proves unsuccessful, the nation may pursue more formal liaison with other members of the WTO, its Director-General, its Council, or any appropriate intergovernmental organization [10]. The nation also may petition the WTO Council to appoint a panel consisting of individuals drawn from member states to investigate the issue and recommend a solution that might involve a directive to the offending nation to change policies and perhaps to compensate the offended nation. The WTO Council must ratify the report and may amend it, or require further negotiation and analysis.

7.3 Limits to What a Trade Forum Can Accomplish

While the WTO provides a forum, now replete with enforcement powers for the resolution of trade disputes, the frequency and the severity of trade disputes have yet to evidence a commensurate reduction. Nations and individual companies still have incentives to game the trade mechanism so they can exploit market access opportunities while restricting access at home to the maximum extent possible. For example, the U.S. Trade Representative in 2000 threatened to seek WTO examination of Mexico's local exchange access charges and other regulatory policies that allegedly have stifled competition with the incumbent carrier TelMex.[14]

The time, effort, complexity, and uncertainty of the WTO dispute resolution process create incentives to retain anticompetitive and market access restrictions. Simply put, some nations' rhetoric has not translated into anything greater than minor, incremental steps toward open market access. Just as nations spread across a continuum insofar as the level of liberalization, privatization, deregulation, and globalization, so too has a continuum developed for plotting where a nation stands on the degree of market accessibility. Nations on the restrictive side of the line risk handicapping their national heros' international marketplace success and blunting whatever comparative advantages they might have.

7.3.1 Tactics for Restricting Telecommunications Equipment Market Access

Transparency in trade should translate into equal opportunities for equipment vendors to know about and to participate in major procurements

14. See "U.S. to Request WTO Consultations with Mexico Regarding Telecommunications Trade Barriers," Press Release No. 00-57 (rel. July 28, 2000) available at http://192.239.92.165/releases/2000/07/00-57.html.

occurring in any WTO member nation. However, nations with a dominant incumbent telephone company still may rely heavily on one domestic equipment manufacturer, often owned by, or affiliated with, the incumbent carrier. The national hero may constitute the single enterprise capable of delivering equipment that meets technical specifications and other requirements that it may have helped prepare.

Nations keen on ensuring a domestic market for the national hero equipment manufacturer may impose costly and time-consuming equipment testing and certification requirements to prequalify equipment vendors. A nation may reject test results from foreign labs, in effect facilitating exclusive market access for domestic enterprises. Rather than support universal standards as promulgated by the ITU, nations or regions may establish their own standards, perhaps with an eye toward relying on domestic manufacturer-supplied technical standards and interfaces.

Few nations fully implement the concept of national treatment; many resort to duties, domestic content requirements, and informatics policies, (requirements that software and other types of intellectual property be created in-country) to create an absolute barrier to entry or a price disadvantage for imports. Additionally, nations may thwart efforts by foreign nations to establish a manufacturing or marketing presence in the nation or refuse to consider such installations as domestic when evaluating domestic content requirements, particularly if assembly rather than manufacturing takes place. Foreign manufacturers and distributors may lack opportunities to dispute discriminatory practices in domestic regulatory agencies and courts.

7.3.2 Tactics for Restricting Telecommunications Service Market Access

Foreign governments and incumbent carriers have no legal or treaty obligation to confer an operating agreement to any and all market entrants. Whether to limit the scope of competition or to limit market entrants to qualified carriers, some nations have restricted the number of facilities-based operators. Indeed many nations may embrace competition slowly and incrementally, initially by privatizing the incumbent but retaining its monopoly and subsequently by authorizing one or two additional competitors.

Without an operating agreement, newcomers cannot access a nation's domestic telephone network to complete calls and, in effect, have no market access. Reciprocity of access is not completely symmetrical in most nations; even where nations accept inbound traffic, few guarantee a commensurate return flow. Fewer still allow foreign carriers to establish a marketing presence within the nation.

Absent extraordinary efforts by regulators to ensure equal-access terms and conditions, the potential exists for incumbent carriers to favor their services and those offered by affiliates. An uncompetitive playing field exists when both domestic and foreign competitors of the incumbent face anticompetitive and unfair access terms and conditions for the same type of network interconnection the network provides affiliates. Some nations, for example, Brazil, persist in denying entrepreneurs the opportunity to resell leased private lines. Other limitations include restrictions on data speed and access to private leased lines, forcing a migration to the usage-sensitive, public-switched service, qualifications on who can install and operate Earth stations, and restrictions on interconnection to the PSTN. In conjunction with equipment testing and certification restrictions, the incumbent may attempt to limit the ability of carriers and customers to attach equipment that provides tailor-made functions unavailable from the incumbent or generated by its central office switches at far higher prices. Unless and until the incumbent has to provide access to its services and facilities on cost-based and nondiscriminatory terms, full and fair competition cannot evolve.

7.4 Leverage in Trade Negotiations

Although they may be reluctant to resort to retaliatory action, nations do have significant domestically authorized and internationally accepted weapons for remedying trade disputes.[15] The reluctance to use tactics involving financial penalties, such as countervailing duties that impose additional, ostensibly parity-fostering costs on imports, stems from the prudent concern that a spiral of sanctions and countersanctions would render all parties worse off.

Accordingly, the most frequently relied-on "weapons" are negotiation and "jawboning." Both tactics involve the forceful advocacy that the parties

15. See Omnibus Trade and Competitiveness Act of 1988, Pub. L. No. 100-418, 1988 U.S. Code Cong. & Adm. News (102 Stat.) 1107 (1988), *codified at* 19 U.S.C. Secs. 3101-3111 (1990); Trade Reform Act of 1974, *codified at* 19 U.S.C. Sec. 2411 *et seq.* Section 301 of the Trade Act of 1974, 19 U.S.C. Sec. 2411 (1990) requires the U.S. Trade Representative to take mandatory remedial action if it is determined that U.S. rights under a trade agreement are being denied. Further amendments to Section 301, 19 U.S.C. Sec. 2420 (1990), commonly known as "Super 301" required the U.S. Trade Representative in 1989 and 1990 to identify U.S. trade liberalization priorities and practices that, if eliminated, would significantly increase exports and to identify priority countries to engage in dialogue to reduce trade barriers.

reach a settlement, lest harsher unilateral actions be legislated or promulgated in a domestic forum like the legislature, which would be far less sympathetic than multinational dispute resolution, like that established under the WTO. In addition to dispute resolution under the aegis of the WTO, nations engage in fact finding and negotiations aimed at addressing the foreign country's industrial structure, for example, whether and how the incumbent carrier might be subject to competition or at least procurement of equipment on an open and nondiscriminatory basis.

7.4.1 Securing Market Access Parity in Telecommunications Service

The United States and other nations that support free and open trade have attempted to achieve closer parity in market access under the auspices of the existing trade and telecommunications policy forums like the WTO and the ITU. However, the WTO has only begun to incorporate its negotiation and trade machinery for services, in addition to goods,[16] and the ITU traditionally has addressed standards and operational rules of the road, and not trade policy per se.[17] Neither organization has an automatic right to serve as the primary forum; accordingly, they may "find themselves in the force field of a jurisdictional battle that may not be resolved in the near future" [3].

The Office of the U.S. Trade Representative was established as an agency in the Executive Office of the President in 1963. Exec. Ord. 11075, Jan. 15, 1963, 28 Fed. Reg. 473, as amended by Exec. Ord. 11106, Apr. 18, 1963, 28 Fed. Reg. 3911, and Exec. Ord. 11113, June 13, 1963, 28 Fed. Reg. 6183. Pursuant to 19 U.S.C. 2171(c)(1) of the Trade Act of 1988, the U.S. Trade Representative, *inter alia*, has primary responsibility to develop and coordinate the implementation of U.S. trade policy.

16. For an analysis of international trade in services and an assessment of whether the existing trade negotiation apparatus can be adjusted to incorporate services, see G. Feketekuty, *International Trade in Services—An Overview and Blueprint for Negotiations* (Cambridge, MA: American Enterprise Inst./Ballinger, 1988); P. Robinson, K. Sauvant, and V. Govitrikar, eds., *Electronic Highways for World Trade Issues in Telecommunication and Data Services* (Boulder, CO: Westview Press, 1989).

17. However, the 1988 World Administrative Telegraph and Telephone Conference (WATTC), convened in Melbourne, Australia, addressed a number of issues with a trade component, for example, establishing a mechanism for "special arrangements" whereby nations can agree on a bilateral basis to permit the provision of new services like value-added networks, even though such services might conflict with conventional rules of the road. WATTC definitely placed the ITU in the mix of legitimate policy-making forums, along

A key impediment to achieving reduced barriers to market entry in telecommunications lies in the almost metaphysical characteristics of the process, that is, the difficulty in defining the transaction or set of transactions in a trade context. Accordingly, both the WTO and the ITU need to make significant adjustments in the way they address telecommunications trade issues. Like broadcasting, which transmits programming into the "ether," telecommunications services can involve a number of transborder transmissions with a variety of intermediate processing.[18] Nations created a multilateral trade forum initially to address concrete, tangible goods, while the ITU has considered telecommunications from a policy or technological perspective, aiming to optimize efficiency, innovation, and social objectives, factors somewhat ancillary in a conventional trade policy assessment.

"The fundamental difficulty with telecommunication and data services as a trade-in-services issue is that it is both a telecommunication-policy issue and a trade-policy issue simultaneously and interactively" [11]. Trade policy officials serve their nation's commercial interests while pressing for a more level competitive playing field elsewhere, including the reduction of barriers or impediments to market access. Telecommunications policy officials ostensibly seek to optimize the rules of the road and standards to promote network connectivity, although industrial policy may color their judgment and make them advocates for national heroes and parochial policies.

A number of trade concepts have a direct translation into concrete telecommunications policy proposals. The multilateral rounds of negotiations in the GATT and the initial rounds of WTO meetings have developed support for several basic principles:

---

with the GATT, the United Nations, and regional organizations like the European Economic Community and the Organization for Economic Co-operation and Development.

18. William Drake and Kalyso Nicolaidis present six problem areas: (1) Not all service transactions fit under the traditional definition of trade as products produced entirely in one country and purchased in another; (2) the boundary between trade and foreign direct investment is fuzzy and if efforts were undertaken to liberalize investment flows, developing nations would seek liberalization of labor flows; (3) the boundary between illegitimate NTBs and legitimate regulations is fuzzy as well; (4) comparative advantage in services sometimes is not based on the same natural factors and endowments as would be the case for goods; (5) applying MFN treatment in telecommunications could erode network integrity, rob incumbents of scale economies, and result in duplication of investment; and (6) most nations would strongly object to simply adding services to the GATT treaty. W. Drake and K. Nicolaidis, "Ideas, Interests and Institutionalization: 'Trade in Services' and the Uruguay Round," *International Organization*, Vol. 46, No. 1, Winter 1992, pp. 37, 62–63.

- Supporting commerce between nations through freely working markets unfettered by discriminatory and nontransparent rules and regulations;
- Reducing barriers to trade on both a bilateral and multilateral basis with due regard to the need to respect national sovereignty, the legitimacy of domestic regulatory goals, and the need to move incrementally to avoid social, political, or economic dislocation;
- Conferring the same market access opportunities to foreign enterprises as those available to domestic ones;
- Accepting a multilateral forum, the WTO, for settling disagreements and for consulting other nations with an eye toward reducing trade barriers.

International telecommunications policy forums may emphasize sovereignty over such trade concepts. Matching half-circuits requires close cooperation, and nations may have greater concerns about the transborder flow of information and entertainment than for the importation of tangible products. Nations relinquish a degree of their sovereignty to accept telecommunications rules with an eye toward promoting beneficial enhancements in efficiency and reduction of transaction costs. While trade policies attempt to achieve similar goals, for example, to enable nations to specialize based on comparative advantage in the production of goods and services, the debate has little to do with relinquishing sovereignty. The sovereignty of a nation is considered a given: The debate focuses on the willingness of nations to allow enterprises of other nations to enjoy MFN treatment, that is, nondiscriminatory market access, including perhaps the right to establish a commercial presence in the foreign country and to benefit from transparent, nondiscriminatory application of regulatory rights and responsibilities.

Nations typically leverage threatened reduction of market access opportunities or increased cost of imported goods to achieve closer parity of access. Such tactics include selective reregulation, delays in further deregulatory initiatives, expanded foreign investment opportunities, imposing expectations for market access reciprocity, and limited deviation from national treatment. The prospective loss of market share creates incentives for foreign nations to agree to language that appears to move them toward easing market access restrictions at home, although the timetable and the procedures may be vague.

On the other hand, nations aggressively pursuing market access initiatives face backlash and retaliation when they insist on too strict a timetable or unilaterally announce what must transpire. Many nations favor the multilateral trade forum system instead of bilateral negotiations for resolving disputes involving telecommunications equipment procurement, network access, interconnection, service pricing, treatment of foreign investment, and regulation of services that add value and enhance basic, leased private lines.

7.4.2 ITU Efforts to Address Telecommunications Trade-in-Service Issues

Although it operates as a multilateral telecommunications policy-making organization, the ITU's portfolio can parallel the WTO and Internet policy-making forums. Telecommunications services clearly fall within both the WTO's and the ITU's jurisdiction, and the potential exists for the two bodies to address similar issues from different perspectives. For example, what might constitute an NTB to trade under the auspices of the WTO might constitute a legitimate matter of national security or otherwise qualify for a market access exemption under the ITU's policies. Bear in mind that the WTO has enforcement powers that the ITU lacks.

In 1988, the ITU convened the World Administrative Telegraph and Telephone Conference (WATTC-88) with an eye toward developing new regulations that could address technological innovations and growing conflict on how best to respond to trade disputes. Some nations viewed WATTC as a last chance for the ITU to carve out a legitimate trade portfolio and to accommodate increasing numbers of commercial enterprises, without traditional common carrier certification or service requirements, seeking access to the facilities of incumbent carriers to provide new enhanced services.[19]

The debate at WATTC-88 centered on the scope of international telecommunications and information-processing services that should be subject to fundamental rules of the road set out in the International Telecommunication Regulations and the instances when incumbent carriers and new enterprises could elect to establish their own "special arrangements." While the ITU convention already contained a section affording administrations

19. For an extensive assessment of WATTC-88, see William Drake, "WATTC-88: Restructuring the International Telecommunication Regulations," *Telecommunications Policy*, Vol. 12, Sept. 1988, pp. 217-234.

that option [12], the United States and the United Kingdom favored expanding the scope of the option and the frequency of its use, while most nations expressed concern about the potential for special arrangements to become a large loophole for avoiding necessary limitations and deviating from the status quo.

"Many of the regulatory issues raised by new interconnection relationships or by [new] hybrid offerings do not fit neatly within the service categories of telephony and telex, which have been the traditional subjects of ITU overseeing and regulations" [13]. Similarly, international value-added networks and other enterprises providing new services often were market entrants, authorized to provide service through the resale of lines leased from incumbent carriers. While the ITU had a category for identifying that type of enterprise, recognized private operating agencies, it did not yet have an acceptable mechanism for allowing them to operate substantially free of the regulations applicable to facilities-based carriers providing the underlying transport functions.

Many nations were content for WATTC-88 to incorporate new services into existing definitions, a decision that would bolster campaigns by incumbent carriers to expand their operations to include such enhanced services. William Drake reports:

> For the USA, the UK and their corporate supporters, a detailed list of services to be covered by the ITU Regulations, would be counterproductive. They argued that any list that might be drawn up would quickly become obsolescent, and could raise regulatory questions about the status of future offerings which would not be included. Moreover, given their sometimes less than muted distrust of PTT ambitions, they feared that the exercise could lead to a list incorporating such items as intracorporate services based on [the use of] leased lines. [14]

WATTC-88 finally adopted a compromise that would apply the regulations to services generally available to the public while exempting specialized services. The ability to reach a compromise reflects enlightened national self-interest in allowing telecommunications to evolve and the ITU to remain effective. But the compromise also reflects how the ITU had to devise solutions to telecommunications trade matters in addressing whether and how nations should accept traffic from new enhanced service providers or allow such enterprises to operate domestically.

7.5 Frustration with Multilateral Dispute Resolution

Many commentators believe that international forums like the WTO and the ITU have declining prospects for resolving disputes and obviating the need for bilateral negotiations.[20] Ineffectual dispute resolution in the GATT motivated the U.S. Congress to legislate self-help procedures implemented on a nation-specific basis. For example, Section 301 of the Omnibus Trade and Competitive Act of 1988 [15][21] provided for expedited investigation and retaliation involving the imposition or increase in tariffs or quotas in response to unfair and injurious trade practices of a foreign government. Such country-specific examinations operate outside the WTO's investigative and judicial function and have the potential for specific retaliatory sanctions outside the WTO's enforcement function. Unilateral action of that sort violates MFN treatment and the commitments the United States made to work within the policies and procedures of the WTO. As such, Section 301 cannot apply to WTO member nations.

Before creation of the WTO, Congress authorized the U.S. Trade Representative to fashion remedies when a foreign government breaches, nullifies, or impairs benefits to the United States from a trade agreement. Domestic legislation creates a unilateral basis for enforcing a trade agreement by authorizing the U.S. Trade Representative to take a more aggressive posture in trade negotiations when the United States perceives an unjustifiable, unreasonable, or discriminatory trade practice that restricts U.S. commerce.

The aggressive use of Section 301 by the United States had the ironic outcome of generating increased support for previous U.S. proposals to make

---

20. The consultative phase could be unproductively long; the party whose measures were challenged could simply drag out the bilateral discussions. Even after parties agreed to the establishment of a panel ... the GATT had no mechanism for appealing it. Even if a panel finally issued a report, a single contracting party—including the disputant to whom it was adverse—could block its adoption by the GATT Council. . . . Finally, no procedure existed to ensure that the GATT Council would monitor the action or inaction of a party whose measures were found to be GATT-inconsistent, unless prompted by the initiative of the party that successfully challenged those measures. J. H. Bello and A. F. Holmer, "U.S. Trade Law and Policy Series No. 24: Dispute Resolution in the New World Trade Organization: Concerns and Net Benefits," *The International Lawyer*, Vol. 28, No. 4, Winter 1994, pp. 1095, 1096–1097.

21. See J. H. Bello and A. F. Holmer, "The Heart of the 1988 Trade Act: A Legislative History of the Amendments to Section 301," *Stanford Journal of International Law*, Vol. 25, 1988, p. 1.

dispute settlement within the WTO more effective and timely. The new rules established at the Uruguay Round of Trade Negotiations[22] and now established in the WTO's governance documents make fundamental changes to the GATT dispute settlement process and have the following provisions:

- Automatic establishment of a panel and automatic adoption of a panel report (unless the Council, by consensus, decides to the contrary);
- An exceptional opportunity for appellate review of panel reports;
- Rigorous surveillance of the implementation of adopted panel reports;
- Compensation or WTO authorization for the suspension of concessions if a report is not implemented in a reasonable period of time;
- Expeditious arbitration in the event of disputes about a reasonable period of time for implementation or the appropriate level of compensation or suspension;
- Recourse to these procedures for practices considered as violating the WTO or nullifying or impairing WTO benefits. [16]

7.5.1 Regional Trade Pacts

The multinational nature of the WTO can result in delayed decision making and a least common denominator ill-fitted for particular nations or regions.[23] Increasingly, nations have bilateral or regional trade negotiations. The North American Free Trade Agreement (NAFTA) provides a 2,000-page case study of what three nations—the United States, Canada, and Mexico—can

22. See General Agreement on Tariffs and Trade: Final Act Embodying the Results of the Uruguay Round of Trade Negotiations, Dec. 15, 1993, 33 I.L.M. 1 (1994).

23. Bello and Holmer do not exclude the United States from the list of nations that contribute to the GATT's ineffectuality: "Like most nations, the United States tends to be schizophrenic; it wants the benefits of free trade, but also the freedom to act on its own without regard to international restraints. It remains the chief champion and cheerleader for the international rule of law, yet it prizes the sovereign right to act in disregard of such law in exceptional circumstances. J. H. Bello and A. F. Holmer, "U.S. Trade Law and Policy Series No. 24: Dispute Resolution in the New World Trade Organization: Concerns and Net Benefits," *The International Lawyer*, Vol. 28, No. 4, Winter 1994, pp. 1103–1104.

produce[24] if they fear the consequences of not responding to free-trade zones in other areas of the world. NAFTA affects $6 trillion of trade among nations totaling 360 million people, making North America the largest free-trade area in the world [17].

NAFTA provides concrete rules for achieving open and free trade between nations, substantially free of distortions and inequality. As stated in the NAFTA preamble, the goal is to "strengthen the special bonds of friendship and cooperation ... contribute to the harmonious development and expansion of world trade ... reduce distortions to trade ... foster creativity and innovation and promote trader in goods and services that are the subject of intellectual property rights ... and preserve their flexibility to safeguard the public welfare."

NAFTA establishes a free-trade area with the objective of eliminating trade barriers, promoting fair competition, increasing investment opportunities, protecting and enforcing intellectual property rights, creating effective procedures to implement and administer the agreement, and establishing a framework to expand and enhance the benefits of the agreement. NAFTA affirms each party's rights under previous agreements (e.g., GATT) to which they are parties. However, to the extent that any inconsistencies exist between NAFTA and other agreements, NAFTA prevails unless it indicates otherwise. Preexisting environmental and conservation agreements and annexes to NAFTA take priority over the basic document to the extent that there are inconsistencies.

NAFTA's primary objective lies in the elimination of trade barriers like import duties on goods that originate in Canada, the United States, and Mexico. The agreement establishes a timetable for the reduction of existing duties and prohibits any party from increasing or adopting new customs or duties. NAFTA also addresses nontariff restrictions such as import licenses and quotas. It specifically requires each party to apply national treatment, in accordance with GATT, to the other parties' goods, but the agreement does allow standards designed to safeguard human, animal or plant health, and the environment.

Chapter 13 of NAFTA contains specific provisions that address telecommunications, ranging from basic public telecommunications transport networks or services to enhanced or value-added services. The agreement

24. The U.S. Congress ratified the NAFTA accord on November 20, 1993. See North American Free Trade Agreement, Dec. 17, 1992, U.S.-Can.-Mex., Pub. L. No. 103-182, 107 Stat. 2057, reprinted in 32 I.L.M. 605 (1993).

includes a requirement that the parties work on establishing open and nondiscriminatory access to such networks by carriers, service providers, and the equipment of end users. The chapter also addresses issues of pricing, standards-related measures, monopolies, transparency, relationships with international organizations and agreements, and technical cooperation for increased compatibility.

Despite progress in promoting free trade, NAFTA also includes a cultural industry exemption clause, which exempts certain industries from application of the agreement. NAFTA does not apply when a nation views its cultural identity and sovereignty at risk as a result of market access. For example, concerns about cultural imperialism have led Canada to restrict both the extent of foreign investment in broadcast enterprises and the scope of foreign programming that can be broadcast.[25]

The telecommunications provisions of NAFTA exempt basic voice telephony services from coverage. That means NAFTA is limited to value-added, or enhanced, services, while basic services must be addressed by a separate bilateral agreement. As to enhanced services, NAFTA requires nondiscriminatory licensing, which means foreign signatories to the agreement should have the opportunity to partial or complete ownership of such an enterprise in another NAFTA nation.

7.6 Ranking Nations in Terms of Market Accessibility

Extensive differences remain as to a nation's political and economic philosophy toward trade as well as its attitudes toward deregulation and liberalization of the telecommunications sector. Accordingly, the pace and the scope of market access initiatives significantly vary across nations and even within regional trading blocs like the European Union. The United States, Canada,

---

25. The European Union has similar restrictions. For a discussion of Europe's "Television Without Frontiers" policies and how they nevertheless still restrict market access, see L. G. C. Kaplan, "The European Community's 'Television Without Frontiers Directive: Stimulating Europe to Regulate Culture," *Emory International Law Review,* Vol. 8, Spring 1994, p. 255; see also Council Directive of 3 October 1989, "Coordination of Certain Provisions Laid Down by Law, Regulation, or Administrative Action in Member States Concerning the Pursuit of Television Broadcasting Activities," 1989 O.J. (L 298) 23, reprinted in 28 I.L.M. 1492 (1989); "Television Without Frontiers: Green Paper on the Establishment of the Common Market for Broadcasting, Especially Satellite and Cable," COM (84) 300 Final (1984).

the United Kingdom, and New Zealand lead in terms of market accessibility. Those nations have erected the legislative changes and implemented the regulatory reforms necessary to foster virtually complete market access opportunities. In practice, those nations have significant resale and facilities-based competition in every telecommunications and information-processing sector.

Not too far behind those four leading nations are a number of countries that have embraced competition and that have some degree of competition, albeit not as robust as possible. Many European Union nations as well as the Scandinavian countries fall into this category. The European Commission has erected a procompetitive, open-access blueprint, and most European Union (EU) nations have followed through with implementing legislation. Additionally, market access leaders in other regions of the world fit into this category, including Chile, Hong Kong, and Australia. Nations that have erected the enabling legislation and completed the structural reform process with some degree of competition in some market segments include Japan, Brazil, Spain, and Argentina. Nations evidencing limited marketplace competition, despite the completion of structural and regulatory reform, include Italy, Mexico, South Africa, India, France, and Germany. Countries pretty much still at the starting line include China, Russia, and most of Africa. Chapter 10, which examines privatization, discusses national deregulation and liberalization strategies in greater detail.

References

[1] Aronson, J., and P. Cowhey, *When Countries Talk—International Trade in Telecommunications Services*, Washington, D.C.: American Enterprise Inst./Ballinger, 1988, p. 218.

[2] General Agreement on Tariffs and Trade, opened for signature Oct. 30, 1947, 61 Stat. A3, 55 U.N.T.S. 187.

[3] Snow, M. S., "Trade in Information Services in Asia, ASEAN, and the Pacific: Conceptual Issues and Policy Examples, *California Western Law Review*, Vol. 28, 1991–1992, p. 329.

[4] Protocol of Provisional Application of the General Agreement on Tariffs and Trade, Oct. 30, 1947, 55 U.N.T.S. 308.

[5] "General Agreement on Trade in Services," *International Legal Materials*, Vol. 33, Sept. 1994, p. 1167.

[6] GATS Article XVII.

[7] GATS Article II.

[8] World Trade Organization, Council for Trade in Services, Decision on Commitments in Basic Telecommunications, S/L/19 (adopted Apr. 30, 1996); see also *Final Report of the Negotiating Group on Basic Telecommunications*, S/NGBT/18.

[9] GATT Article XXIII(1).

[10] GATT Article XXIII(2).

[11] Woodrow, R. B., "Trade in Telecommunication and Data Services," *Electronic Highways for World Trade*, p. 23.

[12] ITU, *International Telecommunication Convention*, Article 42, "Special Arrangements," Kyoto, Japan, 1994.

[13] Renaud, J., "The Role of the International Telecommunication Union: Conflict, Resolution and the Industrialized Countries," in K. Dyson and P. Humphreys, *The Political Economy of Communications—International and European Dimensions*, New York: Routledge, 1990, pp. 33, 43–44.

[14] Drake, W., "WATTC-88: Restructuring the International Telecommunication Regulations," *Telecommunications Policy*, Vol. 12, Sept. 1988, pp. 219.

[15] Pub. L. No. 100-418,. § 1301, 102 Stat. 1107, 1164.

[16] Bello, J. H., and A. F. Holmer, "U.S. Trade Law and Policy Series No. 24: Dispute Resolution in the New World Trade Organization: Concerns and Net Benefits," *The International Lawyer*, Vol. 28, No. 4, Winter 1994, p. 1099.

[17] Koniigsberg, S. R., "Think Globally, Act Locally: North American Free Trade, Canadian Cultural Industry Exemption, and the Liberalization of the Broadcast Ownership," *Cardoza Arts & Entertainment Law Journal*, Vol. 12, 1994, p. 281.

8

Regulatory and Deregulatory Rationales

This chapter examines the reasons for a heritage of substantial government involvement in the telecommunications sector and just the opposite in information processing. The players and the policies that affect U.S. telecommunications and information processing provide a case study and insights on future regulatory quandaries.

International telecommunications policy making in most nations has involved close government participation, if not centralized management from the beginning. Governments have justified such action on national security grounds and on assumptions made about the economics of telecommunications and information. Concerns about national security have justified restrictions on, if not absolute bars to, foreign ownership of telecommunications facilities.

The essentiality of the telecommunications service mission has prompted nations to erect a public utility regulatory framework that not only moderates exposure to downside financial risk but also limits the range of upside profitability. Traditional telecommunications utility regulation provides carriers with financial assurances, for example, insulation from competition and many types of lawsuits in exchange for rate of return regulation, that is, regulator determination of what carriers can earn and how much they can charge for service.

Public utility regulation derives from economic analysis that makes certain conclusions about the viability of competition and the need for government sponsorship of objectives that the marketplace would not achieve.

Regulators conferred exclusive telecommunications franchises, having concluded that a single enterprise could accrue economies of scale and scope while also pursuing public policy objectives such as the extension of service into unprofitable areas like the rural hinterland. Under a rationale of that sort, competition could not exist, and to permit market entry would have the doubly damaging result of harming the ability of the incumbent to meet its public service objectives and duplicating resources and effort in a wasteful and unproductive manner.

The economic term *market failure* refers to a situation in which legislators and regulators do not accept marketplace-driven outcomes, because public policy factors support a different result. For example, governments use tax proceeds to expand the supply of education and other public services, including lighthouses, police and fire departments, armies, and libraries, on the belief that the citizenry should have greater access than what the well-to-do could afford. Presumably, a nation's total welfare increases with the strategic decision to allocate tax proceeds to sectors where marketplace resource allocation has proved insufficient. In the telecommunications sector, governments have articulated a universal service mission and forced carriers and their customers to fund it based on the view that the nation accrues great dividends when all citizens have access to basic services. Governments also have seen the need to underwrite citizens' access to information, including the incubation of the Internet through research and development funding and supporting education, libraries, public broadcasting, and cultural institutions like museums.

The deregulation and liberalization of telecommunications work to dismantle the long-standing role of government. The economic theories of market failure and natural monopoly coupled with national security concerns no longer provide sufficient rationale for insulating the sector from competition. However, the preexisting regulatory regime, based on the old rationale, may relinquish control grudgingly. Accordingly, just as the market must realign to reflect competitive opportunities, so must regulators revamp and reshape their mission.

8.1 The U.S. Telecommunications Regulatory Regime

The U.S. telecommunications regulatory system provides a case study in the sometimes difficult and time-consuming task of responding to changed circumstances. The United States does not have a Ministry of Posts and Telecommunications with a near exclusive portfolio in telecommunications policy making and regulation. Instead, a number of agencies in the executive branch share the portfolio with an independent regulatory agency, the FCC. With a variety of players, each having different expertise, constituencies, perspectives, and agendas, the international telecommunications policy-making process in the United States is complicated and confusing. The great importance of trade policy, coupled with technological and marketplace convergence, compounds the complexity.

U.S. telecommunications policy makers are subject to Congressional oversight hearings and budget authorizations that sometimes result in "regulation by lifted eyebrow," that is, indirect policy making by the legislature without its having enacted or revised a law. However, those agencies generally have significant discretion in assessing what rules and regulations serve the public interest and how to implement legislation. On a day-to-day basis, the agencies have ample discretion to determine, with the assistance of public participation, what initiatives, rules, and regulations will flesh out the general language contained in the Communications Act of 1934 and its amendments.

Such decisions affect industry structure and affect individual company profitability. For example, the FCC used to monitor and regulate the manner in which international carriers activated new submarine cable and satellite circuits. On the view that forcing carriers to activate more expensive satellite circuits would serve the public interest by promoting routing and facility diversity, the FCC promoted the financial well-being of Comsat, the former sole investor in the INTELSAT and Inmarsat cooperatives and may have artificially reduced demand for submarine cables. The FCC also used to prevent AT&T from providing nonvoice data services, in part to promote the financial viability of a small group of international record carriers and to preclude AT&T from dominating the telegram and telex markets.

The judiciary performs a review function to determine whether Congress, the FCC, and executive branch agencies have complied with applicable laws on both procedural and substantive grounds. Occasionally, a federal court may issue an activist, results-oriented decision, for example, ordering the FCC to authorize MCI to provide switched long-distance service in addition to unswitched private-line service, on the grounds that the FCC did not expressly state that long-distance competition should be limited to private line service.[1] Typically, courts limit their role to reviewing administrative

1. See, for example, MCI Telecommunications Corp. v. FCC, 561 F.2d 365 (D.C.Cir.1977), *cert. denied*, 434 U.S. 1040, 98 S.Ct. 780, 54 L.Ed.2d 790 (1978) (Execunet I); MCI Telecommunications Corp. v. FCC, 580 F.2d 590 (D.C.Cir.), *cert. denied*, 439 U.S. 980, 99 S.Ct. 566, 58 L.Ed.2d 651 (1978) (Execunet II).

agency actions to ensure that decisions are rational and that the agency followed proper procedures in fact finding and seeking public participation.[2]

8.1.1 The Legislature

The commerce clause of the U.S. Constitution, Article I, Section 8, gives Congress the power to legislate on matters affecting interstate and foreign commerce. That provision affords Congress exclusive and preemptive power over commercial matters that affect more than one state, or that involve relations with other nations. Congress has ceded day-to-day regulatory authority to the FCC through the powers established in the Communications Act of 1934, the Communications Satellite Act of 1962, and other pieces of legislation. Congressional committees with a budgetary or oversight role keep close track of FCC actions.

The power of the purse and the ability to convene oversight hearings on any topic accord Congress substantial power. That means legislators can stimulate regulatory activity and influence the decision-making process through questions posed in a hearing, correspondence with the FCC, and "unofficial" telephone calls. Congress cannot operate outside the laws it created requiring government to operate "in the sunshine," that is, in public forums that allow all interested parties to observe, participate, and know the positions of other parties. Similarly, it cannot engage in ex parte communications, that is, advocacy on matters before the FCC but outside the conventional process in which all parties know of the communication and have an opportunity to comment on it. Nevertheless, Congress has broad powers to affect the FCC decision-making process, both officially and informally.

8.1.1.1 Primary Telecommunications Legislation

The Communications Act of 1934 (47 U.S.C. Secs. 151–799) and the Communications Satellite Act of 1962 (47 U.S.C. Secs. 701–757) provide the basis for the structure and broad policies in international telecommunications.[3] The Communications Act created the FCC to serve as an expert regulatory agency with a mission to serve the public interest, promote the widespread use and availability of radio, and ensure that the enterprises it

2. See, for example, Bell Atlantic Telephone Companies v. Federal Communications Commission, 206 F.3 1 (D.C. 2000). (The FCC failed to adequately explain why local exchange carriers that hand off calls to ISPs are routing interstate traffic instead of the FCC's determination that the traffic is local and subject to the requirement that local carriers compensate other local carriers when receiving and routing local traffic.)

regulates likewise operate in the public interest. In international telecommunications, the Communications Act contemplates service by common carriers, which operate as public utilities providing nondiscriminatory service on tariffed terms and conditions subject to FCC review.

The Communications Satellite Act extends the common carrier model to satellite service and, until 2000, had specified that only one carrier should serve as the U.S. participant in the INTELSAT global cooperative. The Satellite Act affects the market composition in international telecommunications by creating a three-tiered satellite services marketplace: Comsat as exclusive investor in INTELSAT, an increasing number of international carriers authorized by the FCC, and other resellers and users. Legislation enacted in 2000 amended the Satellite Act to reflect the privatization of INTELSAT and the need to ensure that it competes with other private ventures fairly and without leveraging previously held privileges and immunities granted to it when it operated as a global cooperative [1]. It also established the foundation for U.S. international carriers to secure direct access to INTELSAT, the basis for fair competition between a privatized INTELSAT and other operators and amendments necessary to legalize the Lockheed-Martin acquisition of Comsat.

8.1.1.2 The Telecommunications Act of 1996

Enactment of the Telecommunications Act of 1996 (hereafter referred to as the 1996 Act) substantially changed the regulatory climate in the United States. It created broad deregulatory and market entry opportunities by superseding laws, court decisions, consent decrees, and regulations that had foreclosed or conditioned facilities-based or resale competition. However, the legislation's authors may have overestimated the willingness of incumbent carriers to cooperate with market entrants, thereby accelerating the loss of market share, even though such cooperation constituted a prerequisite for gaining new market entry opportunities, such as authorizing the Regional Bell Operating Companies (RBOCs) to provide long-distance services after these companies open their local networks to competitors.[4]

---

3. In 1996, the U.S. Congress enacted a comprehensive rewrite of the Communications Act. See Telecommunications Act of 1996, P.L. 104-104, 110 Stat. 56 (1999). See also Robert M. Frieden, "The Telecommunications Act of 1996: Predicting the Winners and Losers," *Hastings Communications and Entertainment Law Journal*, Vol. 20, 1997, p. 11.

4. A Bell Operating Company (BOC) may provide inter-LATA, "long-distance" telephone services in areas where it also provides local exchange services (commonly referred to as

The legislation revalidated the concept of common carriage as the primary model for classifying telecommunications carriers. In doing that, however, Congress created another paradox: It provided greater specificity on the duties that status entails[5] but also created a mechanism for the FCC to abandon almost all traditional common carrier requirements if doing so would serve the public interest.[6] Similarly the legislation favors carrier-to-carrier negotiations for facilities access and interconnection instead of the customary application of tariffs that are publicly available and applicable to similarly situated users.

The legislation also created a shared jurisdictional scheme between the FCC and state public utility commissions. The combination of state commissions' concerns over federal meddling with state matters and incumbent local exchange carriers' desire for a judicial conclusion that the 1996 Act violated the Constitution and that the FCC exceeded legislated authority has

---

"in-region inter-LATA service") if (1) it has entered into one or more binding interconnection agreements approved by the state public utility commission having jurisdiction, or (2) in the absence of such a request it has filed a statement of generally available terms and conditions approved by the appropriate state commission. *Id.* 47 U.S.C. §271(c)(1)(A)-(B). LATAs are local access and transport areas, 47 U.S.C. § 153(25), a contiguous geographic area within which a BOC may provide local and toll calling. LATAs were created as part of the AT&T consent decree with an eye toward safeguarding long-distance telephone service competition while also providing the divested BOCs with a sufficiently broad geographic area to ensure ample BOC toll-calling revenues and at least one carrier providing ubiquitous service. Limiting the BOCs to intra-LATA service emphasized the expectation that they would exclusively provide local exchange services. See United States v. American Telephone and Telegraph Co., 552 F. Supp. 131 (D.D.C. 1982), *aff'd sub nom.* Maryland v. United States, 460 U.S. 1001 (1983). The BOCs have resented this line of business restriction, which forecloses access to in-region domestic and international interexchange service. In 1998, the total domestic interexchange service market was $105 billion and the international switched telephone service traffic generated $14 billion in revenues with U.S.-based callers making 4.5 billion international calls. FCC, Industry Analysis Div., *Trends in Telephone Service,* Table 19.2, Telecommunications Revenue Reported by Type of Service and Sec. 7, International Telephone Service (March 2000) available on-line at http://www.fcc.gov/Bureaus/Common_Carrier/Reports/FCC-State_Link/recent.html.

5. "A telecommunications carrier shall be treated as a common carrier under this Act only to the extent that it is engaged in providing telecommunications services." *Id.* 47 U.S.C. §153(44).

6. The Act authorizes the FCC to "forbear from enforcing any regulation or provision of the Act if the Commission determines that such enforcement is unnecessary to guard against discrimination, to ensure just and reasonable services, to safeguard consumers and

resulted in significant delays in implementation of the 1996 Act. This creates yet another paradox: Delay stifles local exchange competition but also precludes RBOCs' entry into long-distance telecommunications markets. Until they receive long-distance authority, the RBOCs have to pursue most international operations through separate subsidiaries not directly linked to their local exchange affiliates.

The 1996 Act overhauls the Communications Act of 1934 to establish a law that expressly favors competition and envisions an Internet-like "network of networks" [2], with localities served by a multiplicity of operators each interconnected with all others.[7] The 1996 Act also provides a greater specificity of what obligations a telecommunications carrier has toward other carriers. For example, the 1996 Act imposes substantial interconnection responsibilities on incumbent local exchange carriers (ILECs), which provide the first-mile and last-mile local loop connection for long-distance carriers. Despite greater specificity, the 1996 Act directs the FCC to implement its new policies, and in doing so the Commission had to initiate dozens of rulemakings, many on an expedited basis, even as stakeholders pursued litigation to stay or overturn one completed aspect of the overall package of implementing dockets.[8]

---

serve the public interest." *Id.* Title IV—Regulatory Reform, Sec. 401(a)(1)–(3), *codified at* 47 U.S.C. § 160, (a)(1)–(3).

7. For a more comprehensive summary of the 1996 Act provisions, see Christopher H. Sterling, "Understanding the Telecommunications Act of 1996," *Federal Communications Law Journal*, Vol. 49, No. 2, Feb. 1997, p. 509; Carl B. Kress, "The 1996 Telekommunikationsgesetz and the Telecommunications Act of 1996: Toward More Competitive Markets in Telecommunications in Germany and the United States," *Federal Communications Law Journal*, Vol. 49, No. 3, Apr. 1997, p. 551; Rob Frieden, "The Telecommunications Act of 1996: Predicting the Winners and Losers," *Hastings Communications & Entertainment Law Journal*, Vol. 20, No. 1, Fall 1997, pp. 11–57.

8. See Implementation of the Local Competition Provisions in the Telecommunications Act of 1996, CC Docket No. 96-98, First Report and Order, 11 FCC Rcd. 15499, 15612 (1996), Order on Reconsideration, CC Docket No. 96-98, 11 FCC Rcd. 13042 (1996), Second Order on Reconsideration, FCC 96-476 (rel. Dec. 16, 1996), *partially reversed and remanded sub nom.* Iowa Utilities Board v. FCC, 118 S.Ct. 879 (1998); Implementation of the Local Competition Provisions of the Telecommunications Act of 1996, CC Docket No. 96-98, Third Report and Order and Fourth Further Notice of Proposed Rulemaking, 15 FCC Rcd 3696 (1999); Promotion of Competitive Networks in Local Telecommunications Markets, Notice of Proposed Rulemaking and Notice of Inquiry in WT Docket No. 99-217 and Third Further Notice of Proposed Rulemaking

The 1996 Act and the FCC's implementation strive to stimulate more resale and facilities-based competition of both local and long-distance telephone service while at the same time fostering a level competitive playing field among incumbents and market entrants. While promoting competition, the drafters of the 1996 Act evidenced an unwillingness to rely solely on marketplace factors, particularly for several preferred constituencies, including schools, libraries, hospitals, clinics, low-income consumers, and telephone subscribers in high-cost, primarily rural and insular[9] areas.[10] The 1996 Act paradoxically seeks to remove market entry barriers to foster competition even as it broadens the concept of universal service in telecommunications, the required undertaking of carriers, and how much consumers must pay to support the mandate.

Nevertheless, the 1996 Act can be characterized as primarily removing prior legislative and judicially imposed barriers to market entry. For example, it eliminates cross-ownership restrictions on telephone company provision of cable television service and vice versa. Likewise, the 1996 Act eliminated the remaining two lines of business restrictions imposed on the RBOCs by the Modification of Final Judgment [3]: (1) long-distance services that cross local access and transport area (LATA)[11] boundaries and (2)

---

in CC Docket No. 96-98, WT Docket No. 99-217 and CC Docket No. 96-98, FCC 99-141 (rel. July 9, 1999).

9. Insular areas include states, commonwealths, and possessions of the United States that are not part of the contiguous 48 states (e.g., Guam and American Samoa).

10. Section 254(b) of the 1996 Act, 47 U.S.C. § 254 (1996) articulates expanded universal service principles, including the mandate that "advanced telecommunications and information services should be provided in all regions of the Nation," 47 U.S.C. § 254(b)(2); that "low-income consumers and those in rural, insular, and high cost areas should have access to telecommunications and information services ... reasonably comparable to those services provided in urban areas," 47 U.S.C. § 254(b)(3) and that "[e]lementary and secondary schools and classrooms, health care providers and libraries should have [subsidized, possibly below cost] access to advanced telecommunications services," 47 U.S.C. § 254(6) under a provisioning scheme set out at § 254(h).

11. Under the 1996 Act of 1996, a local access and transport area (LATA) is "a contiguous geographic area (A) established before the date of enactment of the [1996 Act] by a [BOC] such that no exchange area includes points within more than 1 metropolitan statistical area, consolidated metropolitan statistical area, or State, except as expressly permitted under the AT&T Consent Decree; or (B) established or modified by a [BOC] after such date of enactment and approved by the Commission." 47 U.S.C. § 153(25). LATAs were created as part of the Modification of Final Judgment's (MFJ) "plan of reorganization" under which

telecommunications equipment manufacturing after the companies meet a 14-point competitive check list evidencing the existence of or opportunity for full and fair interconnection and facilities access by other carriers.[12]

Congress leaves to the FCC and state public utility commissions the task of ensuring full and fair interconnection between incumbent and new local exchange carriers (LECs), as well as a host of other issues requiring complex economic analysis. For example, the FCC and state public utility commissions must determine appropriate discounts for resold LEC services and the actual cost of providing switching and routing of traffic generated by competing LECs, interexchange carriers (IXCs), and end users.

Billions of dollars in carrier revenues and cross-subsidies to promote universal service are at stake as the FCC attempts to define a national framework for rational cost-based pricing and to move from implicit to explicit (direct and identifiable) universal service subsidies. For example, until the FCC ordered sizable reductions in 2000, IXCs had to pay LECs an access charge of about 3 cents a minute for the use of local exchange facilities to originate and terminate long-distance toll calls. IXCs estimate that the actual cost of such access lies in a range from 0.4 to 1.2 cents, meaning that the then-current access charges exceeded cost by 250% to 700% and provided large sources of funds for profits or universal service subsidies.[13] The FCC and state commissions have the unenviable task of sorting out complex pricing issues and determining competitively neutral facilities interconnection and access issues for local and long-distance carriers, as well as for end users.

---

the BOCs were divested from AT&T. United States v. Western Elec. Co., 552 F. Supp. 131 (D.D.C. 1982), *aff'd sub nom.* Maryland v. United States, 460 U.S. 1001 (1983); United States v. Western Elec. Co., 569 F. Supp. 1057 (D.D.C. 1983) (Plan of Reorganization), *aff'd sub nom.* California v. United States, 464 U.S. 1013 (1983); see also United States v. Western Elec. Co., No. 82-0192 (D.D.C. Apr. 11, 1996) (vacating the MFJ). Pursuant to the MFJ, "all BOC territory in the continental United States [was] divided into LATAs, generally centering upon a city or other identifiable community of interest." United States v. Western Elec. Co., 569 F. Supp. 990, 993 (D.D.C. 1983).

12. See 47 U.S.C. §§ 271(c)(2)(B)(i)–(xiv).

13. "AT&T asserts, for instance, that the current average per-minute access rates of the BOCs are nearly seven times the forward-looking economic cost of providing that service, and that total interstate access charges collected today from interexchange carriers exceed forward-looking economic cost by $11 billion, or 70 percent of the total." Access Charge Reform Price Cap Performance Review for Local Exchange Carriers, CC Docket No. 96-262, Notice of Proposed Rulemaking, Third Report and Order, and Notice of Inquiry, FCC 96-488, 1996 WL 733469 (F.C.C.) at 11 (rel. Dec. 12, 1996).

8.1.1.3 Telecommunications Laws that Affect Trade

Aside from limitations on foreign ownership in telecommunications, few laws in the United States impose absolute market access reciprocity or financial and access penalties in the face of unequal access. One law that does is the Submarine Cable License Act, enacted in 1921.[14] That law allows foreign enterprises to land an international cable on U.S. soil if and only if U.S. carriers have such a right of access in the carrier's home country. Because national security concerns have prompted all other nations to prohibit the landfall of foreign-owned international cable facilities, the United States imposes a similar prohibition. While a reciprocity requirement may exacerbate trade barriers and access restrictions, it may enable nations to negotiate compromise positions without resorting to further market access restrictions.

The United States officially refrains from more closely scrutinizing foreign manufacturers and carrier activities. It has committed to the trade concept of national treatment, which means that foreign enterprises receive the same rights and responsibilities conferred to domestic companies. Trade commitments notwithstanding, the FCC has considered imposing more burdensome reporting requirements on foreign-dominated companies doing business in the United States. While such records could assist the U.S. government in trade negotiations and provide empirical evidence of unequal market access, the FCC abandoned the strategy in deference to executive branch trade policies and the requirement that MFN treatment apply to WTO member nations.

However, the FCC currently imposes greater regulatory scrutiny of carriers operating in the United States that are affiliated with foreign carriers that have dominant market power in the foreign country.[15]

The FCC has linked a finding of equivalent market access opportunities with evidence of deregulatory and investment opportunities for U.S. enterprises abroad, including market access commitments made under the auspices of the WTO [4]. For nations that provide equivalent market access opportunities to U.S. carriers, the FCC may, on a case-by-case basis, grant waivers to the 20% to 25% foreign ownership cap established in Section 310

14. "An Act relating to the landing and operation of submarine cable in the United States," 47 U.S.C. Secs. 34–39 (1990); see also Executive Order 10530 (May 10, 1954) (delegating to the FCC certain Presidential functions relating to submarine cable landing licenses).

15. See Rules and Policies on Foreign Participation in the U.S. Telecommunications Market, Report and Order and Order on Reconsideration, 12 FCC Rcd 23,891 (1997), *recon. pending.*

of the Communications Act. The Commission also has expressed the desire to use Section 214 of the Communications Act, which requires carriers to apply for its authorization before providing service, as a basis for gauging whether a foreign carrier's home country permits U.S. carriers to provide the very service the foreign carrier seeks authority to provide in the United States.

8.1.2 The Judiciary

Courts interpret, rather than make, law. Their primary role in telecommunications policy and regulation is to resolve disputes by assessing whether the FCC has acted in compliance with applicable laws. In particular, courts determine whether the FCC has complied with the terms of the Communications Act, the Satellite Act, and the Administrative Procedure Act (APA) [5]. The substantive laws that affect telecommunications establish the FCC's powers and mission with broad and flexible terms to account for changed circumstances and technological innovation. The APA establishes procedural standards that agencies like the FCC must satisfy to ensure that the public has a full and fair opportunity to participate in the decision-making process and that the agencies reach reasoned decisions based on the record of evidence and findings generated. The FCC cannot act arbitrarily or capriciously. It cannot abuse the discretion conferred to it by Congress, nor can it otherwise act outside its legal scope of authority.

8.1.3 The Executive Branch

The President and executive branch agencies have powers created by the Constitutional separation of powers, for example, foreign relations and operation of the military. In addition, legislation can create additional executive branch responsibilities, for example, participation in shaping international satellite policy and monitoring the conduct of Comsat in its former capacity as the sole U.S. representative and investor in INTELSAT and Inmarsat.

Executive branch agencies have portfolios in trade policy (the U.S. Trade Representative and the Commerce Department), management of the federal government spectrum (the Commerce Department's National Telecommunications and Information Administration [NTIA]), and telecommunications policy (the Department of State and NTIA). While these agencies typically do not regulate, they can affect regulatory policy primarily through interagency coordination and by long-range planning. That allows them to consider issues in a larger context and presents the opportunity, not always realized, to affect long-term policy making and strategy.

8.1.3.1 Department of State

The Department of State manages the foreign policy and international relations agenda of the United States. Substantive issues like telecommunications and aviation are subsumed within the State Department's broad interpretation of its mission. In application, that means career foreign service officers in the State Department acquire a temporary portfolio in international telecommunications and work with civil servants who are dedicated to careers in telecommunications policy. Such a blend of general foreign relations and specific telecommunications policy skills sometimes does not result in the best work product. A 1993 reorganization within the State Department further diluted in-house expertise in the subject when the separate Bureau of International Communications and Information Policy (CIP) became part of the much larger Bureau of Economic and Social Affairs.

The goals of diplomacy and international comity with foreign nations can conflict with policy initiatives of other executive branch agencies. Depending on the personalities of the leaders at the FCC and the NTIA, turf battles and spirited interagency debates can develop. Some critics allege that the State Department tends to emphasize international comity and cooperation over the nation's sometimes parochial concerns. Wisely or not, foreign policy concerns can dampen the tone and stridency of U.S. advocacy.

The CIP attempts to coordinate international telecommunications and trade policy with all other involved agencies. It assembles the delegations that represent the United States at international forums like the ITU, often conferring the status of ambassador to the delegation head. On matters involving INTELSAT and Inmarsat, it issues instructions to Comsat on what it should advocate and how it must vote on matters of national interest.

Despite perennial calls for better coordination among the CIP, the NTIA, and the FCC, the policy-making process can become disjointed. The potential for such problems is exacerbated by the fact that different Congressional agencies have oversight and budgetary responsibility for each agency. So far, proposals to unify the policy-making process and make it more coherent have failed to gain significant Congressional attention.[16]

---

16. See, for example, Congress of the United States, Office of Technology Assessment, "Jurisdictional Issues in the Formulation and Implementation of National Communication Policy," Chap. 13 in *Critical Connections—Communications for the Future*, 1990.

8.1.3.2 The NTIA

The NTIA, an agency within the Department of Commerce, serves as the executive branch's principal voice in telecommunications policy.[17] The effectiveness of the agency depends in large part on the visibility and stature of its head, because the NTIA has no regulatory or licensing portfolio. Likewise, it has a small staff of only 15 to 20 involved in aspects of international policy making.

Initially, the NTIA constituted a department within the Office of the President. As the Office of Telecommunications Policy (OTP), it had a closer reporting line to the President, perhaps commensurate with the importance of telecommunications to the national economy. The NTIA was spun off in 1978, to become a part of the Commerce Department. In 1993, the NTIA was threatened with a further downgrading in visibility and proximity to the President by a proposal to make it a part of the National Institute of Science and Technology, formerly known as the National Bureau of Standards.

The NTIA participates in the policy-making process principally through advocacy in FCC proceedings and by issuing reports on issues like the National Information Infrastructure.[18] Its statutorily conferred responsibilities address coordination and registration of the federal government's use of frequency spectrum,[19] including the identification of government-used

17. "The [Commerce] Department's telecommunications policymaking functions are centered in the National Telecommunications and Information Administration (NTIA), which was created pursuant to Reorganization Plan No. 1 of 1977, with the responsibilities of the Secretary of Commerce under Executive Order 12046 [Fed. Reg. 13349-13357 (Mar. 29, 1978)] delegated to it. "NTIA Telecom 2000: Charting the Course for a New Century," Chap. 9 in *International and Domestic Policymaking in the Year 2000*, NTIA Spec. Pub. 88-21 (Washington, D.C.: Government Printing Office, Oct. 1988), p. 168.

18. See Information Infrastructure Task Force, Sec. Ronald H. Brown, Chairman, *National Information Infrastructure: Agenda for Action* (Washington, D.C.: Government Printing Office, Sept. 15, 1993); *National Information Infrastructure: Progress Report*, Sept. 1993–1994 (Washington, D.C.: Government Printing Office, Sept. 1994).

19. Under Section 305 of the Communications Act of 1934, as amended, the President retains the authority to assign frequencies to all radio stations belonging to the federal government. The President has delegated authority to the Secretary of Commerce, who in turn has delegated it to NTIA. See Exec. Order No. 12046, as amended, 3 C.F.R. 158 (1978), reprinted in 47 U.S.C. Sec. 305 app. at 127 (1989); U.S. Dept. of Commerce, Dept. Organization Orders 10-10 and 25-7. The applicable executive order and

spectrum that can be transferred to private use as ordered by Congress[20] and participation in the oversight of Comsat's activities in INTELSAT and Inmarsat.[21]

Most of the NTIA's responsibilities are shared with other agencies. The FCC has greater technical expertise and a larger staff. The State Department leads on foreign policy issues, and other trade offices in the Department of Commerce and the U.S. Trade Representative (USTR) hold more direct authority to formulate trade policy. Nevertheless, the NTIA's studies and advocacy documents help shape the international telecommunications and trade policy agenda. Particularly in the trade area, staff technical expertise augments the more generalist USTR on telecommunications facilities, services and equipment matters.

If the FCC's constituency constitutes its licensees and the public that uses licensee services, the NTIA's natural allies are manufacturers and carriers concerned with market access and competition issues. However, in view of the fact that it confers no licenses and has no regulatory function, the NTIA has a less direct impact on any single enterprise. It accrues clout when

---

an Office of Management and Budget Circular No. A-qq "provide NTIA with the power to assign frequencies and approve the spectrum needs for new systems." U.S. Dept. of Commerce, National Telecommunications and Information Administration, *U.S. Spectrum Management Policy: Agenda for the Future,* 20 NTIA Spec. Pub. 91-23 (Washington, D.C.: Government Printing Office, Feb. 1991).

20. In 1993, Congress directed the Secretary of Commerce to identify at least 200 MHz of spectrum currently allocated on a primary basis for the federal government use that is not required for present or identifiable future use by the federal government and that is most likely to have the greatest potential for productive uses and public benefit if allocated for non-Federal use. Omnibus Budget Reconciliation Act of 1993, Pub. L. No. 103-66, Title VI, § 6001(a)(3), 107 Stat. 312 (approved Aug. 10, 1993); see also H.R. Rep. No. 103-213, 103rd Cong., 1st Sess. (1993). On February 10, 1994, the Secretary of Commerce released *Preliminary Spectrum Reallocation Report (Preliminary Report),* NTIA Special Publication 94-27, identifying spectrum for reallocation from federal government use to private sector, including local government, use. See also *Report to Ronald H. Brown, Secretary, U.S. Department of Commerce, Regarding the Preliminary Spectrum Reallocation Report,* 9 FCC Rcd. 6793 (1994).

21. "Functions relating to international communications satellite systems, vested in the President by the Communications Satellite Act of 1962, and the Inmarsat Act of 1978 have also been delegated to NTIA." "NTIA Telecom 2000: Charting the Course for a New Century," Chap. 9 in *International and Domestic Policymaking in the Year 2000,* NTIA Spec. Pub. 88-21 (Washington, D.C.: Government Printing Office, Oct. 1988), p. 169.

pocketbook and industrial leadership issues predominate concerns about foreign policy, national security, and interagency "turf."[22]

8.1.3.3 U.S. Trade Representative

The USTR serves as the lead agency for coordinating negotiations with foreign nations on trade matters.[23] Telecommunications has become a significant component in the USTR's portfolio.[24] Because of the technical complexity of the issues and the generalist nature of its staff, the USTR regularly consults with other agencies and includes representatives of those agencies on its negotiating delegations. However, USTR staff have expressed concern when the FCC attempts unilaterally to carve out a trade portfolio. For example, in the 1980s, the FCC proposed to monitor U.S. carrier procurement of foreign-manufactured equipment and to scrutinize closely the operations of foreign-owned carriers in the United States. The Commission articulated a model view of the international telecommunications and trade marketplace [6] and strongly signaled its desire to use regulatory oversight as leverage to achieve its goals. To promote market access[25] opportunities for U.S. carriers and equipment manufacturers, the FCC proposed more extensive regulatory burdens on foreign carriers operating in the United States,

---

22. See Department of Commerce, National Telecommunications and Information Administration, Office of Spectrum Management homepage, available on-line at http://www.ntia.doc.gov/osmhome/osmhome.html.

23. See Trade and Tariff Act of 1984, Pub. L. No. 98-573, 98 Stat. 2948 (directing the President to negotiate reductions in foreign NTBs) *codified at*, 19 U.S.C. § 2122 (1985); see also The Omnibus Trade and Competitiveness Act of 1988, Pub. L. No. 100-418, 1988 U.S. Code Cong. & Admin. News 1107 (1988).

24. See T. Howell, et al., "International Competition in the Information Technologies: Foreign Government Intervention and the U.S. Response," *Stanford Journal of International Law*, Vol. 22, 1986, p. 215; Note, "Opening Doors in Foreign Market Trade Through the Telecommunications Trade Bill," *Brigham Young University Law Review*, 1989, p. 639.

25. "[W]e believe that it is our duty to assess whether the regulations or practices of certain foreign administrations, such as restrictions on the availability of telecommunications equipment or the use of the telecommunications network, may have an adverse impact on our ability to meet the public interest goals set forth in Section 1 of the [Communications] Act." Regulatory Policies and International Telecommunications, CC Docket No. 86-494, Notice of Inquiry and Proposed Rulemaking, 2 FCC Rcd. 1022, 1023 (1987). The Commission sought to establish an "ideal" international telecommunications model based on four key elements: open entry, nondiscrimination, technological innovation, and international comity. *Id.* at 1022.

thereby making them subject to FCC jurisdiction if U.S. enterprises were unable to access markets in the foreign carriers' home country.

The USTR strongly opposed the FCC's trade initiative as one that deviated from the commitment to national treatment. The FCC's final order backed off from a confrontation with the USTR on trade policy, opting instead to require reports only from the handful of foreign government–owned and operated telecommunications carriers operating in the United States. In effect, the Commission agreed to refrain from using regulation to leverage market access opportunities for U.S. carriers.

In 1995, the FCC again proposed to use its regulatory authority to secure parity of market access between U.S. and foreign carriers operating within the United States.[26] While the agency appears to defer to the USTR's lead on trade policy, it has sought to establish its own portfolio, particularly for leveraging market access opportunities. In the context of establishing a national trade policy, the USTR must consult with other agencies. But in doing so, it must come to terms with the different constituencies represented. When the State Department participates, foreign policy concerns and appreciation for international comity are incorporated. When two agencies of the Department of Commerce participate (the NTIA and the International Trade Administration), domestic industry concerns are voiced, sometimes not necessarily by the same person or with the same perspective. When the FCC's technical expertise is tapped, the trade delegation or task force runs the risk of bolstering the Commission's interest in pursuing its own agenda.

The potential for turf battles, multiple constituencies and positions, and uncoordinated initiatives rises with the number of agencies and individuals involved. In the worst-case scenario, foreign governments with a better ability to reign in disparate agencies and players can divide and conquer the U.S. interagency policy-making process. At best, foreign governments have a number of agencies to visit and convince.

8.1.3.4 The FCC

The Communications Act of 1934 vests the FCC with authority to regulate interstate and foreign communications by wire and radio. The Commission's public interest mandate propels a regulatory portfolio into every aspect of international telecommunications and, at least peripherally, into matters of international trade policy as well. A separate International Bureau was

26. See 47 U.S.C. §§ 271(c)(2)(B)(i)–(xiv).

created in 1994 to consolidate international functions that were distributed throughout the Commission. Although the FCC may reorganize on functional grounds rather than jurisdictional or type of medium, the International Bureau currently handles all satellite issues, spectrum coordination with other nations, preparations for Commission participation at ITU, and other bilateral and multilateral conferences and international policy matters.

The FCC operates as an independent regulatory agency, separate from the executive branch and the legislature. However, the President appoints the agency's five commissioners (three from the President's political party) subject to Senate approval. Congress votes operating funds and conducts regular oversight hearings to ensure that the agency accommodates the legislature's interest even in the absence of new laws.

FCC rules, regulations, and policies substantially affect the terms, conditions, and profitability under which international telecommunications service providers operate. The Communications Act confers ample flexibility for the FCC to fashion new rules, regulations, and policies as public interest so dictates. While subject to judicial review, the actions of this expert agency usually pass muster unless the Commission has violated procedural and fairness requirements [7].

The Communications Act of 1934 authorizes the FCC to license facilities and service providers and to ensure that they operate in the public interest. The power to license and regulate also means that the FCC has ongoing responsibilities, under Title II of the Communications Act, to ensure that service providers do not operate in an unreasonable or discriminatory manner. Likewise, the Commission may require a licensee to operate as a common carrier,[27] thereby imposing a duty to hold itself indifferently to the public and to serve all who seek service, usually under a public contract known as a tariff.

8.2 U.S. Policies Prior to Deregulation

Until the mid-1970s, U.S. international telecommunications policy making closely paralleled the philosophy and approach of most nations. It placed a

27. The Communications Act of 1934, as amended, defines a common carrier as "any person engaged as a common carrier for hire, in interstate or foreign communication by wire or radio" 47 U.S.C. Sec. 153(h) (1990). The Communications Act requires common carriers to provide service "upon reasonable request," at tariffed rates that are "just and rea-

premium on industrial stability and universal, sometimes subsidized, service. Prior to aggressive efforts to replicate domestic marketplace deregulatory successes, the United States shared the view that natural monopolies would optimally serve the major international telecommunications market segments. Although privately owned, U.S. international service carriers (USISCs) were limited in the type of markets they could enter and the terms and conditions under which they could offer services.

U.S. international telecommunications policy established a number of service, market, technological, and geographic dichotomies that collectively restricted competition and market access. The FCC imposed restrictions on who could provide international services and where service could originate and terminate. The Commission created a voice/data dichotomy whereby only AT&T could provide switched voice services, and only a small group of international record carriers (IRCs) could provide text services like telegrams and telexes.

The United States was divided into locations in which international record traffic could originate—gateway cities—with the rest of the nation considered the hinterland, where customers would have to secure services of a domestic carrier to route traffic to a gateway. That "end-on-end" routing arrangement generated higher costs to consumers. For record services, users in the hinterland had to pay one carrier, The Western Union Telegraph Company, a separate domestic carriage charge and another IRC a separate international carriage charge. As part of an antitrust court case settlement, Western Union agreed to limit the scope of its services to domestic record traffic, including the carriage of traffic to and from the international gateways of other carriers.

The FCC also imposed policies that blunted the potential for competition by multiple service providers using the same transmission technology (intramodal competition) or different transmission media (intermodal competition). Until the mid-1980s, the FCC dutifully adhered to the international consensus of consortium and cooperative planning, deployment, circuit activation,[28] pricing, and maintenance of international facilities. The Commission

sonable" and "without unjust or unreasonable discrimination." 47 U.S.C. Sec. 201(a), (b) and 202(a). See Note, "Redefining 'Common Carrier': The FCC's Attempt at Deregulation by Redefinition," *Duke Law Journal*, 1987, p. 501.

28. "The Commission has made decisions affecting the distribution of circuits among available international facilities nearly since the advent of communications satellites in 1965." Policy for the Distribution of United States International Carrier Circuits Among Avail-

accepted the premise that nations need to coordinate on facilities planning and investment to achieve scale economies, avoid wasteful duplication of investment, and secure the benefits of averaging costs over a larger user base.

Only the former Soviet Union's Intersputnik system provided some degree of intramodal, satellite competition with INTELSAT for transoceanic routes. The spillover signals of domestic satellites did provide a limited, regional service alternative to INTELSAT, but only if a nation agreed to permit market access. The nearly universal decision to blend the cost of cable and satellite facilities into a single composite rate meant that users had no opportunity to select the cheaper of the two media or perhaps to pay a premium for all cable service if satellite signal echoing presented a problem.

U.S. avoidance of intramodal competition included legislation that created an international satellite monopoly. The Communications Act of 1962 created Comsat and authorized it to serve as the nation's sole investor and participant in INTELSAT. Comsat received a legislative franchise to serve as a "carrier's carrier" (i.e., intermediary) for INTELSAT international satellite capacity. No other enterprise could deal directly with INTELSAT and carriers and only the federal government could qualify as "authorized users" who could acquire wholesale capacity from Comsat. Other users acquired marked-up retail capacity further down the distribution chain.

That structure supported the creation of multiple market tiers with a single facilities operator (INTELSAT), a single U.S. intermediary (Comsat), and a select group of wholesalers (the USISCs). Channeling satellite capacity under such a system resulted in a number of pricing markups by intermediaries who provided little enhancement to the basic transmission capacity and whose administrative functions involved primarily order taking.

Until the mid- and late 1970s, the FCC accepted the world view that resale of leased lines presented an unreasonable threat to carrier profitability

---

able Facilities During the Post-1988 Period, CC Docket No. 87-67, Notice of Proposed Rulemaking. 2 FCC Rcd. 2109 (1987), policy abandoned, 3 FCC Rcd. 2156 (1988). For an example of how the balanced loading policy was applied, see ITT Cable and Radio, Inc.—Puerto Rico, 5 FCC 2d 823 (1966); AT&T, 7 FCC 2d 959 (1967) (FCC determination of the proper mix of submarine cable and satellite facilities for Puerto Rico); Comsat, 29 FCC 2d 252 (1971) (proportional loading ordered at the rate of five Atlantic Ocean Region INTELSAT satellites circuits for every one TAT-5 submarine cable circuit activated). But see also Communications Satellite Corp., 5 FCC Rcd. 5952 (1990) (evidencing reduced scrutiny and greater sensitivity to amount of time for consideration of a Comsat investment obligation in INTELSAT prior to scheduled vote by the cooperative's signatories); *accord*, Communications Satellite Corp., DA 92-955 (rel. July 23, 1992) (authorizing Comsat investment in the INTELSAT K satellite).

and universal service. Resale is an exercise in arbitrage: securing bulk capacity intended for use by a single large-volume user and subdividing it for use by a number of customers with lower individual capacity requirements. The FCC grew to believe that such a function reduced the potential for facilities-based carriers to discriminate against small-volume users and could stimulate the benefits of competition without the delay and expense in parallel facilities construction.

8.2.1 Undoing a Heritage of Pervasive Government Regulation and Oversight

Before embracing a deregulatory philosophy, starting in the late 1970s and early 1980s, the FCC established industrial policies that had the effect of segmenting the international telecommunications marketplace and subjecting all segments to pervasive regulation. At least initially, such division may have had a technological justification: Until the onset of high-frequency radio, which could transmit voice and record signals, the submarine cable medium could transmit only textual messages via Morse code. However, any technological justification for segregating carriers into only voice and only record service markets ended well before 1980, when the FCC abandoned the policy.[29] The Commission finally permitted AT&T to provide record services [8] and the IRCs to provide voice services [9]. Actions like that were the result of an increasingly prevalent view that government need not intrude on the interplay of market forces, because such involvement probably did more harm than good. Segmenting markets and foreclosing market access, for example, denying AT&T the opportunity to provide record services, ostensibly "protected" users from monopolization by a carrier dominant in another market (e.g., AT&T's near monopoly in voice services). But instead of foreclosing additional monopolization, market segmentation policies prevented competition and allowed carriers to maintain high rates.

It took an act of Congress to reauthorize Western Union to provide international services,[30] a market it had relinquished to secure government

---

29. See American Telephone and Telegraph Co., 37 FCC 2d 1151 (1964) (TAT-4 voice/record dichotomy).

30. See Record Carrier Competition Act of 1981, Pub. L. No. 97-130, 95 Stat. 1687 (1981), implemented in Interconnection Arrangements Between and Among the Domestic and Int'l Record Carriers, Interim Order, 89 FCC 2d 928 (1982), implementing tariffs rejected, 91 FCC 2d 483 (1982), *modified on recon.*, 93 FCC 2d 845 (1983). For an analysis of the costs incurred and charges by Western Union to interconnect with

consent to acquisition of a key competitor.[31] Limiting a dominant carrier like Western Union also served an FCC policy that sought to maintain a dichotomy between hinterland and gateway localities,[32] but in application it financially disadvantaged users in localities not authorized to make direct international connections, regardless of demand.

Dichotomies based on service (voice/record) and location (gateway/hinterland) outlasted any technological justification. For example, new generations of submarine cables[33] and radio facilities made it possible for

---

international record carriers, see Western Union Telegraph Co., 95 FCC 2d 881 (1983), *aff'd sub nom.* FTC Communications, Inc. v. FCC, 750 F.2d 226 (2d Cir. 1984); Western Union Telegraph Co., FCC 86-190 (rel. May 2, 1986) (ordering the international record carriers to reimburse Western Union for service whose 15% discount was later deemed unjustified), *on recon.*, 2 FCC Rcd. 2999 (1987), *on further recon.*, 3 FCC Rcd. 2597 (1988). See also R. Frieden, "International Telecommunications and the Federal Communications Commission," *Columbia Journal of Transnational Law*, Vol. 21, No. 3, 1983, pp. 423, 466–485.

31. In 1943, Congress enacted Section 222 of the Communications Act to permit Western Union to acquire its major competitor, the Postal Telegraph Co. Western Union was required to divest itself of international record carrier operations in exchange for Congressionally sanctioned antitrust immunity. For a history of Section 222 prior to enactment of the Record Carrier Competition Act, see Western Union Tel. Co. New Telex Serv. Arrangements via Mexico and Canada, 75 FCC 2d 461 (1979), *vacated*, ITT World Comms., Inc. v. FCC, 635 F.2d 32 (2d Cir. 1980) (FCC authorization of Western Union to provide international service via Mexico and Canada was deemed a violation of Sec. 222).

32. International Record Carriers' Scope of Operations, 38 FCC 2d 543, 545 (1972) (describing the gateway concept), relaxed to include 21 new gateway cities, 76 FCC 2d 115 (1980), *on recon.*, 80 FCC 2d 303 (1980), *aff'd sub nom*, Western Union Tel. Co. v. FCC, 665 F.2d 1112 (D.C. Cir. 1981). Until enactment of the Record Carrier Competition Act of 1981, the FCC had established a geographical dichotomy between "gateway" cities, where international traffic could originate, and the "hinterland," where foreign-destined traffic had to first access a gateway via a domestic carrier for subsequent retransmission via an international record carrier. See Domestic Public Message Serv., 71 FCC 2d 471 (1979), *aff'd sub nom.* Western Union Tel. Co. v. FCC, 665 F.2d 1126 (D.C. Cir. 1981) (expanding the domestic carrier set to include Graphnet, Inc., but maintaining the domestic/international service dichotomy).

33. AT&T, the dominant domestic telephone carrier, initiated transatlantic telephone message service in 1927 by high-frequency radio. In 1956, AT&T developed a reliable underwater repeater and, together with the British government, laid the first transatlantic cable (TAT-1). Overseas Communications Services, CC Docket No. 80-632, 84 FCC 2d 622, 623 (1980).

international carriers to provide both voice and record services. The IRCs invested in the new submarine cables, thereby providing those companies with the technological wherewithal to provide voice services in competition with AT&T. Yet the FCC perpetuated the voice/record dichotomy, presumably to ensure that AT&T did not dominate record markets as it had voice markets. While designed to guard against AT&T leveraging of market power into record services, the policy substantially reduced intramodal competition among carriers using the same transmission medium they collectively owned. Even without such a policy, the action by all carriers to form cable consortia and satellite cooperatives discouraged price competition, primarily because each carrier had roughly the same transmission capacity costs. AT&T's disqualification from record service markets made it that much easier for the handful of IRCs to form an oligopoly and agree not to engage in price competition. Likewise, the absence of IRC voice services in effect guaranteed that AT&T could maintain its monopoly.

8.2.2 The Deregulatory Campaign (1980 to Present)

In the late 1970s and early 1980s, the United States began to abandon support for market segmentation and regulated monopolies. In 1980, the FCC resolved to foster "an improved international communications system with more choices for consumers, more diverse service offerings, and lower rates" [10]. The FCC embraced the view that reduced government oversight and marketplace intervention would enhance consumer welfare, reduce rates, promote competition, spur innovation, foster service diversity, and achieve a more efficient telecommunications marketplace.

Many U.S. officials viewed international telecommunications policy as unnecessarily lagging and diverging from already completed domestic deregulatory initiatives. They hoped that unilateral action could set a deregulatory foundation that foreign counterparts would embrace. The industry and consumer welfare–enhancing dividends achieved in the United States over the last 25 years[34] presumably could apply internationally.

---

34. See, for example, Competitive Carrier, 77 F.C.C.2d 308 (1979), First Report and Order, 85 F.C.C.2d 1 (1980), Further Notice, 84 F.C.C.2d 445 (1981), Second Report and Order, 91 F.C.C.2d 59, Third Report and Order, 48 Fed. Reg. 46791 (1983), Fourth Report and Order, 95 F.C.C.2d 554, Fifth Report and Order, 98 F.C.C.2d 1191, Sixth Report and Order, 99 F.C.C.2d 1020 (1985), *rev'd and remanded sub nom.* MCI Telecom. Corp. v. FCC, 765 F.2d 1186 (D.C. Cir. 1985); Hush-A-Phone v. United States,

However, international policy making driven from domestic experience fails to consider different and unavoidable forces and philosophies. Prior to its philosophical transformation, the FCC ignored evidence that marketplace forces could work in international telecommunications. Its newfound deregulatory zeal may have compounded problems, because efforts to make up for lost time typically backfired because the regulatory counterparts and legislators in other nations did not match the Commission's initiatives. Representatives from other nations refused to be rushed or threatened by initiatives that they viewed as requiring study to confirm the absence of harm to incumbent carriers and their essential service mission.

Even now, few governments have opted to diminish substantially the scope of oversight responsibilities in international telecommunications. While contemplating the free interplay of marketplace and technological forces, even the FCC has recognized the need to consider countervailing factors that work against facilities-based carrier competition and new regulatory classifications, such as private, noncommon carriers[35] that promote market entry by enterprises substantially free of government oversight. Many U.S. deregulatory initiatives initially failed to achieve the support from other

---

239 F.2d 266 (D.C. Cir. 1956), *on remand sub nom.* Hush-A-Phone v. AT&T, 22 FCC 112 (1957) (establishing right of customers to attach acoustical devices to their telephones); Use of the Carterfone Device in Message Toll Service, 13 FCC 2d 420 (1968) (establishing public's right to attach electronic devices to the telephone); Allocation of Frequencies in the Bands Above 890 Mc. 27 FCC 359 (1959), *on recon.* 29 FCC 825 (allocation of frequency spectrum for competing microwave radio services); Microwave Communication, Inc., 18 FCC 2d 953 (1969) *on recon.*, 21 FCC 2d 190 (authorizing common carrier competition); Regulatory Policies Concerning Shared Use and Resale of Common Carrier Services Facilities, 47 FCC 2d 644 (1974), 48 FCC 2d 1077 (1974), 60 FCC 2d 261 (1976), modified, 61 FCC 2d 70 (1976), *further modification*, 62 FCC 2d 588, *aff'd sub nom.* AT&T v. FCC, 572 F.2d 17 (2nd Cir. 1977), *cert. den.*, 439 U.S. 875 (1978) (authorization for shared use and resale of long-distance lines, which created the opportunity for arbitrage); MCI Telecom. Corp. v. FCC, 561 F.2d 365 (D.C. Cir. 1977), *cert. den.*, 434 U.S. 1040 (1978) mandate enforced, 580 F.2d 590 (D.C. Cir. 1978), *cert. den.*, 439 U.S. 980 (1978) (creating right of alternative carriers to access the PSTN and thereby provide alternative long-distance telephone service).

35. See, for example, Tel-Optik, Limited, 100 FCC 2d 1033 (1985) (authorization of a private, noncommon carrier international fiber optic submarine cable operator); Pan American Satellite Corp., 101 FCC 2d 1318 (1985) (authorization of a private, noncommon carrier international satellite operator). The Record Carrier Competition Act of 1981, Pub. L. No. 97-130, 95 Stat. 1687 (1981) authorized Western Union Telegraph Company's return to international services after years of prohibited access.

nations necessary to achieve new investment and market access opportunities. While theoretically supportable and often bolstered by empirical proof of enhanced consumer welfare in the United States, many FCC initiatives initially were deemed too bold, risky, and threatening by other governments. Over time and with a growing consensus, many nations now share at least some of the FCC's deregulatory philosophy.

Officials at the FCC have recognized belatedly the futility in applying unilaterally and unconditionally a domestic market–oriented deregulatory model to the international telecommunications environment. In the 1980s, the Commission initiated a number of rulemakings to undo the complex web of protectionist policies it designed to create market segments, service dichotomies, and limit competition. While the FCC could undo domestic policies, which retarded competition, the Commission generally failed to convince other nations of the need for speedy and dramatic change. Until nations perceived the benefit of privatization, liberalization, and deregulation, FCC edicts could do little to foster international telecommunications policies based on theoretical premises or even domestic experiences.

For example, the FCC could provide concrete evidence of public dividends accruing from authorizing enterprises to resell bulk private-line services.[36] The Commission's pro-resale domestic policy included large-volume calling services like WATS lines, initially made available only to individual companies generating high volumes of long-distance calls [11]. Private-line resale involves the aggregation of small-volume users whose collective demand qualifies for a bulk-volume discount otherwise unavailable to any single small-volume user. The FCC viewed that option as putting pressure on "underlying [facilities-based] carrier[s] … to realign the relationship between unit and bulk prices to make that relationship wholly cost-based" [12].

When the FCC first attempted to apply the resale concept internationally [13], PTTs universally rejected it. They viewed the Commission as authorizing pirates who would jeopardize PTT revenue streams and threaten sovereignty. U.S. resellers could enter the international market without PTT authorization simply by leasing the facilities of other carriers that had foreign operating agreements. Foreign carriers threatened to eliminate international private-line tariff options used by resellers and forcefully objected to the

36. See Regulatory Policies Concerning Resale and Shared Use of Common Carrier Services and Facilities, 60 FCC 2d 261, modified, 61 FCC 2d 70, further modifications, 62 FCC 2d 588 (1977), *aff'd sub nom.*, AT&T v. FCC, 572 F.2d 17 (2nd Cir. 1978), *cert. denied*, 439 U.S. 875 (1978).

concept. Over the last 15 years, a small but growing number of nations have recognized the public benefits in having market entrants that lease lines and either add value and enhancements to them (e.g., international value-added networks) or provide a brokerage (arbitrage) function for basic services. Even nations with less than cutting-edge regulatory policies (e.g., Argentina, Hong Kong, Japan, Korea, Singapore, and South Africa) support market entry by resellers that enhance basic capacity. Canada, the United Kingdom, Sweden, Australia, and New Zealand were the first to join the United States in supporting basic service resale, and the number now exceeds 25.[37]

8.3 Themes in Current U.S. International Telecommunications and Trade Policy

Perhaps even at the risk of overbearingness, the United States has pursued international telecommunications policies that foster competition through market entry and decreased regulatory oversight. While most foreign counterparts now recognize the benefits accruing from recalibrating the scope of regulation to stimulate innovation and efficiency, they may continue to have some misgivings about promoting broader consumer choices and downward rate pressure if such action threatens the incumbent carrier's ability to generate revenues needed for subsidizing deliberately underpriced services. Unlike the FCC's campaign to foster cost-based pricing, other nations establish clear or implicit obligations on incumbent carriers to underwrite certain offerings like POTS with profits from other services. Low-cost and ubiquitous POTS supports national cohesiveness and security. Many U.S. initiatives have achieved less than universal support from foreign counterparts, because of the real or perceived challenge to the ability of incumbents to carry out their mission of universal service.

8.3.1 Facilities-Based Competition

The United States has encouraged market entry by facilities-based carriers as a way to expedite the introduction of technological innovations [14]. Many nations only recently have eliminated the incumbent carrier's monopoly. Those nations did not purposefully block innovation, nor did they want to stifle competition throughout the industry. However, they believed that

37. See FCC, International Bureau, ISR Approved Countries, available on-line at http://www.fcc.gov/ib/isr.html.

basic services constitute a natural monopoly; in other words, economic analysis supports the view that a single enterprise can operate most efficiently, because it can reduce costs by serving a large user base. Such nations believed that limits on the scope of permissible competition are necessary to ensure that the incumbent carrier can achieve social goals like universal service and below-cost provision of POTS.

8.3.2 Cost Deaveraging

The United States has sought to promote transmission facility pricing as a function of the particular medium used, without averaging or forced use of more costly or less efficient facilities. The FCC no longer requires balanced loading of cable and satellite circuits by U.S. international service carriers. Likewise, the Commission has abandoned a policy that required carriers to average costs of different facilities into a single, composite rate [15]. Some nations support cost averaging between high- and low-density routes and between cable and satellite facilities. That decision favors users who generate comparatively less traffic or who operate in high-cost areas. Other nations, including ones generating high volumes, believe that averaging supports global connectivity and equity between urban and rural residents.

8.3.3 Private Networking and Access to the PSTN

The United States strongly supports the opportunity for users to engineer their own private network solutions to particular needs, including access to foreign PSTNs. Nations have grave concerns about the extent to which end users and entrepreneurs can engineer networks that access the PSTN. Except in the central business district of most developed nations, the PSTN typically constitutes a bottleneck, that is, an essential facility over which most traffic traverses with no widely available alternative. Carriers controlling the PSTN have great leverage in setting charges for access. If an incumbent carrier has a duty to provide POTS at less than fully compensatory rates, then it must generate higher revenues from other services. The ability to charge rates well in excess of costs in large part depends on the incumbent's control of PSTN access.

Incumbent carriers fear that new service providers, like resellers, and users will find ways to access the PSTN without detection or at lower rates than designated for the type of service provided. They also worry that market entrants and users will migrate switched service traffic off high-cost public networks onto cheaper private ones. If providers of highly profitable international message telephone traffic, that is, resellers and private-line operators, can avoid paying high rates for access to the PSTN, then the incumbent may have less revenue available to subsidize POTS. Accordingly, many nations still refrain from supporting any market entry or liberalization initiative that jeopardizes captive traffic streams allegedly necessary to maintain a system that makes it possible to provide some services at subsidized levels. Notwithstanding those restrictive policies, technological innovations and new ventures willing to push the legal envelope make the PSTN more accessible and porous. Current substantial volumes of international traffic now leak into the PSTN. While the local carrier does receive some compensation, the rate typically is substantially less than if the traffic was identified as part of an international routing.

8.3.4 Alternatives to Satellite Cooperatives and Cable Consortia

The United States has aggressively sought to promote market entry by these methods:

- The transborder services of domestic satellites;[38]
- Facilities-based satellite alternatives to international cooperatives [16];
- Private cable competitors to common carrier consortia facilities [17].

Soon after the development of a domestic satellite industry, the FCC authorized the limited international, transborder use of domestic satellites whose footprints spill over into adjacent nations.[39] In 1985, the Commission authorized the first dedicated international satellite system separate from INTELSAT [18]. To promote competition in the provision of international services delivered via satellite, the FCC also unbundled regulatory consideration of the space segment from the Earth station services and determined that the latter could be provided independently by entities instead of an FCC-

---

38. See, for example, Transborder Video Services, 88 FCC 2d 258 (1891). But in Communications Satellite Corp. v. FCC, 836 F.2d 623 (D.C. Cir. 1988), the D.C. Circuit vacated an FCC order granting the use of a domestic satellite for transborder services to and from Jamaica on the grounds that the FCC failed to articulate why it was uneconomical or impractical to use the INTELSAT cooperative. The Earth station in Jamaica was subsequently sold and reconfigured to provide international service via INTELSAT.

39. See Transborder Satellite Video Services, 88 FCC 2d 258 (1981).

organized consortium of international carriers.[40] The FCC also allowed limited two-way international Earth station operations by private carriers.[41] After rejecting the option for end users and international carriers to achieve direct, cost-based access to INTELSAT capacity,[42] the Commission in 2000 reversed its view and expressed its intent to permit direct access.

---

40. Modification of Policy on Ownership and Operation of U.S. Earth Stations That Operate with the INTELSAT Global Communications Satellite System, 100 FCC 2d 250 (1984), *aff'd sub nom.*, Western Union Int'l v. FCC, 804 F.2d 1280 (D.C. Cir. 1986). Previously, the FCC had ordered the bundling of satellite capacity with Earth station services, with the latter available only from a Modification of Policy on Ownership and Operation of U.S. Earth Stations That Operate with the INTELSAT Global Communications Satellite System, Comsat-managed facility owned by a consortium of international carriers. See *Ownership and Operation of Earth Stations*, Second Report and Order, 5 FCC 2d 812 (1966); see also R. Frieden, "Getting Closer to the Source: New Policies for International Satellite Access," *Federal Communications Law Journal*, Vol. 37, 1985, p. 293.

    Previously, the FCC designated Comsat as a 50% owner and manager of all continental U.S. Earth stations. Amendment of Pt. 25 of the Commission's Rules and Regulations with Respect to Ownership and Operation of International Earth Stations in the United States for Use in Connection with the Proposed Global Commercial Communications Satellite System, Docket No. 15735, Second Report and Order, 5 FCC 2d 812 (1966). The other 50% was allocated to all other USISCs on the basis of their relative use of the Earth stations. AT&T subsequently acquired complete ownership of the ESOC facilities in the continental United States, Am. Tel. & Tel. Co., File No. I-T-C-87-109, 4 FCC Rcd. 2327 (1989).

41 Licensing Under Title II of the Communications Act of 1934, as amended, of Private Transmit/Receive Earth Stations, 3 FCC Rcd. 1585 (1988) (declaratory ruling authorizing licensure of noncommon carrier Earth stations under Title III of the Communications Act notwithstanding Section 201(c)(7) of the Communications Satellite Act that only expressly recognized Comsat and other authorized user/carriers), *aff'd sub nom.* TRT Telecommunications, Inc. v. FCC, 876 F.2d 135 (D.C. Cir. 1989); *policy applied in* Reuters Information Services, Inc., 4 FCC Rcd. 5982 (1989).

42. The Communications Satellite Act of 1962 established Comsat as the sole U.S. signatory to INTELSAT and specified that only certain special "Authorized Users," for example, the military and international telecommunications common carriers, shall have the right to deal directly with Comsat to procure capacity on a wholesale basis. Authorized User Policy, 4 FCC 2d 421 (1966) reconsideration granted in part, 6 FCC 2d 593 (1967) (Authorized User-I). Nonauthorized users, such as end users, must acquire INTELSAT capacity on a retail basis from authorized users. The Authorized User-I policy specified that Comsat shall serve as the "carriers' carrier," on a common carrier basis.

    In 1982, the FCC modified Authorized User-I to permit Comsat to provide international satellite capacity directly to a broader set of carriers and users who could arrange

When it first refused to allow a variety of carriers to access INTELSAT directly in the 1980s [19], the FCC did allow a larger group of carriers to deal directly as authorized users with Comsat rather than face a double markup of INTELSAT capacity by both Comsat and the previously limited consortium of international carriers with Comsat as Earth station facility manager and majority owner. The Commission also took a hard look at the corporate structure and performance of the COMSAT Corporation, the sole U.S. participant in the INTELSAT and Inmarsat international satellite cooperatives.[43] In 1984, the FCC also authorized the first private international cable system, Tel-Optik, Ltd., now owned by Sprint International and Cable & Wireless of the United Kingdom, whose fiber optic cable PTAT-1 links the United States, Bermuda, Ireland, and the United Kingdom and serves as an alternative to the conventional multiple-carrier consortia system [20].

Facilities-based competition can achieve a fundamental change in the industrial structure of international telecommunications. Although the

---

with Comsat to have service made available at a carrier's Earth station, and to offer retail, end-to-end service through a separate subsidiary. Authorized User Policy, 90 FCC 2d 1394 (1982) (Authorized User II). That decision was vacated and remanded to the FCC, because the Commission failed to consider contemporaneously two related issues: (1) whether any carrier should have the right of direct access to INTELSAT capacity, thereby eliminating the Comsat intermediary markup; and (2) whether carriers could operate their own Earth stations independent of the Comsat-managed facilities owned by a few major carriers. Consideration of modifications to its Authorized User Policy. ITT World Communications, Inc. v. FCC, 725 F.2d 732 (D.C. Cir. 1984). After further proceedings, the Commission in effect readopted its Authorized User II policy, Authorized User Policy, 97 FCC 2d 296 (1984), reaffirmed, 99 FCC 2d 177 (1985) (Authorized User-III). Contemporaneously, the Commission rejected direct access for carriers, 97 FCC 2d 296 (1984), but authorized independent Earth station ownership, 100 FCC 2d 250 (1984). This time, the D.C. Circuit Court of Appeals affirmed the Commission's action. Western Union International, Inc. v. FCC, 804 F.2d 1280 (D.C. Cir. 1986).

43. See Comsat Study—Implementation of Sec. 505 of the International Maritime Telecommunications Act, 77 FCC 2d 564 (1980); Changes in the Corporate Structure and Operations of the Communications Satellite Corp., Notice of Proposed Rulemaking, 81 FCC 2d 287 (1980), First Report and Order, 90 FCC 2d 1159 (1982), *reconsideration denied*, 93 FCC 2d 701 (1983), Second Report and Order, 97 FCC 2d 145 (1984), *on recon.*, 99 FCC 2d 1040 (1984). The FCC concluded that Comsat should be permitted to engage in nonjurisdictional, that is, competitive, non-INTELSAT/Inmarsat signatory activities if such ventures were "not inconsistent with its statutory mission" and provided it modified its corporate structure to establish a separate retail services subsidiary and revised cost allocation formulas.

United States was instrumental in forming international cooperatives and supporting their exclusivity, it now believes that, with adequate safeguards, the marketplace can support competition even for some switched services that access the PSTN. Most nations initially objected to that scenario and only with the passage of time and the onset of limited separate system competition did they accept the view that conditional competition would not prevent incumbents from achieving their service mission.

Facilities-based competition and access to the PSTN constitute key U.S. policy predicates that even now do not meet with global support. Many nations have yet to promote robust facilities-based competition and unlimited opportunities to access the PSTN. However, an increasing number of nations have privatized the incumbent carrier. The EU established a 1998 deadline for elimination of facilities-based telecommunications monopolies. Enhanced and even basic resale of leased lines have become a more acceptable vehicle for promoting some degree of competition. Depending on one's viewpoint, the FCC deserves credit for persevering in a deregulatory campaign that has achieved success, and that credit should be shared with economists, advisors, World Bank officials, commercial lenders, and a larger cast that collectively have embraced competition, deregulation, and liberalization as essential predicates for an effective telecommunications infrastructure.

8.3.5 Permitting Self-Help Strategies Like Callback

In the late 1990s, the FCC evidenced an even greater willingness to rely on entrepreneurship and technological innovations to achieve procompetitive objectives. That reliance appeared to endorse tactics that may violate ITU Recommendations and the general consensus of how far one nation can deviate from uniform rules of the road. Specifically, the FCC did not object to individual or carrier strategies that achieved immediate cost savings by evading artificial barriers to market entry and other impediments to competition established by foreign regulatory agencies. Absent proof that a U.S. carrier was violating the laws of another nation, the FCC refused to revoke licenses when carriers exported U.S. dial tone to users in foreign locales. Callback, also known as call-reorigination, provides significantly lower-cost international dialing opportunities to callers located in nations with high retail international calling rates. The caller in such a high-cost area can avoid having to pay the prevailing retail rate by establishing a virtual presence in the United States and "importing" dial tone.

At the risk of encouraging illegal behavior, the FCC chose to concentrate on the immediate consumer benefits in terms of lower rates than whether such tactics might harm incumbent carriers and violate the spirit, if not the letter, of the law.

8.3.6 Prescribing Settlement Rates

Even as the FCC refrained from foreclosing the export of U.S. dial tone abroad, the Commission embarked on an aggressive extraterritorial campaign to lower international calling rates everywhere. Frustrated with the pace of retail rate reductions, the FCC decided to calculate and prescribe what it thought carriers should charge each other for terminating international calls. The FCC interposed itself in what had been private carrier-to-carrier negotiations. It justified such involvement with the belief that without external prodding international carriers would not negotiate cost-based settlement rates and that such high rates imposed a multibillion-dollar burden on U.S. international callers. Chapter 9 more extensively documents that new, proactive, and aggressive FCC strategy.

8.3.7 Relaxing Foreign Ownership and Other Restrictions

Even before completion of successful market access negotiations under the auspices of the WTO, the FCC sought to lead by example. Its 1995 Foreign Carrier Entry Order [21] aimed to enhance consumer welfare and to promote competition by reducing barriers to U.S. market access by foreign carriers and investors, but with adequate safeguards to ensure that the United States did not open its telecommunications markets without commensurate access opportunities abroad. The Commission's primary safeguard involved a case-by-case assessment of whether foreign carriers and their affiliates operated in nations affording U.S. ventures effective competitive opportunities (ECOs) in the foreign carrier's home nation. The FCC proposed to make an ECO assessment whenever it received Section 214 applications to operate in the United States, including proposals to exceed the foreign ownership limitations set forth in Section 310(b)(4) of the Communications Act of 1934, as amended.

Upon successful formation of the WTO Basic Agreement on Telecommunications,[44] the FCC abandoned the ECO test for WTO member countries [22]. In light of market opening commitments, the FCC concluded that

44. The results of the WTO basic telecommunications services negotiations were incorporated into the GATS by the Fourth Protocol to the GATS, Apr. 30, 1996, 36 I.L.M. 366 (1997).

WTO member nations deserved virtually open market access[45] absent evidence of an unequal market access risk to U.S. ventures. The FCC imposed only three major safeguards:

1. For telecommunications routes from and to the United States served by carriers affiliated with a foreign carrier, the Commission conditioned facilities-based authorizations on the application of settlements with U.S. carriers for the route at or below the benchmark standard rate it calculated as the cost for terminating an international call [23].
2. The Commission required quarterly annual traffic and revenue reporting requirements for switched resale carriers affiliated with foreign carriers having market power in the home (foreign) country [24].
3. The Commission classified as dominant, subject to more burdensome regulatory scrutiny, any carrier affiliated with a foreign carrier having market power in the foreign end of the route [25].

8.4 Fair Trade in Telecommunications

As part of its campaign to foster competition, the United States has engaged other nations in a dialogue with an eye toward encouraging market entry by multiple carriers and equipment manufacturers. Only through market entry can end-user rates drop significantly, in part because monopoly carriers may have no incentive to negotiate lower accounting rates, the level of competition set by carrier correspondents to reflect the cost of providing service. Likewise, market entry is essential to reduce a growing trade deficit in telecommunications equipment. The United States pursues that goal through greater advocacy to eliminate closed procurements, discriminatory testing and certification requirements, nondisclosure of technical interface details, discriminatory duties on imports, and so forth.[46]

45. The new rules apply to Sec. 214 applications, applications to exceed foreign ownership limitations in Section 310(b)(4) of the Communications Act, as amended, and applications for cable landing licenses.

46. See, for example, *U.S. Telecommunications in a Global Economy: Competitiveness at a Crossroads,* Report from the Secretary of Commerce to the Congress and the President of the United States (Aug. 1990). Similarly, the European Economic Community and other

To achieve a more receptive hearing for any of the above policies, the United States has attempted to appeal to the growing business orientation of many incumbent carriers and the evolving responsiveness to consumer requirements by other nations' regulatory and policy-making bodies. While it remains true that no policy initiative may achieve acceptance without thorough ventilation of the issues and foreign carrier cooperation, financial matters increasingly predominate over loftier notions of sovereignty and international comity. Foreign carriers and government ministries increasingly realize that a robust and fairly priced telecommunications network is a sine qua non to commercial development. Both existing and prospective employers that heavily rely on telecommunications facilities and services can more readily vote with their feet to other, more accommodating nations.

Nations increasingly question the rationale for favoring national heros and for insulating incumbent carriers from competition. Protectionist policies may make it possible for them to remain profitable in the short term even as their long-term viability may become doubtful. Without pressure to innovate and streamline, such enterprises may become inefficient and noncompetitive outside their closed markets. Such policies leave the nation vulnerable to trade disputes, customer migration, and comparatively inadequate service.

On the other hand, nations must find ways to underwrite substantial investments to upgrade rather than maintain existing networks. The allure of broadband and Integrated Services Digital Networks,[47] new intelligent systems,[48] and the Internet results in part from incumbent carriers' view that a wider services wingspan will spread costs over a broader base and enable it to exploit economies of scale and scope. However, a too visible incumbent carrier campaign to incorporate, if not monopolize, information services, may backfire as large-volume users of private networks authorized by government regulators vigorously object to any migration to one-size-fits-all public networks.

---

nations appear inclined to reduce barriers to efficiently operating markets. See "Towards a Dynamic European Economy—Green Paper on the Development of the Common Market for Telecommunications Services and Equipment," COM(87) 290 final, June 30, 1987; see also Frieden, "Open Telecommunications Network Policies for Europe: Promises and Pitfalls," *Jurimetrics*, Vol. 31, Nov. 1990, pp. 319–328.

47. See Integrated Services Digital Networks (ISDN), Notice of Inquiry, 94 FCC 2d 1289 (1983), First Report, 98 FCC 2d 249 (1984).

48. See Intelligent Networks, CC Docket No. 91-346, Notice of Inquiry, 6 FCC 7256 (1991).

8.4.1 FCC Strategies

The FCC has pursued a number of different strategies to promote open and competitive international telecommunications markets. Its greatest success occurs when it can dismantle policies, rules, or regulations that insulate U.S. carriers and manufacturers from competing. It has achieved far less success in securing parallel deregulatory initiatives in other nations, particularly when it threatens retaliatory sanctions on foreign ventures operating in the United States.

For example, the FCC heightened its attention to the net outflow of over $5 billion per year in international toll revenue settlements with foreign carriers.[49] Mindful of the contribution telecommunications services and equipment sales can have on the national economy, the Commission believed that it could leverage its power to monitor and comparatively regulate more heavily foreign companies operating in the United States to secure lower accounting rates. The Commission proposed to scrutinize foreign carriers' and manufacturers' U.S. operations but substantially curbed the scope of its oversight and the reporting requirements it imposed[50] in response to concerns voiced by the executive branch that such selective scrutiny would violate the U.S. commitment to national treatment.

The FCC also reduced regulatory oversight of international carriers and service providers that operate in a competitive marketplace as evidenced

49. See *International Accounting Rates and the Balance of Payments Deficit in Telecommunications Services, Report of the Common Carrier Bureau to the Federal Communications Commission* (Dec. 12, 1988); See also Uniform Settlement Rates, 84 FCC 2d 121 (1980); Implementation and Scope of the International Settlements Policy for Parallel International Routes, Notice of Proposed Rulemaking, 50 Fed. Reg. 28,418 (1985), Report and Order, 59 Rad. Reg. 2d (P&F) 982 (1986) *on partial recon.*, 2 FCC Rcd. 1118 (1987), *on further recon.* 3 FCC Rcd. 1614 (1988); Regulation of International Accounting Rates, CC Docket No. 90-337, Report and Order, Phase I, 6 FCC Rcd. 3552 (1991), Further Notice of Proposed Rulemaking, Phase II, 6 FCC Rcd. 3434 (1991), First Report and Order, Phase II, 7 FCC Rcd. 559 (1992); Regulation of International Accounting Rates, CC Docket No. 90-337, Phase II, Fourth Report and Order, 11 FCC Rcd 20,063 (1996); 1998 Biennial Regulatory Review Reform of the International Settlements Policy and Associated Filing Requirements, IB Docket No. 98-148, Report and Order and Order on Reconsideration, 14 FCC Rcd. 7963 (1999).

50. See Rules and Policies on Foreign Participation in the U.S. Telecommunications Market, Report and Order and Order on Reconsideration, 12 FCC Rcd 23,891 (1997), *recon. pending*; see also Amendment of the Commission's Regulatory Policies to Allow Non-U.S. Licensed Space Stations to Provide Domestic and International Satellite Service in the United States, Report and Order, 12 FCC Rcd 24,094 (1997), First Order, *on*

by market access commitments made under the auspices of the WTO.[51] The Commission narrowed the scope of conventional common carrier regulation by creating a new regulatory category for nondominant carriers, that is, carriers lacking the power to affect the price or supply of a service. Such carriers qualify for streamlined regulation, including a shorter time period between filing for a new service or price change and FCC authorization.

The FCC may have overestimated the potential for other nations to adopt a similar strategy for reducing or eliminating regulation. Absent parallel deregulatory action abroad, FCC initiatives have the effect of promoting domestic market access without securing commensurate opportunities in other nations. The Commission recognized that outcome by conditioning the opportunity for foreign-owned carriers operating in the United States to qualify for nondominant carrier status. Carriers with majority foreign ownership can qualify for nondominant status on a route-by-route basis only after demonstrating that their parent carrier does not have dominant, bottleneck control in the foreign home country.

8.4.2 Resistance to U.S. Procompetitive Initiatives

Many of the international telecommunications and trade policy initiatives in the United States result from interest in broader global market access and the belief that enhanced domestic consumer welfare from deregulation can apply internationally.[52] However, international telecommunications and trade

---

recon., FCC 99-325, W.L. 983820 (F.C.C.) (rel. Oct. 29, 1999) *appeals pending* (DISCO II Order); 1998 Biennial Regulatory Review—Review of International Common Carrier Regulations, IB Docket No. 98-118, Notice of Proposed Rulemaking, 13 FCC Rcd. 13713 (1998), Report and Order, 14 FCC Rcd. 4909 (1999); Extension of Lines, New Lines, and Discontinuance, Reduction, Outage and Impairment of Service by Common Carriers; and Grants of Recognized Private Operating Agency Status Extensions and Supplements, 47 C.F.R. § 63.12 (1999).

51. Rules and Policies on Foreign Participation in the U.S. Telecommunications Market, Report and Order and Order on Reconsideration, 12 FCC Rcd 23,891 (1997), *recon. pending*. The results of the WTO basic telecommunications services negotiations were incorporated into the GATS by the Fourth Protocol to the GATS, Apr. 30, 1996, 36 I.L.M. 366 (1997).

52. The FCC believes its procompetitive, open-entry policies will foster "an improved international communications system with more choices for consumers, more diverse service offerings and lower rates." Am. Tel. & Tel. Co., 75 FCC 2d 682, 288 (1980) (removal of restriction on the use of international telephone lines for data applications), *aff'd sub nom.*, Western Union International, Inc. v. FCC, 673 F.2d 539 (D.C. Cir. 1982).

matters do not have symmetry with any one nation's domestic policies. Despite empirical success in the United States, deregulation does not necessarily discredit the traditional monopoly carrier model, nor does it trigger liberalization and dismantling incumbent carriers. Furthermore, the multilateral and cooperative nature of international telecommunications and trade policy militate against simply extending one nation's policies abroad. Accordingly, the pace of international policy liberalization may lack the speed U.S. deregulators and entrepreneurs would prefer.

8.5 Recalibrating Market Access Opportunities in the United States

In view of mounting trade deficits in telecommunications and no apparent progress in fostering parity of access, national treatment may have become a red herring. The FCC and other federal agencies have focused their attention on parity of treatment between and among national and foreign enterprises, without any linkage to how the matter is addressed in foreign countries. In turn, the Commission and other agencies are reminded of their moral duty not to discriminate, something many foreign governments persist in doing.

It appears that the FCC, in conjunction with other agencies with trade portfolios, now will consider using regulation as market access leverage and selective deregulation as a reward for progressive nations in the following areas:

- Establishment of an operating presence in foreign locales, particularly for ventures that do not use frequency spectrum (e.g., installation of switches, resale of leased lines, construction of fiber optic networks);
- Acquiring whole circuits and the ability to operate a network largely independent of domestic carriers;
- Providing cost-based and nondiscriminatory interconnection rights, particularly for international value-added networks and market entrants;
- Reducing accounting rates to align them closer to actual costs;
- Relaxing alien ownership restrictions.

8.5.1 Action Necessitated by Users' Resorting to Self-Help

Accounting-rate evasion, bypassing incumbent carrier facilities, creative traffic routing, and a host of technology-aided user schemes mean that national governments have limited time remaining to manage timely and incremental trade reforms. Without expedited consideration by federal agencies and their counterparts on a bilateral and multilateral process, user-directed self-help schemes will override the ability of the ITU and other forums to manage change.

The international accounting rate issue provides an instructive case study. For years, the United States seemingly could afford to ignore a process increasingly out of synch with changing circumstances. But the pace at which the United States financially suffers from neglect has substantially increased as the net settlement deficit grew almost $2 billion, to more than $3 billion in just four years and peaking in 1998 at $5.4 billion.[53] The population of the United States, its multicultural composition, and multinational corporate enterprises contribute to approximately half of the accounting rate settlement deficit incurred by U.S. carriers. However, the other half reflects the fact that foreign countries retard demand for switched telephone services into the United States. They do that by imposing excessively high collection charges, promoting outbound private lines instead of switched services, refusing to authorize facilities-based competition and inbound reseller access, and insisting on retaining high accounting rates despite technological innovations that can substantially reduce the per-unit cost of service.[54] Absent conscientious efforts by carriers and governments to reduce accounting rates, users will take matters into their own hands by using a variety of legal, illegal, and gray-area tactics to avoid accounting rate liability while still securing access to the PSTN.

8.6 Conclusion

The track record of the United States in international telecommunications policy shows a long period of pervasive regulation that segmented markets, tolerated or promoted monopolies, and used public utility ratemaking. In the

---

53. See Regulation of International Accounting Rates, CC Docket No. 90-337, Phase I, Order on Reconsideration, 7 FCC Rcd. 8049, 8051 (1992).

54. See R. Frieden, "International Toll Revenue Division: Tackling the Inequities and Inefficiencies," *Telecommunications Policy,* Vol. 17, Apr. 1993, pp. 221–233.

late 1970s and early 1980s, deregulatory initiatives gathered momentum, but for some issues the U.S. government seemed to have allowed anachronistic policies to remain, for example, service dichotomies that artificially segmented the market and limited the number of carriers authorized to provide service.

Absent coordinated efforts by governments to foster cost-based services, diversity, and competition, high-volume users will have incentives to find ways to avoid excessive rates, service limitations, and carrier unresponsiveness. Already, both legitimate and questionable enterprises provide new and creative solutions to service or price barriers erected by incumbents. That market will grow as more users and entrepreneurs direct their attention to bypassing the current regime.

Similar kinds of domestic bypass threats forced the FCC to revamp the pricing of local facilities' interconnection to avoid encouraging users to pursue less efficient but cheaper routing options. The international telecommunications marketplace has the same kind of tariff and pricing anomalies. Yet in this instance the FCC cannot unilaterally act to remedy an international matter involving sovereign nations. Accordingly, national regulatory authorities need to revamp the terms and conditions for market access. The failure to do so threatens the very foundation by which sovereign carrier correspondents match half-circuits. Absent immediate and conscientious action to foster competition or at least to eliminate unreasonable service limitations (e.g., the inability to use a leased line to route both voice and data traffic), the sovereign customer will resort to self-help that in time could divest carriers of network control and the ability to fully control market access and network interconnection. Unequal market access has enabled some carriers and governments to extract a larger share of the financial benefits accruing from deregulation. Heightened efforts to foster parity of market access will balance the flow of benefits and also reduce incentives for users to resort to self-help through black-market or gray-market options.

References

[1] Open-Market Reorganization for the Betterment of International Telecommunications Act, Pub. L. 106-180, 114 Stat. 48 (2000). See also Applications of INTELSAT LLC for Authority to Operate, and to Further Construct, Launch, and Operate C-band and Ku-band Satellites that Form a Global Communications System in Geostationary Orbit, FCC 00-287, 2000 WL 1129036 (F.C.C.) (Rel. Aug. 8, 2000).

[2] Noam, E. M., "Beyond Liberalization—From the Network of Networks to the System of Systems," *Telecommunications Policy*, Vol. 18, No. 4, 1994, pp. 286–294.

[3] United States v. Western Elec. Co., 552 F. Supp. 131 (D.D.C. 1982), *aff'd sub nom.* Maryland v. United States, 460 U.S. 1001 (1983).

[4] Market Entry and Regulation of Foreign-affiliated Entities, IB Docket No. 95-22, 11 FCC Rcd 3873 (1995); Report and Order and Order on Reconsideration, 12 FCC Rcd 23891 (1997).

[5] Administrative Procedure Act, as amended, 5 U.S.C. Secs. 701–706 (1990).

[6] Regulatory Policies and International Telecommunications, CC Docket No. 86-4994, Notice of Inquiry and Proposed Rulemaking, 2 FCC Rcd. 1022 (1987); Report and Order and Supplemental Notice of Inquiry, FCC 88-71 (rel. Mar. 25, 1988); *on recon.*, 4 FCC Rcd. 323 (1989).

[7] 47 U.S.C. Sec. 706(2)(a).

[8] American Telephone and Telegraph Co., 75 FCC 2d 682 (1980); Overseas Communications Services, 84 FCC 2d 622 (1980), modified, 92 FCC 2d 641 (1982) *aff'd sub nom.*, Western Union Int'l v. FCC, 673 F.2d 539 (D.C. Cir. 1982).

[9] Western Union Int'l, Inc., 76 FCC 2d 166 (1980), *aff'd sub nom.*, Western Union Int'l v. FCC, 673 F. 2d 539 (D.C. Cir. 1982).

[10] American Telephone and Telegraph Co., 74 FCC 2d 682, 688 (1980) *aff'd sub nom.*, Western Union International, Inc. v. FCC, 673 F.2d 539 (D.C. Cir. 1982) (removal of restrictions on the use of international telephone lines for data applications).

[11] Regulatory Policies Concerning Resale and Shared Use of Common Carrier Domestic Public Switched Network Services, 83 FCC 2d 167 (1980), *on recon.*, 86 FCC 2d 820 (1981).

[12] Domestic Resale Policy, 60 FCC 2d at 298–99.

[13] International Resale Policy, 77 FCC 2d 831 (1980), proceeding terminated in Regulation of International Accounting Rates, CC Docket No. 90-337, Phase II, First Report and Order, 7 FCC Rcd. 559 (1992).

[14] Graphnet Sys., Inc., 63 FCC 2d 402 (1977), *aff'd sub nom.* ITT World Communications, Inc. v. FCC, 595 F.2d 597 (2d Cir.), *on remand*, 71 FCC 2d 1066 (1979); International Relay, Inc., 77 FCC 2d 819 (1980), *on recon.*, 82 FCC 2d 41 (1980).

[15] Policy for the Distribution of United States International Carrier Circuits Among Available Facilities During the Post-1988 Period, CC Docket No. 87-67, Notice of Proposed Rulemaking, 2 FCC Rcd. 2109 (1987), Report and Order, 3 FCC Rcd. 2156 (1988) (abandoning circuit distribution guidelines).

[16] Pan American Satellite Corp., 101 FCC 2d 1318 (1985) (conditional grant), 2 FCC Rcd. 7011 (1987) (final C-band authority), 3 FCC Rcd. 677 (1988) (final Ku-band authority).

[17] Tel-Optik, Ltd., 100 FCC 2d 1033 (1985).

[18] Application of Pan American Satellite Corporation for Authority to Construct, Launch and Operate a North Atlantic Region Satellite System, 101 FCC 2d 1201 (1985); see also Establishment of Satellite Systems Providing International Communications, 101 FCC 2d 1046 (1985).

[19] Authorized User Policy, 97 FCC 2d 296 (1984), *reaffirmed*, 99 FCC 2d 177 (1985), *aff'd sub nom.*, Western Union Int'l v. FCC, 804 F.2d 1280 (D.C. Cir. 1986). In 2000, the FCC revisited the direct access issue and signaled its intention to permit it.

[20] Tel-Optik, Ltd. (Private Submarine Cable Decision), 100 FCC 2d 1033 (1985).

[21] Market Entry and Regulation of Foreign-Affiliated Entities, 11 FCC Rcd 3873 (1995).

[22] Rules and Policies on Foreign Participation in the U.S. Telecommunications Market, Report and Order and Order on Reconsideration, 12 FCC Rcd 23,891 (1997), *recon. pending*. The ECO test continues to apply for non-WTO nations. *Id.* 12 FCC Rcd. 23949.

[23] *Id.* 12 FCC Rcd. at 23976-77.

[24] *Id.* 12 FCC Rcd. at 24013.

[25] *Id.* 12 FCC Rcd. at 23991.

9

Case Studies in Change

In view of the changes affecting telecommunications and information processing, a book of this sort can offer only a snapshot of the current state of affairs. Simply documenting changes, while helpful, does not provide insights and perspective nor would it detect trends that may help provide readers with some ability to predict future changes and opportunities. While doing so is ambitious, we ought to consider how recent case studies contribute to the development of trends. Set out here are accounts of several recent policy debates offered with an eye toward providing examples of how converging technologies and markets affect how policy makers make decisions in this still new and unpredictable climate.

9.1 The International Telecommunications Toll Revenue Division Process

International telecommunications service requires two or more carriers to enter into a commercial arrangement for the reciprocal delivery of traffic. Carriers negotiate operating agreements that set out the terms and conditions for the complete routing of calls, including the domestic "tail" circuits running to and from international gateway facilities. Operating agreements specify an accounting rate to be used as the amount of compensation shared by the carriers for their joint provision of service. For nations with competitive international calling options, the accounting rate may exceed the retail collection rate charged to end users.

A high accounting rate adversely affects the ability of carriers to reduce rates, because it specifies an often steep amount owed a participating carrier for terminating calls. Carriers facing competition often have to rely on return traffic settled at the high accounting rate to offset a deficit created by charging users a collection rate below the accounting rate. Accounting rates distort the international telecommunications marketplace by creating incentives for carriers not to reduce rates and for governments to insist that carriers route traffic on a proportionate basis, that is, requiring carriers to route outbound traffic to carriers proportionate to the percentage of total inbound traffic generated by each carrier.

The accounting rate regime limits the nature and scope of direct competition in international telecommunications. By creating a large margin between what carriers charge and their actual costs, the accounting rate regime generates a big incentive for both carriers and customers to avoid the system through surreptitious routing schemes that evade the accounting rate settlement process. Over time, the incentive to avoid the accounting rate regime may trigger a collapse, but most carriers have a perverse incentive to maintain the system, because it accrues above market revenues and creates artificial barriers to more robust price competition.

9.1.1 Overview

Carriers theoretically specify accounting rates to represent the approximate costs they incur in the complete routing of a telephone call from call originator to call recipient.[1] For most routes, however, correspondents have failed to reduce the accounting rate to reflect lower costs in new generations of

1. For extensive background on accounting rate regulation by the FCC, see Paul W. Kenefick, "A Step in The Right Direction: The FCC Provides Regulatory Relief in International Settlements and International Services Licensing," *CommLaw Conspectus*, Vol. 8, Winter 2000, p. 43; Rob Frieden, "Falling Through the Cracks: International Accounting Rate Reform at the ITU and WTO," *Telecommunications Policy*, Vol. 22, No. 11, Dec. 1998, pp. 963–975; "The Impact of Call-Back and Arbitrage on the Accounting Rate Regime," *Telecommunications Policy*, Vol. 21, No. 9/10, 1997, pp. 819–827; "International Toll Revenue Division: Tackling the Inequities and Inefficiencies," *Telecommunications Policy*, Vol. 17, No. 3, Apr. 1993, pp. 221–233; "Accounting Rates: The Business of International Telecommunications and the Incentive to Cheat," *Federal Communications Law Journal*, Vol. 43, No. 2, 1991, pp. 111–139.

transmission facilities. If two carriers have equal traffic volumes, a high accounting rate presents no problem because the carriers do not transfer funds from the toll revenues they collect. But many telephone service routes, including most originating in the United States, generate an imbalance in traffic flows, that is, more traffic originates from the United States than terminates in the United States. An imbalance in traffic flows requires a transfer of funds when carriers settle accounts: The carrier that generates more outbound traffic must compensate its correspondent carrier for accepting more traffic that traverses the correspondent's international and domestic facilities.

In 1997, U.S. carrier settlements with foreign counterparts resulted in a \$5.45-billion deficit.[2] Application of excessive accounting rates has a direct and adverse impact on the rates charged to users. While technological innovations like digitization and compression have reduced dramatically the cost of transmission facilities, not all retail rates have dropped commensurately, because the frame of reference for tariffing service is not actual transmission facility costs but what the carriers agree should be the per-minute level of compensation they will share. For example, the actual cost of providing a minute of telephone service between the United States and many points in Europe is a few cents per minute, but both the accounting and collection rates can exceed that rate by several hundred percent.

Seven major factors explain the substantial transfer of funds:

- United States International Service Carrier (USISC) outbound traffic volumes greatly exceed the return, inbound flow, because of a larger population, widespread multinational business activity, often significantly lower end-user charges that stimulate outbound calling, relative per-capita wealth, and a large immigrant population that seeks to maintain ties with family members abroad.[3]
- Foreign international telecommunications rates greatly exceed what USISCs charge, thereby retarding demand for inbound U.S. calling.

2. See *Trends in the U.S. International Telecommunications Industry*, Table 24, Measures of International Settlement Deficit, Sept. 1999, available on-line at http://www.fcc.gov/Bureaus/Common_Carrier/Reports/FCC-State_Link/trends.html.

3. A study by the FCC's Common Carrier Bureau revealed that charges for calls to the United States can exceed by as much as 80% the outbound rate. FCC, Common Carrier Bureau, *Calling Prices for International Message Telephone Service Between the United States and Other Countries*, Aug. 1992. The rate has narrowed considerably for many routes, particularly where resale and facilities-based competition thrive.

- Most current accounting rates substantially exceed service cost, because they fail to reflect technological innovations that have reduced the per-unit cost of providing service.
- While recognizing that they lose some revenues in having to incur an accounting rate settlement, most USISCs appear content to offset the outbound traffic deficit with inbound settlements at rates well in excess of their actual costs to terminate calls.
- With increasing frequency, high-volume international telecommunications users, particularly in nations with high toll charges, have devised traffic-routing arrangements exempt from the accounting rate settlement process, For example, they use private lines that "leak" into the PSTN and callback services that deliver dial tone for services in a country with low rates to users in a nation with high rates, thereby reducing the amount of inbound U.S. traffic that would have offset some of the accounting rate surplus generated.[4]
- The ITU has only begun to address the issue of whether international telecommunications services should be priced on the basis of actual costs or on other factors that would build in a subsidy, for example, keep accounting rates high or flow a larger portion of a lower, cost-based rate for lesser developed countries to support infrastructure construction.
- While willing to discount rates when competitively necessary, international carriers seek to reduce cannibalization of retail services.

A 1988 report from the Federal Communications Commission's Common Carrier Bureau acknowledged that higher outbound U.S. traffic volumes account for half of the growing accounting rate deficit.[5] Low

---

4. See G. Staple, "Winning the Global Telecoms Market: The Old Service Paradigm and the Next One," in G. Staple, ed., *TeleGeography 1992 Global Telecommunications Traffic Statistics and Commentary* (London: International Institute of Communications, 1992), pp. 32–53.

5. "[I]t is clear that the price of calling the United States has not declined as rapidly as the price of calling from the United States. Because American prices were lower to begin with, the differential in prices has increased. This has undoubtedly contributed to the continued and increasing imbalance in traffic flows and in turn contributed to the rapidly growing balance of payments deficit." FCC, International, Common Carrier Bureau, *Accounting Rates and the Balance of Payments Deficit in Telecommunications Services,* Dec. 12, 1988, p. 29.

charges to end users in the United States stimulate traffic growth, while much higher rates in foreign countries retard demand and reduce return traffic flows [1, p. 29]. The FCC objected to such "direct underwriting by U.S. consumers of foreign telecommunications administrations" [2], but a solution remains elusive.

On one hand, the FCC has engaged in an activist posture to force both domestic and foreign carriers to reduce accounting rates and has gone so far as to prescribe what it thinks cost-based accounting rates should be. But on the other hand, the FCC wants to rely on marketplace forces[6] and entrepreneurial creativity to dismantle the accounting rate regime without regulatory intervention. The Commission has focused on the apparent lack of cooperation by foreign carriers, but has failed to note that U.S. carriers have not aggressively lowered retail rates when accounting rates fall and passed through the savings in terminating calls for foreign carriers at settlement rates well in excess of their costs.

Throughout the 1980s and early 1990s, the FCC appeared ambivalent on how to remedy the accounting rate issue. When operating with an activist posture, the FCC may impermissibly intrude on the sovereignty and jurisdiction of other nations, particularly because the ITU did not substantially consider the issue until 1992.[7] On the other hand, a less aggressive posture may rely too heavily on the carriers to remedy the problem by negotiating accounting rates downward even when they may lack a commercial incentive

6. The FCC recognizes that accounting rates result from business negotiations and that the Commission should support international comity that it defines as "the mutual recognition and accommodation by nations of their differing philosophies, policies, and laws." Regulatory Policies and International Telecommunications, CC Docket No. 86-494, Notice of Inquiry and Proposed Rulemaking, 2 FCC Rcd., 1986, p. 1022, para. 52.

7. Article 6.1.1 of the *International Telecommunication Regulations* recognizes that the level of toll charges is a "national matter" but recommends that carriers avoid "too great a dissymmetry between the charges applicable." *The International Telecommunication Regulations, Final Acts of the World Administrative Telegraph and Telephone Conference,* Art. 6.1.1 (Melbourne, Australia, 1988). Study Groups of the ITU's Consultative Committee on International Telegraph and Telephone (CCITT) frequently have considered accounting rate reforms. However, 1992 marked the first time any CCITT Study Group has suggested on the record that administrations consider actual costs when setting accounting rates. See Regulation of International Accounting Rates, Docket No. 90-337, Phase II, Second Report and Order and Second Further Notice of Proposed Rulemaking, 7 FCC Rcd. 8040, para. 36 (rel. Nov. 27, 1992). The ITU continues to examine the issue without achieving consensus on how to achieve reform. See Rob Frieden, "Falling Through the Cracks" (note 1).

to do so. Only recently has the FCC undertaken a balanced and increasingly successful campaign to create incentives for carriers to act. AT&T in particular has recognized the value in having lower accounting rates, because in the current regulatory environment it does not have to flow through all savings and efficiency gains to consumers.[8] The FCC hopes to avoid direct confrontation with foreign governments on the accounting rate issue by linking access to U.S. markets with an assessment of whether equivalent opportunities exist for U.S. companies, in light of market access commitments made under the auspices of the WTO and with regard to opportunities for private-line resale.

9.1.2 Accounting Rate Fundamentals

Routing international telecommunications traffic involves a contract negotiated between carriers for each type of service between each pair of nations. Such bilateral arrangements promote multilateral collaboration in transmission facility investment. International carriers jointly own, operate, and maintain international submarine cables through consortia and international satellites through cooperatives like INTELSAT and private ownership or consortium participation. Typically, ownership interests are allocated as a function of anticipated use. Joint ownership means that most carriers incur roughly the same cost per unit of international capacity. It follows that most parallel routes to the same region of the world would have roughly the same total costs, even though expenses for the domestic haul portion of a complete route can vary as a function of equipment vintage, traffic volumes, and regulatory policies that may impose access surcharges to support universal service.

Key components of international traffic routing are the financial terms and conditions under which one carrier compensates another carrier for agreeing to match international circuits and to provide switching and domestic routing needed to deliver calls to the intended recipient. An accounting rate serves as the basis for dividing the toll charges[9] collected for the joint provision of an international service.

---

8. See Strategic Policy Research, *The U.S. Stake in Competitive Global Telecommunications Services: The Economic Case for Tough Bargaining*, Dec. 16, 1993 (study conducted for AT&T).

9. The accounting rate does not necessarily equal the rate charged end users. The collection rate may be lower, as is the case for some outbound U.S. calls, but typically exceeds the accounting rate, particularly in foreign nations.

Accounting rates are negotiated by the participating carriers and have two components:

1. The unit of currency used and the applicable rate per unit of traffic carried. For example, in 1999, USISCs and French carriers divided \$0.20 per minute (down from \$1.71 in 1990) of full-rate international message telephone service traffic.
2. The settlement process, that is, how the accounting rate amount will be divided between correspondents, usually 50-50 when two carriers participate.

Carriers establish accounting rates to represent the total cost generally incurred to establish a complete international circuit, that is, both international half-circuits and both domestic-tail circuits to and from the international gateway. However, the FCC has concluded that "accounting rates for international message telephone service (IMTS) are significantly greater than the current costs of providing service" [3]. Disparity between accounting rates and actual costs incurred by carriers would have little significance if there were parity in traffic volumes. Accounting rates have great significance when an imbalance of traffic streams exists and substantial funds must be transferred, as is the case for IMTS between the United States and many foreign nations, both developed and developing.

Comparatively low IMTS rates in the United States have stimulated outbound IMTS calling, resulting in traffic volumes far in excess of inbound flows. Typically, the IMTS rates in foreign nations vastly exceed carrier cost and the equivalent outbound U.S. rate to callers, thereby dampening demand for inbound U.S. calling.[10] Foreign carriers have viewed accounting rate surpluses as a painless way to subsidize below-cost provisioning of postal and certain telecommunications services.

Artificially high accounting rates (rates have not dropped commensurately with reduced costs per unit of capacity) require net debtor carriers to make even higher settlement payments to compensate for the disparity in traffic volumes. It appears that despite increasing outpayments, USISCs have been less than vigorous in their advocacy for accounting rate reductions. Such high accounting rates do not necessarily reflect foreign carrier "whipsawing" (discussed in Section 9.1.3.1) of USISCs by leveraging inbound U.S.

10. See Regulation of International Accounting Rates, CC Docket No. 90-337, Phase II, Further Notice of Proposed Rulemaking, 6 FCC Rcd, 1992, pp. 3434, 3436.

traffic to secure concessions. However, they do represent a surcharge to U.S. ratepayers, because USISCs incur financial burdens that prevent them from offering further rate reductions.

Perhaps most important, reliance on the accounting rate system distorts the marketplace for international telecommunications services by institutionalizing false pricing signals and creating incentives for carriers to mask their costs and the ways in which they respond to competition. For example, while USISCs can evidence a net settlement deficit in terms of what they have to pay out to foreign carriers, they can largely offset that payment by receiving settlement payments for terminating inbound traffic at a fraction of the cost attributed by the accounting rate regime to that function.[11] Additionally, USISC reluctance to press for lower accounting rates stems in part from concern that such advocacy might prompt foreign-carrier retaliation in ways that could further hurt profitability. The FCC's Common Carrier Bureau identified two other reasons:

- International calling volumes are rapidly growing. Despite the massive outpayment in accounting rates, IMTS typically "generates at least four times as much revenue as the average domestic interstate call."
- A higher percentage of international telephone calls are metered and charged on a usage-sensitive basis, contrary to usage-insensitive private lines, which are more prevalent for intercity links in domestic markets. [1, p. 9]

With robust traffic growth and high margins, USISCs may not have perceived a need to push accounting rates downward. In fact, an aggressive campaign might trigger a backlash by foreign carriers, particularly because the FCC, until recently, required all USISCs to apply a uniform accounting rate to secure a commitment from foreign-carrier correspondents to route inbound U.S. traffic to USISCs in proportion to the inbound traffic the foreign carrier received from each U.S. carrier. Higher than necessary accounting rates also may help preserve the appearance of IMTS as an expensive and unattractive business to enter, thereby reducing the scope of competition. If accounting rates were to drop toward cost, market entry would become more

11. See William H. Melody, "Telecom Myths: The International Revenue Settlements Subsidy," *Telecommunications Policy*, Vol. 24, No. 1, Feb. 2000, p. 51.

attractive, and incumbent USISCs would have to reduce both retail and wholesale international rates significantly, thereby reducing or eliminating a source of revenues to offset losses or low margins in the more competitive domestic markets. Figure 9.1 identifies the cost items and revenue flows in an international telephone call. Figure 9.2 emphasizes that unless otherwise permitted by regulators, international carriers must route international traffic on a proportionate basis, that is, routing outbound traffic in a direct relationship to the volume received inbound.

9.1.3 Previous FCC Approaches to the Problem

After years of neglecting the issue, the FCC in 1985 began to consider high accounting rates a trade issue and a major irritant. The Commission's International Settlements Policy sought to prevent foreign carriers from playing one USISC against the others[12] to extract financial concessions [4, 5]. The International Settlements Policy, like its predecessor, the Uniform Settlements Policy,[13] required "all [U.S. International Service] carriers providing the same service to the same foreign point to have the same accounting, settlement and division of tolls arrangement with the foreign administration," [4], including an expectation that carriers will negotiate an operating agreement requiring the proportionate routing of return traffic. In other words, if MCI generated 60% of British Telecom's inbound traffic from the United States, then MCI should secure from British Telecom the commitment to route 60% of its United States–destined traffic to MCI.

The Commission mandated uniform accounting rates ostensibly to prevent monopoly foreign carriers from whipsawing multiple USISCs.[14]

12. The international telecommunications marketplace in 1985 could be characterized as "becoming increasingly competitive on the U.S. side, in contrast to the foreign side which continues to be dominated by monopoly post, telephone and telegraph (PTT) carriers." L. Johnson, "Dealing With Monopoly in International Telephone Service: A U.S. Perspective," *Information Economics and Policy*, Vol. 4, 1991, p. 225.

13. See Uniform Settlement Rates on Parallel International Communications Routes, 84 FCC 2d 121 (1980). "The policy ... was first developed in the 1930's and had its roots in antitrust law." International Settlements Policy Reconsideration, 2 FCC Rcd., p. 1118. See also Mackay Radio and Telegraph Co., 2 FCC 592 (Telegraph Committee 1936), *affirmed*, 97 F.2d 641 (D.C. Cir. 1938).

14. The term whipsawing refers to the potential for a monopoly PTT to have a superior bargaining position vis-à-vis multiple USISCs. As long as the PTT has one operating agree-

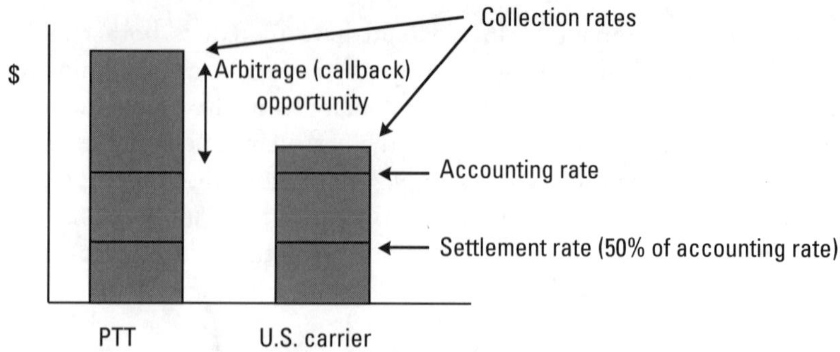

Cost to carrier = true cost/half-circuit + settlement cost

Example: (Assume true cost of half-circuit = \$0.05 per minute,
Accounting Rate: \$0.20 per minute)

Five-minute call to Germany via AT&T @ \$ 0.29 per minute

Cost to Subscriber:	5 × \$0.29 = \$1.45
Settlement Rate:	\$0.10 (half accounting rate)
Cost to AT&T:	5 × (\$0.09 + \$0.10) = \$0.95
Profit to AT&T:	\$1.45 – \$0.95 = \$0.50
Cost to Deutsche Telekom:	5 × .05 = \$0.25
Profit to Deutsche Telekom:	(5 × 10) – .25 = \$0.25

Figure 9.1 The collection rate model.

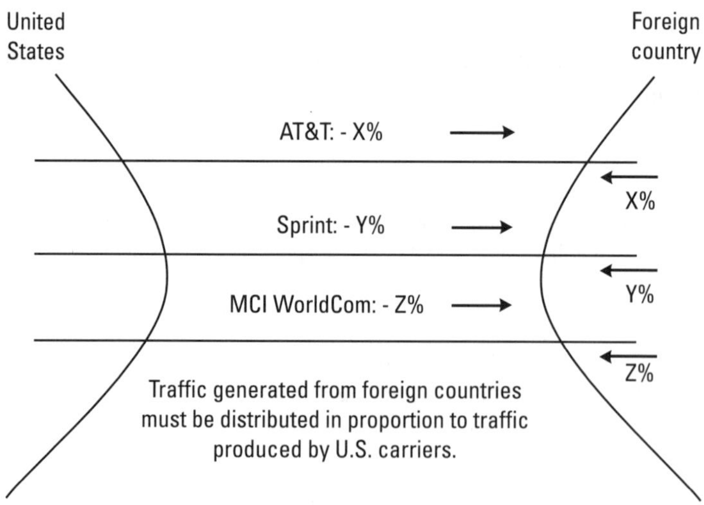

Figure 9.2 The proportionality rule.

That mandatory uniformity may have exacerbated the accounting rate settlement problem for these reasons:

- A foreign carrier would be less likely to agree to any accounting rate reduction if instead of rewarding one high-volume carrier, the lower rate had to apply for settlements with all USISCs.
- A single USISC cannot opt for a strategy of stimulating traffic volume growth by offering foreign carriers a lower accounting rate.
- Any single USISC considering hardball negotiations with foreign carriers to lower the accounting rate had to balance the potential for increased outbound traffic and higher revenues against the prospect for the foreign carriers to retaliate by routing less return traffic, regardless of FCC-articulated concerns that foreign carriers should not discriminate against USISCs.

9.1.3.1 Whipsawing

Notwithstanding a preference for avoiding marketplace intervention and oversight, the FCC believed that it had to intervene and respond to superior foreign-carrier negotiating leverage, particularly for routes where only one foreign carrier controlled market access. Monopoly carriers may enjoy bargaining leverage, because they have several USISCs available to handle traffic to all U.S. destinations, but all USISCs have to secure an operating agreement with a single carrier. Foreign carriers have few immediate incentives to confer additional operating agreements unless a USISC can generate additional traffic volumes to compensate for the additional transaction costs incurred to provide interconnection. That disincentive grows more pronounced where new USISCs seeking access could not propose a temporary deal sweetener, that is, an offer to pay a premium over the applicable accounting rate until a threshold volume of traffic is generated, because the

ment with a USISC, it has national access. Because more than one USISC seeks access to the foreign locale, the PTT may play one carrier against the others by auctioning off access to the carrier willing to accept the lowest settlement rate. "Generally, this has involved a PTT seeking a modification of an accounting rate in a manner more favorable to the PTT—that is, a reduction in the rate paid by the PTT to the U.S. carrier for delivery of traffic in the U.S., or an increase in the rate paid by U.S. carriers to the PTT for delivery of traffic in that foreign country." International Settlements Policy Reconsideration Report and Order, 59 Rad. Reg. 2d 982 (1986), *on reconsideration*, 2 FCC Rcd. at 1118.

International Settlements Policy required uniformity in the terms and conditions for settling accounts.

Absent a regulatory mandate of accounting rate uniformity, the Commission theorized that foreign carriers could auction off inbound U.S. traffic flows to the highest bidder, that is, the USISC willing to part with the greatest share of the toll revenues. In application, the Commission's policy appeared to have mandated accounting rate uniformity stuck at artificially high levels, far exceeding actual service costs.

New USISCs, as potential whipsawing victims, have no way to pressure foreign correspondents for lower accounting rates, and established USISCs have little incentive to force the issue. The initiating carrier would risk good corporate relations with the foreign carrier whose managers typically have great latitude in determining how to route return, inbound U.S. traffic, despite FCC concerns about the potential for discrimination. Even if one USISC cared to take the risk, with an eye toward generating even higher returns, the FCC's accounting rate uniformity requirement would have conferred the benefit to all USISCs.

9.1.4 The Accounting Rate Regime and Bypass Incentives

Because accounting rates remain at artificially high levels for many routes, carriers and their customers strategize how to route traffic exempt from the settlement process. The vehicles for avoiding high accounting rates include the use of callback services, which provide dial tone to end users physically situated in another country, and linking international private lines with a switch that secures access to the PSTN. Those options may violate ITU recommendations[15] and carrier tariffs, because they enable end users to secure services in a manner that the carrier did not intend on providing. While such bypass may expedite reforms, it flouts uniform rules of the road. For example, the ITU Recommendations on leased international private lines contemplate consultation and agreement on the scope of service. Private lines, by

15. Recommendation D.1, Sec. 7.1.1 of the ITU's International Telegraph and Telephone Consultative Committee Blue Book, Vol. II, Fascicle II.1, General Tariff Principles, Charging and Accounting in International Telecommunications Services, suggests that administrations can condition, consult, and agree to the scope of access to public networks provided to users of international private leased circuits. To the extent that a private line reseller or end user does not engage in such consultation and erects a system for evading the accounting rate, the host country may deny access to the PSTN. However, in many instances, accounting rate avoidance schemes may go undetected by the carrier providing interconnection.

definition, provide closed, intracorporate networking capabilities, not the functional equivalent to switched public, long-distance services.

What is occurring in international telecommunications parallels the gray market in international commercial aviation, in which carriers look the other way or clandestinely collaborate with ticket resellers, consolidators, and brokers who offer seats at rates well below the published tariff.[16] In international telecommunications, sophisticated users and system integrators design private-line networks that avoid accounting rate liability. Carriers originally offered unmetered private lines as a way to fill excess capacity and satisfy large-volume user requirements for closed, internal networks. Private-branch exchanges and other customer-controlled equipment have enabled users to interconnect unmetered international private lines with local PSTNs. Such "leakiness" enables the private-line subscriber to access users outside the internal network. Expanded access to a private-line network means that users who otherwise would have to use IMTS circuits can opt for specially configured private-line access for functionally equivalent service.

Resellers can expand the reach of leaky private lines with higher capacity switches. Some carriers and their regulatory overseers do not object to that type of pure resale, which does not enhance leased lines but simply repackages for sale to many the discounted bulk calling capacity designed for a single user. Resale stimulates overall capacity demand, and it can reduce outbound IMTS accounting rate liability, particularly where regulatory policies block or limit inbound resale. Some carriers, intent on capturing larger market shares by aggregating and routing regional traffic through a hub, may engineer a complex array of private lines and acquire both half-circuits on routes to handle accounting rate–exempt traffic. Transiting, the routing of traffic destined for another country across the domestic facilities of an intermediary carrier, presents another opportunity for carriers and new international telephone entrepreneurs alike to engineer innovative arrangements for users.[17]

---

16. International carriers do provide discounted rates to high-volume users, for example, as an incentive to migrate from unmetered private lines to metered virtual (software-defined) private lines using the public switched network. The carriers avoid application of artificially high accounting rates by creating a new service category and applying a different—and lower—accounting rate. Foreign carriers typically have no obligation to justify how the new rate does not discriminate against users paying higher charges for existing offerings subject to accounting rates.

17. Even companies with limited budgets can get into the international telecommunications business and exploit high accounting rate and end-user charge differentials. A "boomerang

Since the early 1990s, the FCC has taken a more proactive role in accounting rate oversight, with an eye toward encouraging carrier and end user self-help, that is, routing strategies that collectively make high accounting rates unsustainable. The FCC adopted a get-tough policy by prescribing what it calculated to be cost-based accounting rates,[18] because it had grown impatient with the pace of progress. While the FCC can properly condition grants of regulatory authorizations and prescribe rates for the carriers it regulates, attempts to affect the behavior and the financial performance of other carriers generated vocal opposition, at home and abroad, that the Commission had failed to appreciate international comity and national sovereignty.[19]

Similarly, an FCC proposal to impose reporting requirements and other means for overseeing the extent of participation in the U.S. telecommunications market by foreign-owned firms [6][20] generated arguments that it would violate the commitment to national treatment of foreign enterprises, that is, applying identical regulatory rights, responsibilities, and opportunities to both

---

box" enables callers in high-cost foreign locations to place a call to the United States, hang up, and soon receive a call from the United States with the intended call recipient on the line. At the microlevel, the foreign caller avoids having to pay the significantly higher charge for originating an international call, the foreign carrier loses some toll revenues, and the USISC handling the international call accrues some additional toll revenues. At the macrolevel, the transaction contributes to the expanding U.S. accounting rate deficit, thereby blunting the foreign carrier's revenue losses and the USISC's revenue gains.

18. The Commission proposed to "establish...determine and prescribe just and reasonable accounting rates" if USISCs and their foreign counterparts failed to negotiate rates downward to an FCC-determined benchmark range. Regulation of International Accounting Rates, Notice of Proposed Rulemaking, 5 FCC Rcd. at 4950.

19. When the FCC attempted to influence the timetable for construction and activation of the TAT-7 overseas cable through direct negotiations with foreign governments, foreign carriers deemed such activism intrusive of national sovereignty, and the U.S. Court of Appeals for the District of Columbia deemed it a violation of the Government in the Sunshine Act. ITT World Communications, Inc., 77 FCC 2d 877 (1980) (order denying petition for rulemaking on permissible scope of FCC contacts with foreign administrations to negotiate delayed deployment of a transatlantic submarine communications cable), *reversed*, ITT World Communications v. FCC, 699 F.2d 1219 (D.C. Cir. 1983), *reversed on other grounds*, 466 U.S. 463 (1984).

20. The FCC has modified its policies that impose more extensive oversight of foreign-owned carriers providing international services from the United States. See Regulation of International Common Carrier Services, CC Docket No. 91-360, Notice of Proposed Rulemaking, 7 FCC Rcd. 577 (1992), Report and Order, FCC 92-463 (rel. Nov. 6, 1992) (retaining more burdensome "dominant carrier" oversight only where the foreign

foreign-owned carriers and domestic carriers. The FCC subsequently decided to calibrate the scope of regulatory oversight of foreign carriers to the degree of market access accorded U.S. carriers, particularly the extent to which U.S. service providers may use leased international private lines to access the PSTN in foreign locales. That mechanism provides strong leverage for achieving market access parity, by linking the scope of inbound U.S. market access with reciprocal opportunities for outbound traffic.[21]

Reliance on proliferating resale of private lines repositioned the FCC from direct confrontation with foreign carriers over their sovereign right to negotiate accounting rates, to "procedural reforms that remove any U.S. regulatory impediments to lower, more economically efficient, cost-based accounting rates" [7]. The Commission assumed that if resale were available on an equivalent basis, inbound and outbound, then the incumbent facilities-based carriers would perceive new incentives to negotiate lower accounting rates to dissuade customers from migrating to private-line and resale options. Facilities-based USISCs, facing competition from resellers,[22] unencumbered by accounting rate liability, may view high accounting rates as imposing a floor on how low they can price end-user rates "to prevent diversion of ... customers to a reseller" [8].[23] Presumably, resellers providing

---

affiliate of a USISC has the ability to discriminate against unaffiliated carriers through control of bottleneck services and facilities in the foreign market).

21. See Cable & Wireless, Inc. DA-1344, Tele. Div. (rel. Dec. 8, 1994); Cable & Wireless, Inc. 8 FCC Rcd. 1664 (Com. Car. Bur. 1993); fONOROLA Corp. and EMI Corp., 7 FCC Rcd. 7312 (1992), *on recon.*, 9 FCC Rcd. 4066 (1994) (authorizing British and Canadian resellers to provide international service upon finding that the foreign country on the other end of the circuit provides equivalent opportunities to U.S. carriers to resell interconnected private lines).

22. "Resale [of leased private lines] would bypass the accounting rate mechanism—a major cost to the traditional carrier mode of operation—and increase the feasibility of creating unidirectional traffic channels." K. Cheong and M. Mullins, "International Telephone Service Imbalances," *Telecommunications Policy*, Vol. 15, Apr. 1991, pp. 107, 116. If resale remains unidirectional, U.S. facilities-based carriers and consumers will not benefit: Resale occurring only in the inbound U.S. direction would increase the U.S. accounting rate deficit. Resale must be bidirectional to have the effect of "expos[ing] the differential between tariffs and accounting rates and ultimately force traditional carriers to renegotiate accounting rates closer to service costs." *Ibid.*, pp. 116–117.

23. "To the extent that the accounting rate is above cost, the underlying carrier will face a constraint on how much of a reduction in its revenues it can tolerate." Accounting Rate Phase II, First Report and Order, 7 FCC Rcd., p. 561, para 561.

outbound services from the United States will acquire market share, thereby reducing the number of IMTS outbound minutes subject to accounting rate settlements. A facilities-based carrier that refuses to negotiate accounting rates closer to cost would "receive fewer revenues from its IMTS customers and, thus, would wind up with fewer revenues overall" [8]. Figure 9.3 outlines how international callback works. Callers in locations with high retail rates can import dial tones from nations with lower international calling rates.

9.1.5 Linking Inbound Private-Line Access with WTO Market Access Commitments

The FCC decided to link resale opportunities into the United States with an evaluation of whether an equivalent outbound opportunity exists, first by a country-specific examination and later with reference to what market access commitments the nation had made under the auspices of the WTO. If the Commission had failed to ensure that U.S. carriers have pure resale access like that available to foreign carriers (so-called two-way resale), it would have given "overseas administrations the incentive and the power to use such [inbound U.S.] resale to evade the international settlements process or otherwise discriminate against competing U.S. carriers" [8].

Two-way resale, instead of the even better facilities-based competition, offers the next best incentive for carriers to reduce accounting rates. Currently, the FCC has certified that equivalent, two-way resale opportunities exist in 25 nations, 14 of which have been certified since 1998.[24] The FCC's two-way resale requirement means that foreign regulatory authorities have to authorize expanded or first-time inbound resale opportunities to ensure continued outbound access for their resellers, including the incumbent, facilities-based carrier interested in routing traffic exempt from accounting rate settlements. For resale to become bidirectional, many nations will have to change prevailing regulatory and licensing policies to permit market entry by both facilities-based carriers and pure resellers, who simply provide IMTS-type service via leased lines. Heretofore, most nations have limited resale opportunities to international value-added networks that provide enhanced services over leased lines.

24. See FCC, International Bureau, International Simple Resale Approved Countries (March 6, 2000), available on-line at http://www.fcc.gov/ib/isr.html.

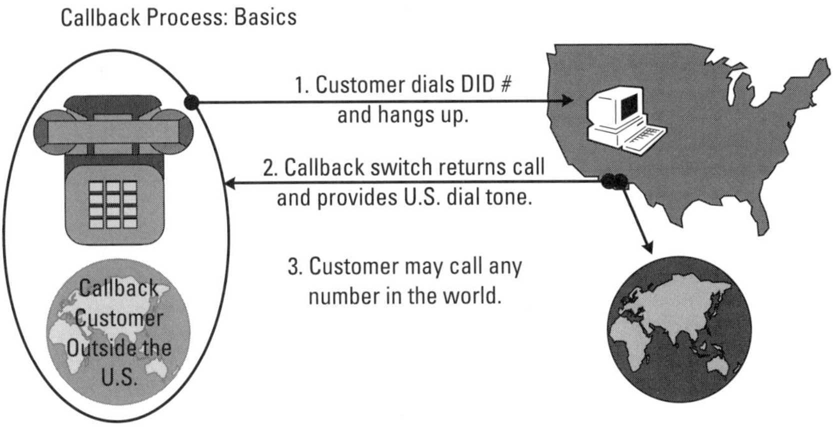

Figure 9.3 International callback applications.

9.1.6 New Strategies

By the late 1990s, the FCC again changed its strategy to one that combined unilateral action, for example, prescribing benchmark settlement rates, with multilateral market access initiatives under the auspices of the WTO. In 1999, the FCC substantially dismantled its International Settlements Policy by exempting U.S. carrier contractual arrangements with nondominant foreign carriers and for all arrangements with foreign carriers operating in competitive foreign markets [9]. In application, that means U.S. carriers can negotiate on a one-to-one basis with foreign carriers without having the terms and conditions made applicable to all other U.S. carriers. The FCC permits such complete flexibility for settlement arrangements between U.S. carriers and foreign telecommunications carriers that lack market power and for all settlement arrangements on routes where U.S. carriers are able to terminate at least 50% of their U.S. billed traffic in the foreign market at rates that are at least 25% below the applicable FCC-prescribed benchmark settlement rate [9, para. 6].

The FCC initiated that deregulatory initiative based on the mandate contained in the Telecommunications Act of 1996 that the Commission remove regulations no longer in the public interest.[25] The Commission

25. See 47 U.S.C. Sec. 161 (1999).

justified its move on the existence of more facilities-based carriers globally[26] and the prospect for even greater competition in view of existing and future market access commitments made at the WTO.[27] With that action, the FCC substantially jettisoned its International Settlement Policy and previous orders that afforded only some degree of flexibility.[28]

In the final analysis "the real problem is not the accounting rate system per se, ... [but] regulatory and licensing policies ... [that] have inhibited competition ... [thereby] buttress[ing] high and static accounting rates" [10]. The incentive to retain high accounting rates can last only as long as incumbent carriers face no competition at home. Should a government authorize competition for basic international telephone services, either facilities-based or resale, the second carrier might project revenue gains by reducing end-user charges and accounting rates to stimulate demand and possibly to encourage USISCs to route more inbound traffic its way. Despite the keen attention to privatization and liberalization, most countries persist in reserving a monopoly for switched international telephone services. Only after market entry by additional carriers, particularly ones that can build and operate their own facilities, will market-driven carrier access charges replace the accounting rate regime. Table 9.1 tracks the accounting rate used by U.S. carriers over time with a number of nations. Table 9.2 tracks the growing international settlements deficit incurred by U.S. carriers.

---

26. "New entrants are being established in regions throughout the world and are rapidly gaining substantial market share in many markets. For example, in Europe, over 50 new facilities-based carriers have entered the market and are providing service in competition with incumbent operators in nearly all countries of the European Union. In the past year, companies have committed to investing over $3 billion to build independent intra-European fiber-optic networks. In Japan, Hong Kong, Australia, and many other countries, similar developments are occurring as U.S. and other domestic carriers are entering the market to compete with incumbent carriers." 47 U.S.C. Sec. 161 (1999), para. 10.

27. "Today, over 30 countries are committed to open and competitive telecommunications markets, and 22 other countries have committed to open their markets in the future as a part of the WTO Agreement on Basic Telecommunications." 47 U.S.C. Sec. 161 (1999), para. 10.

28. See Policy Statement on International Accounting Rate Reform, 11 FCC Rcd. 3146 (1996) (Policy Statement); Regulation of International Accounting Rates, CC Docket No. 90-337, Phase II, Fourth Report and Order, 11 FCC Rcd. 20,063 (1996).

Table 9.1
Telephone Service Accounting Rates for Selected Countries

Peak Period Accounting Rates

Year-End Data

	1980	1981	1982	1983	1984	1985	1986	1987	1988	1989	1990	1991	1992	1993	1994	1995	1996	1997	1998	1999
Australia	$2.00	$2.00	$2.00	$2.00	$2.00	1.5 SDR	1.2 SDR	1.2 SDR	1.2 SDR	1.2 SDR	0.8 SDR	0.68 SDR	0.6 SDR	0.55 SDR	0.4 SDR	0.4 SDR	0.308 SDR	0.308 SDR	0.22 SDR	0.22 SDR
Brazil	$3.00	$3.00	$3.00	$3.00	$2.50	$2.50	$2.50	$2.50	$2.50	$2.50	$2.50	$1.60	$1.50	$1.40	$1.27	$1.14	$1.03	$0.85	$0.65	$0.60
Canada[A]	N/A	N/A	N/A	N/A	N/A	$0.42	$0.42	$0.42	$0.42	$0.42	$0.28	$0.28	$0.28	$0.26	$0.24	$0.22	$0.20	$0.20	$0.20	$0.20
China	N/A	N/A	N/A	N/A	N/A	9.0 GF	9.0 GF	9.0 GF	8.0 GF	8.0 GF	7.5 GF	7.5 GF	6.5 GF	6.0 GF	5.5 GF	4.5 GF	3.8 GF	3.1 GF	0.85 SDR	
Colombia	$2.20	$2.10	$2.10	$2.10	$2.00	$2.00	$2.00	$1.90	$1.85	$1.85	$1.75	$1.60	$1.55	$1.50	$1.40	$1.30	$1.25	$1.00	$0.80	$0.75
Dominican Republic	N/A	N/A	N/A	N/A	N/A	$1.45	$1.45	$1.45	$1.45	$1.45	$1.42	$1.39	$1.36	$1.29	$1.30	$1.10	$0.90	$0.70	$0.60	$0.52
Egypt	N/A	N/A	N/A	N/A	N/A	$2.10	$2.10	$2.00	$2.00	$1.95	$1.90	$1.80	$1.70	$1.60	$1.50	$1.40	$1.40	$1.30	$1.30	$1.10
El Salvador	N/A	N/A	N/A	N/A	N/A	$1.55	$1.55	$1.50	$1.50	$1.50	$1.50	$1.40	$1.30	$1.25	$1.20	$1.20	$1.10	$0.88	$0.77	$0.62
France[A]	1.6 SDR	1.6 SDR	1.6 SDR	1.6 SDR	1.6 SDR	1.6 SDR	1.4 SDR	1.4 SDR	1.4 SDR	1.2 SDR	1.2 SDR	1.0 SDR	0.7 SDR	0.7 SDR	0.42 SDR	0.36 SDR	0.24 SDR	0.19 SDR	0.15 SDR	0.15 SDR
Germany	1.6 SDR	1.6 SDR	1.3 SDR	1.2 SDR	1.2 SDR	1.2 SDR	1.2 SDR	1.2 SDR	1.2 SDR	1.2 SDR	1.2 SDR	1.0 SDR	0.8 SDR	0.6 SDR	0.34 SDR	0.26 SDR	0.16 SDR	0.15 SDR	0.15 SDR	0.15 SDR
Greece	N/A	N/A	N/A	N/A	N/A	5.0 GF	5.0 GF	5.0 GF	5.0 GF	5.0 GF	5.0 GF	1.53 SDR	1.21 SDR	1.12 SDR	0.95 SDR	0.85 SDR	0.7 SDR	0.63 SDR	0.4 SDR	0.25 SDR
Guatemala	N/A	N/A	N/A	N/A	N/A	$1.50	$1.50	$1.50	$1.50	$1.50	$1.50	$1.50	$1.40	$1.30	$1.20	$1.18	$1.00	$0.90	$0.77	$0.68
Hong Kong	$8.00	$8.00	$8.00	$8.00	$8.00	$2.60	$2.50	$2.50	$2.50	$2.35	$2.20	$1.90	$1.60	$1.20	$1.00	$1.00	.65 SDR	.58 SDR	.52 SDR	.1 SDR
India[A]	N/A	N/A	N/A	N/A	N/A	$2.70	$2.50	$2.50	$2.25	$2.25	$2.25	$2.00	$1.90	$1.80	$1.80	$1.60	$1.42	$1.28	$1.28	$1.28
Israel	$3.00	$3.00	$2.50	$2.40	$2.40	$2.40	$2.40	$2.40	$2.40	$2.40	$2.40	$2.40	$2.28	$2.16	$2.16	$1.90	$1.18	$0.70	$0.59	$0.30
Italy	5.5 GF	5.0 GF	5.0 GF	5.0 GF	5.0 GF	5.0 GF	5.0 GF	5.0 GF	5.0 GF	5.0 GF	4.6 GF	4.38 GF	1.2 SDR	1.1 SDR	0.82 SDR	0.48 SDR	0.36 SDR	0.24 SDR	0.16 SDR	0.16 SDR
Jamaica	N/A	N/A	N/A	N/A	N/A	$1.75	$1.75	$1.75	$1.75	$1.75	$1.65	$1.60	$1.55	$1.50	$1.40	$1.40	$1.30	$1.25	$1.25	$1.15

Table 9.1 (continued)

	1980	1981	1982	1983	1984	1985	1986	1987	1988	1989	1990	1991	1992	1993	1994	1995	1996	1997	1998	1999
Japan[A]	$3.00	$3.00	$2.66	$2.66	$2.66	$2.35	$2.35	$2.35	1.34 SDR	1.34 SDR	1.34 SDR	1.13 SDR	0.95 SDR	0.75 SDR	0.63 SDR	0.63 SDR	0.63 SDR	0.63 SDR	0.21 SDR	0.21 SDR
Korea, Rep.	$3.00	$3.00	$3.00	$3.00	$2.80	$2.80	$2.60	$2.52	$2.44	$2.36	$2.10	$1.90	$1.60	$1.44	0.95 SDR	0.85 SDR	0.85 SDR	0.72 SDR	$0.85	$0.71
Mexico[B]	N/A	N/A	N/A	N/A	N/A	N/A	N/A	N/A	N/A	N/A	$1.45	$1.32	$1.16	$1.10	$1.00	$0.91	$0.67	$0.68	$0.70	$0.74
Netherlands	1.6 SDR	1.6 SDR	1.6 SDR	1.2 SDR	1.2 SDR	1.2 SDR	1.2 SDR	1.2 SDR	1.2 SDR	1.1 SDR	1.1 SDR	0.9 SDR	0.5 SDR	0.5 SDR	0.4 SDR	0.25 SDR	0.25 SDR	0.2 SDR	0.14 SDR	0.1 SDR
Nigeria	N/A	N/A	N/A	N/A	N/A	$1.50	$1.50	$1.50	$1.50	$1.50	$1.50	$1.50	$1.50	$1.50	$1.50	$1.50	$1.50	$1.50	$1.40	$1.15
Pakistan	N/A	N/A	N/A	N/A	N/A	$2.55	$2.55	$2.55	$2.55	$2.55	$2.30	$2.30	$2.30	$2.30	$2.30	$2.30	$2.20	$2.00	$1.20	$1.20
Peru	N/A	N/A	N/A	N/A	N/A	$2.00	$2.00	$2.00	$2.00	$2.00	$2.00	$2.00	$1.50	$1.40	$1.30	$1.30	$1.23	$1.00	$0.85	$0.66
Philippines	$2.70	$2.70	$2.70	$2.70	$2.50	$2.30	$2.30	$2.15	$2.00	$1.95	$1.92	$1.85	$1.68	$1.68	$1.34	$1.23	$1.00	$0.82	$0.72	$0.57
Poland	N/A	N/A	N/A	N/A	N/A	$2.00	$1.75	$1.75	$1.65	$1.50	$1.50	$1.35	$1.30	$1.25	$1.20	$1.15	$0.95	$0.70	$0.55	$0.42
Russia	N/A	N/A	N/A	N/A	N/A	$9.00	$9.00	$9.00	$9.00	$3.00	$3.00	$2.60	$2.60	$2.60	$2.60	$2.60	$2.60	$2.60	$2.60	
Saudi Arabia	$2.36	$2.36	$2.36	$2.36	$2.36	$2.36	$2.36	$2.36	$2.36	$2.36	$2.36	$2.20	$2.20	$2.20	$2.20	$2.20	$2.20	1.5 SDR	1.25 SDR	1.0 SDR
South Africa	N/A	N/A	N/A	N/A	N/A	$2.00	$2.00	$2.00	$2.00	$2.00	$2.00	$1.80	$1.80	$1.50	$1.20	$1.20	$1.00	$1.00	$0.80	$0.70
Spain[A]	5.3 GF	5.3 GF	5.3 GF	5.3 GF	5.3 GF	5.3 GF	5.3 GF	1.7 SDR	1.6 SDR	1.6 SDR	1.5 SDR	1.5 SDR	1.3 SDR	1.2 SDR	1.0 SDR	0.44 SDR	0.35 SDR	0.2 SDR	0.2 SDR	
Switzerland	5.5 GF	1.6 SDR	1.6 SDR	1.6 SDR	1.6 SDR	1.6 SDR	1.4 SDR	1.4 SDR	1.25 SDR	1.12 SDR	1.12 SDR	1.12 SDR	0.808 SDR	0.808 SDR	0.61 SDR	0.61 SDR	0.35 SDR	0.25 SDR	0.2 SDR	0.2 SDR
Taiwan	$2.25	$2.25	$2.25	$2.25	$2.25	$2.00	$2.00	$2.00	$2.00	$1.90	$1.80	$1.60	$1.40	$1.20	$1.20	$1.20	$1.20	$1.14	$0.45	
Thailand	N/A	N/A	N/A	N/A	N/A	$2.50	$2.50	$2.50	$2.50	$2.30	$2.30	$2.00	$1.75	$1.60	$1.60	$1.55	$1.50	$1.20	$0.90	$0.70
United Kingdom[A]	$1.20	$1.20	$1.06	$1.06	$1.06	$1.06	$1.06	$1.06	$1.06	$1.06	$1.06	0.68 SDR	0.54 SDR	0.44 SDR	$0.33 SDR	0.25 SDR	0.15 SDR	0.15 SDR		

[A] Where more than one PTT connects with U.S. carriers, the table shows rates for the first carrier providing service. These include Korea Telecom in Korea, Kokusai Denshin Denwa in Japan, PLDT in the Philippines, CODETEL in the Dominican Republic, and British Telecom in the United Kingdom. Accounting rates may vary by time of day, or by volume of traffic.

[B] Accounting rates for service between Mexico and the United States vary by service classification and the originating or terminating locations. Each rate shown in the table represents a simple average for all accounting rates in effect for service between the United States and Mexico.

Source: FCC.

Table 9.2
Measures of the International Settlements Deficit (in Millions of Dollars)

	43.61 Settlements Data[A]			BEA Estimate of International Sales and Purchases of Private Services[B]										
	Receipts	Payments	Deficit	Receipts					Payments					Deficit
				Europe	South & Central America	Asia & Pacific	Other	All Countries	Europe	South & Central America	Asia & Pacific	Other	All Countries	
1980	912	1,277	365											
1981	992	1,553	561											
1982	1,168	1,901	733											
1983	1,298	2,270	972											
1984	1,296	2,497	1,201											
1985	1,471	2,623	1,152											
1986	1,606	3,076	1,470	653	356	388	430	1,827	946	920	732	655	3,253	1,426
1987	1,859	3,613	1,753	763	407	468	473	2,111	1,122	1,064	824	726	3,736	1,625
1988	2,051	4,189	2,137	822	400	519	455	2,196	1,410	1,283	1,035	848	4,576	2,380
1989	2,325	4,849	2,524	911	449	643	516	2,519	1,540	1,539	1,127	966	5,172	2,653
1990	2,515	5,389	2,873	973	531	687	544	2,735	1,616	1,783	1,245	939	5,583	2,848
1991	2,767	6,063	3,296	1,093	651	956	591	3,291	1,960	1,971	1,523	1,154	6,608	3,317
1992	2,892	6,229	3,336	1,007	569	776	533	2,885	1,660	2,028	1,293	1,071	6,052	3,167
1993	2,938	6,645	3,707	928	589	756	512	2,785	1,593	2,146	1,404	1,222	6,365	3,580
1994	3,003	7,294	4,291	924	669	756	516	2,865	1,603	2,361	1,684	1,280	6,928	4,063
1995	3,073	8,016	4,943	950	750	942	586	3,228	1,561	2,527	1,876	1,341	7,305	4,077
1996	2,829	8,484	5,655	876	855	698	571	3,270	1,378	2,981	2,445	1,500	8,304	5,034
1997	2,865	8,315	5,450	1,092	986	1,087	606	3,771	1,252	2,920	2,570	1,371	8,113	4,342

[A]The totals include settlement payments and settlement receipts for international telephone, telex, and telegraph. Settlements associated with transiting traffic are first included for 1991.

[B]1997 data from the *Survey of Current Business*, October 1998, Table 5.4, pp. 104–105. Data for some prior years were revised.

Source: FCC.

9.2 Internet Telephony

The Internet has evolved into a vibrant medium for communications, entertainment, education, and commerce. One of the primary drivers for the growing consumer reliance on Internet mediation involves the ability of the Internet to offer instant real-time delivery of digital packets in addition to the store-and-forward, nonreal-time delivery of packets in applications like electronic mail. Real-time streaming of information packets means that the Internet can serve as a medium for audio and video programming and also for telephone services.

In the accelerated pace of product and service life cycles common to the Internet, telephone-type services have quickly evolved from an awkward personal computer–mediated curiosity to a commercial service available from not just computers, but conventional telephones as well. Internet telephony has the potential to serve as a major threat to the international accounting rate regime and possibly to how telecommunications carriers price retail long-distance services for two primary reasons:

- The Internet architecture provides for efficient facilities loading, including the ability of telecommunications networks dedicated for the data services to handle voice traffic at near zero cost, absent congestion.
- Regulatory policies throughout the world largely exempt providers of Internet-services from having to divide toll revenues based on an accounting rate settlement, to pay interconnection charges, and to contribute to universal telecommunications service funding imposed on telecommunications carriers.

Internet telephony constitutes a formidable threat to conventional telecommunications carrier profit margins. ISPs easily can add telephony traffic onto their data lines, and technological innovations provide ways to inject Internet voice traffic into the PSTN for the last-mile delivery to call recipients. Given the large difference between ISPs' costs incurred in providing Internet telephony and the retail charges for conventional telephone services, especially international rates, ISPs can profit handsomely by pricing service well below the preexisting retail toll charge. Such exploitation of a wide pricing differential constitutes a type of arbitrage. The ISP can make a business case for delivering services to consumers at significantly lower costs, because the ISP has plenty of margin with which to work, that is, the difference between its actual costs and the imputed cost established by route-specific accounting rates based on conventional telephony. Figure 9.4 is an estimate of the growth in Internet-mediated telephone calls.

9.2.1 The Technology of Internet Telephony

Internet telephony uses the digital, packet-switched nature of the Internet along with its routing and addressing standards to provide real-time, two-way audio conferencing.[29] Internet switching and routing technology manages the transmission and processing of text, graphics, data, audio, and video. The Internet's TCP/IP protocol[30] provides a standard vehicle for subdividing content (e.g., a voice conversation) into a stream of packets that are routed

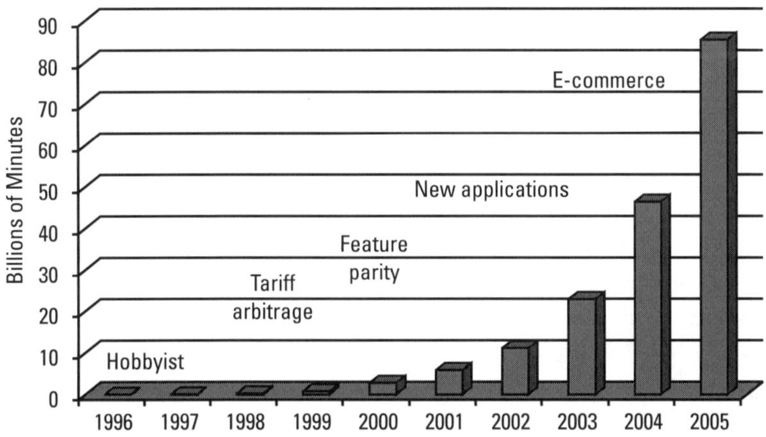

Figure 9.4 Global voice/fax over IP traffic and market drivers.

---

29. For a helpful nontechnical introduction to Internet telephony, see Internet Telephony Consortium, "Frequently Asked Questions: How Can I Use the Internet as a Telephone?" available on-line at http://itel.mit.edu/voice_faq.html.

30. "The common denominator for e-mail communications is the use of a standard programming protocol, TCP/IP-Transmission Control Protocol/Internet Protocol-upon which inter-computer communications are based. The TCP protocol divides messages into packets which are marked with a sequence number and the address of the recipient. TCP also inserts error control information. The packets are then sent over the network to the addressee. The routing of the individual packets varies, with IP controlling the transport of the packets to the remote host computer. At the remote host, TCP receives the packets and checks for errors. When an error occurs, TCP asks for the particular

via any available path between the sender and the intended call recipient. Each packet has space reserved for destination information so that intermediary routing facilities can read the header data to determine how and where to send the packets onward toward their intended destination. Headers include a sequence of digits that correspond to an Internet address, much like the numbering sequence in direct-distance dialing via telephone.

Packet switching efficiently uses available switching and routing capacity. Likewise, it operates despite outages, blockages, and busy conditions, because the TCP/IP scheme makes it possible for multiple efforts to route traffic onward in the event that initial efforts fail. Resending misdelivered or unreceived packets and routing them via different and possibly circuitous links require software processing to reassemble the packets in proper order. For traffic and services that do not require immediate, real-time delivery (e.g., electronic mail), possible delays and reassembly present few problems. However, Internet telephony requires immediate, real-time delivery of the packets in their proper order. Any delay, loss, or improper sequencing of packets will result in distortion or the temporary loss of the audio stream.

Heretofore, Internet telephony has lacked the quality, reliability, and security to be considered comparable to conventional telephone services. Traditional telephone services use circuit switching that sets up a dedicated link between call originator and call recipient. That technology provides high-quality service and reliability, because a dedicated pathway exists, as opposed to the virtual, on-the-fly links provided via the Internet. A dedicated pathway may be technologically wasteful in the sense that switching, routing, and transmission capacity lies dormant during pauses in a conversation. Packet-switching technology efficiently fills in gaps with other traffic so that traffic can traverse different routes and arrive at different times at the same destination. In circuit switching, all parts of a traffic stream traverse the same pathway, providing greater quality assurance.

What Internet telephony lacks in quality of service and reliability it makes up in lower costs. However, some users may care more about reliability of service and less about savings. Currently, Internet traffic cannot easily be classified by priority of service or by type of application. Best-efforts

---

packet to be re-sent. Once all the packets have been received, TCP will then use the sequence number to reconstruct the original message. It is the job of IP to get the packets from one place to another; it is the job of TCP to manage the flow and insure that the data are correct." Richard Allan Horning, "Has Hal Signed a Contract: The Statute of Frauds in Cyberspace," *Santa Clara Computer and High Technology Law Journal*, Vol. 12, Aug. 1996, pp. 253, 258.

routing of traffic may not provide the security, safety, and reliability a user requires. For those willing to take the qualitative risk, the financial savings are significant. However, Internet telephony consumers have to incur some initial, up-front costs. Unlike conventional telephone service, the cheapest types of Internet-mediated telephony require a significant initial capital outlay of about $2,000 for a personal computer, modem, sound card, speakers, microphone, software, and Internet access. Conventional telephone services use an inexpensive, "dumb" terminal, the telephone handset, but users incur per-minute charges that can exceed $1 a minute for many international destinations. Internet telephony provided on a conventional dial-up basis (e.g., a toll-free access number) requires an ISP to install devices that can convert circuit-switched telephone traffic into packets and vice versa. Additionally, those devices must provide a routing function, using the IP to bring traffic to a facility, commonly referred to as a point of presence, in the vicinity of the call recipient. While estimates of the financial impact of Internet telephony may vary, one can extrapolate the potential impact by considering the traffic and revenue diversion impact of pure, or simple, resale (i.e., the use of leased international private lines to handle voice traffic). Table 9.3, compiled by the FCC, shows significant growth in such traffic.

9.2.2 Financial and Regulatory Arbitrage and the Potential Impact on Telecommunications Pricing

Internet telephony provides profitable opportunities for incumbents and newcomers alike to offer services functionally equivalent to conventional telephony but treated in a manner that subjects the service to little or no regulation and accrues lower operational costs. Entrepreneurs savor the opportunity to exploit financial and regulatory anomalies and asymmetries in telecommunications, for example, the ability to lease private lines, link them with the PSTN, and offer a long-distance telephone service to individual consumers who otherwise would not qualify for the bulk discounts previously offered only to high-volume private-line users.

Over the years, a number of such anomalies and asymmetries have existed. For example, the price (not necessarily the cost) of a minute of telecommunications use has depended on factors like these:

- The perceived value of the service;[31]

31. Both the FCC and state regulatory commissions have allowed carriers to price some services on the perceived value consumers accrue. For example, some local exchange tele-

Table 9.3
Pure Resale Compared with Facilities-Based and Facilities-Resale International Telephone Service*
Revenues and Minutes Shown in Thousands

Company	1991	1992	1993	1994	1995	1996	1997
AT&T							
Facilities-based and facilities-resale							
U.S. minutes	6,569,014	7,038,969	7,200,907	8,039,868	8,831,143	9,546,094	10,331,159
U.S. billed revenue	$6,962,147	$7,313,799	$7,482,250	$7,984,452	$8,424,874	$8,558,833	$8,351,335
Pure Resale							
U.S. minutes							61,220
U.S. billed revenue							$42,732
Percent minutes handled by pure resale							0.6
MCI							
Facilities-based and facilities-resale							
U.S. minutes	1,599,786	2,101,422	2,856,890	3,529,197	4,485,504	5,371,998	5,912,943
U.S. billed revenue	$1,486,972	$2,065,113	$2,779,355	$2,952,375	$3,967,807	$3,549,540	$4,243,318
Pure resale							
U.S. minutes	64,013	37,142	30,745	28,637	50,559	45,958	34,796
U.S. billed revenue	$65,366	$47,967	$34,307	$55,757	$51,553	$42,140	$39,012
Percent minutes handled by pure resale	3.8	1.7	1.1	0.8	1.1	0.8	0.6
Sprint							
Facilities-based and facilities-resale							

Table 9.3 (continued)

Company	1991	1992	1993	1994	1995	1996	1997
U.S. minutes	728,442	946,173	1,181,074	1,489,696	1,772,347	2,744,892	2,794,250
U.S. billed revenue	$603,785	$785,940	$1,047,749	$1,228,667	$1,288,845	$1,493,416	$1,478,284
Pure resale							
U.S. minutes	72,861	42,326	32,441	32,880	38,425	97,141	124,420
U.S. billed revenue	$88,149	$63,438	$28,110	$56,685	$68,222	$87,178	$94,956
Percent minutes handled by pure resale	9.1	4.3	2.7	2.2	2.1	3.4	4.3
Carriers other than AT&T, MCI, and Sprint							
Facilities-based and facilities-resale							
U.S. minutes	147,616	207,352	274,404	556,885	968,075	1,605,975	3,773,521
U.S. billed revenue	$166,267	$166,637	$195,192	$265,361	$478,265	$632,293	$1,195,378
Pure resale							
U.S. minutes	358,087	485,257	665,119	1,260,389	2,452,755	6,981,704	8,352,290
U.S. billed revenue	$286,925	$399,660	$528,686	$1,007,750	$1,634,614	$3,506,753	$3,934,092
Percent minutes handled by pure resale	70.8	70.1	70.8	69.4	71.7	81.3	68.9
Combined totals							
Facilities-based and facilities-resale							
U.S. minutes	9,071,858	10,293,916	11,513,275	13,615,646	16,057,069	19,268,959	22,811,873
U.S. billed revenue	$9,219,171	$10,331,489	$11,504,546	$12,430,855	$14,159,791	$14,234,082	$15,268,315
Revenue per minute	$1.02	$1.00	$1.00	$0.91	$0.88	$0.74	$0.67

Table 9.3 (continued)

Company	1991	1992	1993	1994	1995	1996	1997
Carriers other than AT&T, MCI, and Sprint as a percentage of total facilities-based and facilities-resale traffic							
U.S. minutes	1.6	2.0	2.4	4.1	6.0	8.3	16.5
U.S. billed revenue	1.8	1.6	1.7	2.1	3.4	4.4	7.8
Pure resale							
Number of reporting carriers	69	86	130	180	230	313	317
U.S. minutes	494,961	564,725	728,305	1,321,906	2,541,739	7,124,803	8,572,726
U.S. billed revenue	$440,440	$511,065	$591,103	$1,120,192	$1,754,389	$3,636,071	$4,110,792
Revenue per minute	$0.89	$0.90	$0.81	$0.85	$0.69	$0.51	$0.48
Pure resale minutes as percentage of facilities-based and facilities-resale minutes	5.5	5.5	6.3	9.7	15.8	37.0	37.6

Carriers handle pure resale traffic by routing calls to other carriers and reselling those carriers' switched services. Facilities-based calls are carried over international circuits in which the carrier has an ownership interest. Facilities-resale calls are carried over international circuits that the carrier leases from other carriers. The data in this table were taken from Section 43.61 Traffic Data reports. The figures are world totals for international message telephone service and include international traffic between domestic and offshore U.S. Points.

Source: FCC.

- Which regulatory agency (state or federal) has jurisdiction over cost allocation and tariffing;[32]
- Whether the service is domestic or international;[33]
- Whether another carrier or an end user seeks facilities interconnection;[34]
- The type of carrier[35] or enterprise[36] providing service;[37]

---

phone service rates have increased when the number of accessible subscribers reaches a benchmark. "In most states, the Bell Operating Companies and larger independents charge higher rates in metropolitan areas than in rural areas—a pricing practice that dates back to the turn of the century and is traditionally justified in the belief that the value of the service provided is higher for subscribers with larger local calling areas." Federal Communications Commission, "FCC Releases Semiannual Study on Telephone Trends," 1991 FCC LEXIS 4305 at *10 (Aug. 7, 1991).

32. Typically, an intrastate long-distance minute of use significantly exceeds the price of an interstate long-distance minute of use. Ironically, an intrastate state call originated via a cellular telephone may be significantly cheaper than the corresponding rate for a call originated over wireline facilities. The rate differential results, in part, from rate-making policies, which may include cross-subsidies to local-exchange service, as opposed to the actual cost of service differences.

33. International message telephone service substantially exceeds domestic rates on a per-minute and mileage-band basis, primarily because international carriers have negotiated toll revenue division agreements that have failed to drop commensurately with cost reductions.

34. The Telecommunications Act of 1996 and preexisting FCC regulations differentiate the terms and conditions for interconnection between carriers as opposed to customer-carrier interconnection. The Telecommunications Act orders favorable and potentially zero-cost interconnection between certain types of carriers. For example, Section 251 requires all local exchange carriers "to establish reciprocal compensation arrangements for the transport and termination of telecommunications." 47 U.S.C. § 251(b)(5). End users and interexchange (long-distance) carriers must pay higher "access charges."

35. During a time when interexchange carrier competitors of AT&T received inferior access to the PSTN, the Commission authorized discounted access charges. However, the Commission never stated that the discounts were cost based as opposed to a rough-justice solution designed to reflect both inferior access and the Commission's desire that carriers like MCI acquire market share. See, for example, Exchange Network Facilities for Interstate Access (ENFIA), CC Docket No. 78-371, Report and Order, 71 FCC 2d 440 (1979); *on recon.*, 93 FCC 2d 739 (1983), *aff'd in part and remanded in part sub nom.*, MCI Telecomm. Corp. v. FCC, 712 F.2d 517 (D.C. Cir. 1983). Currently, the FCC is considering whether wireless mobile service providers like cellular radio operators should have to compensate wireline local exchange carriers for terminating calls while such wireline carriers do not have to compensate the wireless operators for similar call termina-

- The type of line or facility providing service[38] and whether the service can access the PSTN.[39]

---

tions. See Interconnection Between Local Exchange Carriers and Commercial Mobile Radio Service Providers, Notice of Proposed Rulemaking, CC Docket No. 95-185, 11 FCC Rcd. 5020 (1996) (proposing reciprocal termination between wireline and wireless carriers, including the possibility of an interim zero termination charge between carriers); First Report and Order and Implementation of the Local Competition Provisions in the Telecommunications Act of 1996, Interconnection Between Local Exchange Carriers and Commercial Mobile Radio Service Providers, CC Docket No. 96-98,11 FCC Rcd. 15499 (1996).

36. "Captive" long-distance callers from hotel rooms and callers not familiar with dial-around options for avoiding price gouging for pay phone service recognize the vast price differences for long-distance telephone service.

37. Certain types of services have qualified for exemption from regulatory burdens that impose extra costs. For example, enhanced services qualify for noncommon carrier status, and its users are exempt from having to pay an access charge payment otherwise applicable to basic service subscribers. A 1987 FCC initiative to eliminate the exemption generated substantial opposition by users who claimed the Commission had proposed to impose a "modem tax." "In 1983 we adopted a comprehensive 'access charge' plan for the recovery by local exchange carriers (LECs) of the costs associated with the origination and termination of interstate calls [citing MTS and WATS Market Structure, Memorandum Opinion and Order, 97 FCC 2d 682 (1983)]. At that time, we concluded that the immediate application of this plan to certain providers of interstate services might unduly burden their operations and cause disruptions in provision of service to the public. Therefore, we granted temporary exemptions from payment of access charges to certain classes of exchange access users, including enhanced service providers." Matter of Amendments of Part 69 of the Commission's Rules Relating to Enhanced Service Providers, CC Docket No. 87-215, Notice of Proposed Rulemaking, 2 FCC Rcd. 4305 (1987) (proposing to impose access charges on enhanced service lines), *terminated*, 3 FCC Rcd. 2631 (1988) (proposal abandoned on the ground that despite the apparent discrimination in charges "a period of change and uncertainty" besetting the enhanced services industry justified ongoing exemption from access charge payments). Currently, the FCC requires users of ISDN services to pay only one subscriber line charge, an access payment, despite the fact that ISDN circuits can derive more than one voice-grade equivalent channel.

38. The FCC's access charge regime established a different pricing structure for switched access and special access. The former includes regular dial-up services and requires end users to pay a monthly flat-rated subscriber line charge, currently \$3.50 for residential and small-business users and \$6.00 for other business users. The latter includes leased, private-line users who certify that the line does not "leak" into the PSTN through the use of, for example, an on-premises switch like a PBX that could couple the private line with trunks that access the PSTN provided by LECs ostensibly for local switched services. See MTS/WATS Market Structure (Phase I), 93 FCC2d 241 (1983), *modified on recon.*, 97

Internet telephony has the potential to migrate substantial traffic volumes from conventional telecommunications networks. Incumbent carriers surely do not want to encourage such a migration, because it will create downward pressure on all telephone toll rates and cannibalize retail rates. On the other hand, incumbent carriers probably will determine that they are financially better off providing the transmission capacity for Internet telephony, albeit at lower margins, than if they lose customers' traffic entirely. The massive increase in domestic and international broadband telecommunication capacity reflects the view that carriers can make up in volume what they lose in margin. While estimates of the financial impact of Internet telephony vary, one can extrapolate the potential impact by considering the traffic and revenue diversion impact of pure, or simple resale, that is, the use of leased international private lines to handle voice traffic. Table 9.4 tracks the various market strategies of Internet telephony entrepreneurs.

9.3 Satellites in the Global Information Infrastructure

The vastly greater bandwidth requirements of the new world information order will rely heavily on fiber optic networks. However, satellites will complement terrestrial networks with emphasis on providing Internet access to remote areas, satisfying consumers' insatiable desire for more video programming, and perhaps offering ubiquitous mobile telecommunications services. The broad geographic footprint of geostationary orbiting satellites can

FCC2d 682, *further modification on recon.*, 97 FCC2d 834, *partially aff'd and partially remanded sub nom.*, Nat'l Ass'n Regl. Util. Comm'rs v. FCC, 737 F.2d 1095 (1984), *cert. den.*, 105 S.Ct. 1224; *further modification*, 99 FCC2d 708 (1984), 100 FCC2d 1222, *further recon. den.*, 102 FCC 2d 899 (1985). See also Investigation of Access and Divestiture Related Tariffs, 101 FCC2d 911 (1985) *recon. denied*, 102 FCC2d 503 (1985) Investigation of Access and Divestiture Related Tariffs, 101 FCC2d 935 (1985).

39. International private-line services, which do not access the PSTN, are exempt from the accounting rate regime. Their per-minute costs are significantly lower than switched services. Undetected private-line leakage has become commonplace, making it possible for resellers to provide a service functionally equivalent to international message telephone service at a fraction of the cost. See Rob Frieden, "The Impact of Boomerang Boxes and Callback Services on the Accounting Rate Regime," in D. Wedemeyer and R. Nickelson, eds., *Proc. Pacific Telecommunications Council Eighteenth Annual Conference* (Honolulu: Pacific Telecommunications Council, 1996), pp. 781–790.

Table 9.4
IP Telephony Market Strategies

Strategy	Time Frame	Users	Players
Tariff arbitrage	1995–2003	Hobbyists, cost conscious	Delta Three, IDT, VolcalTec
Simplified entry	1997–????	Cost conscious, users unaware	ICG Netcom, RMI VIP Calling
Converged networking	1999–????	Users unaware, cost conscious	AT&T, Sprint, MCI WorldCom, Qwest, Level 3
Niche marketing	1998–2003	Technologically savvy	AT&T, Sprint
Value-added apps	2000–????	Mainstream users, technologically savvy	AT&T, MCI WorldCom

Source: The Strategis Group.

partially abate the dichotomy of access to information infrastructures between rural and urban locales.

The term *digital divide* refers to the potential that new information technologies will exacerbate rather than narrow differences in accessibility and affordability between the elite and the poor, between elderly and rural residents. Satellites might provide a cost-effective way to bridge the gap, particularly given the low incremental cost to serve an additional point of communications under a broad footprint. Efforts to make satellites an integral part of the new world information order include a revamping of the industrial structure from one dominated by cooperatives formed by intergovernmental agreement to robust competition among private ventures. As well, entrepreneurs will have to risk billions of dollars to test the market viability of satellite roles in the Internet and mobile telecommunications services.

Visions of a GII have become front-page news, the fancy of elected officials and just about every general-interest magazine. Broadband digital telecommunications and information processing have become a reality. However, most of the attention has centered on terrestrial, wireline delivery. Fiber optic cables do provide the most efficient and cost-effective method for transmitting large volumes of digital bitstreams on point-to-point routes. However, that preferred wireline distribution method requires relatively high traffic density, and a business case cannot yet be made for extending fiber optic cable ubiquitously. Likewise, some GII applications involve multicasting, point-to-multipoint services for which satellites historically have accrued a comparative advantage over terrestrial links.

Satellites will be a key ingredient in the development of a broadband telecommunications and information-processing infrastructure.[40] Satellites extend the reach of terrestrial networks, thereby complementing wireline technologies by filling gaps. But to secure market share, equipment manufacturers and operators must redefine the industry by marketing satellites as more than passive conduits or "bent pipes." Marketplace success for satellites in the current and future commercial environment will require efforts to shape the technology to provide intelligent, value-added services in addition to basic transmission capacity. Many ventures already have begun to do so, despite numerous marketplace, political, regulatory, and technological challenges. This section briefly examines the developing trends that generate enhancements and handicaps to the satellite industry's access to GII markets.

9.3.1 What Is the Global Information Infrastructure?

The GII is a vision of a future telecommunications and information-processing environment that has greater robustness, accessibility, intelligent services, and functionality than what is currently available. One major application of the GII is the Internet[41] and in particular the World Wide Web, the interactive, multimedia, and user-friendly "network of networks."[42] Ideally, access to both basic and enhanced telecommunications should be ubiquitous and affordable, but even as technological advances promise a broadband information superhighway, narrowband plain old telephone service (POTS)[43] is beyond the reach of a majority of the world's population.

---

40. For background on the role of satellite in the distribution of video programming, see Patrick R. Parsons and Robert M. Frieden, *The Cable and Satellite Television Industries* (Boston: Allyn and Bacon, 1998).

41. "The Internet today is a worldwide entity whose nature cannot be easily or simply defined. From a technical definition, the Internet is the 'set of all interconnected IP networks'—the collection of several thousand local, regional, and global computer networks interconnected in real time via the TCP/IP Internetworking Protocol suite." Religious Technology Center v. Netcom On-Line Communication Services, Inc., 907 F. Supp. 1361 (N.D. Cal. 1995), quoting Daniel P. Dern, *The Internet Guide for New Users,* (New York: McGraw-Hill, 1994), p. 16.

42. See Eli M. Noam, "Beyond Liberalization—From the Network of Networks to the System of Systems," *Telecommunications Policy,* Vol. 18, No. 4, 1994, pp. 286–294.

43. Even in the United States, a multibillion-dollar universal service initiative has taken years to achieve a comparatively high telephone line density but one far from achieving ubiquitous access.

Accordingly, policy makers must balance the GII vision of high speed, advanced Internet applications with long-standing basic telecommunications infrastructure needs that have yet to be completely fulfilled.[44]

The GII is a global adaptation of infrastructure development initiatives in several countries.[45] In the United States, the National Information Infrastructure (NII) initiative seeks to promote ubiquitous access to a robust telecommunication infrastructure.[46] An upgraded infrastructure should be able to support narrowband services like electronic mail plus broadband, digital services like access to the Internet, large file transfer, remote access to data bases, video teleconferencing, and telemedicine (on-line access to patient

44. Until enactment of the Telecommunications Act of 1996, P.L. 104-104, 110 Stat. 56, signed into law Feb. 8, 1996, *codified at* 47 U.S.C. § 151 *et seq.* (1996), the FCC and state public utility commissions pursued infrastructure development primarily through subsidization of access by rural and low-income consumers. Section 254 of the Communications Act, as amended, requires the formation of a federal-state joint board on universal service to recommend changes to any universal service policy. 47 U.S.C. 254 (1996). That section sets out several guiding principles: (1) access to quality services at just, reasonable, and affordable rates; (2) access to advanced services throughout the nation now defined to include low-income consumers and those in rural, insular, and high-cost areas as well as advanced telecommunications service access for schools, health care providers, and libraries; (3) equitable and nondiscriminatory contributions by all providers of interstate telecommunications services to universal service funding; and (4) specific and predictable support mechanisms. See 47 U.S.C. Sec. 254(b); see also Federal-State Joint Board on Universal Service, Notice of Proposed Rulemaking and Order Establishing Joint Board, FCC 96-93 (rel. March 8, 1996); reprinted in 61 Fed. Reg. 10499 (March 14, 1996).

45. See, for example, Information Policy Advisory Council, *A National Policy Framework for Structural Adjustment Within the new Commonwealth of Information, A Report to the Minister for Communications and the Arts* (Commonwealth of Australia, 1997); Dora Mozes, *Measuring the Global Information Infrastructure for a Global Information Society* (Industry Canada, Sept. 1996); Survey Telecommunications, *The Economist*, Sept. 13, 1997.

46. See Ronald H. Brown, et al., *The Global Information Infrastructure: Agenda for Cooperation* (U.S. Information Infrastructure Task Force, 1996); *U.S. Information Infrastructure Task Force's Committee on Applications and Technology: Putting the Information Infrastructure to Work*, Special Publication 857 (Gaithersburg, MD: National Institute of Standards and Technology, 1994); and *The Information Infrastructure: Reaching Society's Goals*, Special Publication 868 (Gaithersburg, MD: National Institute of Standards and Technology, 1994); see also the *National Information Infrastructure: Agenda for Action* available on-line at http://sunsite.unc.edu/nii/NII-Executive-Summary.html; access to most Information Infrastructure Task Force publications is available at: http://www.iitf.nist.gov/iitf-pub.html.

records, remote diagnosis, and transmission of x-rays and other diagnostic tools).

The U.S. initiative seeks to promote private-sector investment through appropriate tax and regulatory policies. Through passage of the Telecommunications Act of 1996 [2], Congress has concretely articulated what constitutes universal service and what objectives the United States should pursue to achieve widespread and equitable access to advanced telecommunications services.[47] The NII seeks to promote technological innovation that can achieve a seamless, interactive, and fully integrated network.

9.3.2 Satellite Roles in Video Programming Delivery and the Internet

Satellites promote the possibility of television without borders and frontiers, in other words, widespread access to diverse video programming throughout the world. Most of the transmission capacity of satellites in geostationary orbit deliver television programs with fixed satellites providing distribution to broadcast and cable television facilities and broadcast satellites providing direct distribution to consumers. The point-to-multipoint feature of satellites supports widespread delivery of content at low cost, the expense of installing an additional receive-only Earth station. Satellites accrue positive networking externalities when additional points of communication tap into the content delivery function.

Satellites can accrue similar positive networking externalities for Internet-mediated content, particularly for distribution of content to remote locales that lack access to nearby fiber optic cables. Additionally, satellites can provide ways to balance traffic loads, handle peak demand, and conserve bandwidth and switching resources. For example, satellites can deliver frequently viewed World Wide Web pages to a storage facility closer to

47. Section 254 of the Communications Act expands the concept of universal service to include insular areas, such as Pacific Island territories, low-income consumers, health care providers for rural areas, elementary and secondary school classrooms, and libraries. Rates for rural health care services shall be "reasonably comparable" to charges for similar service in urban locales, and service provided to meet an educational purpose must be discounted with the difference offsetting the carrier's universal service payments or qualifying it for reimbursement from the universal service fund. The section also requires the formation of a new federal-state joint board to review existing universal service support mechanisms, including financial subsidies, with an eye toward recommending new procedures. All telecommunications carriers providing interstate telecommunications must contribute, on an equitable and nondiscriminatory basis, to a universal service funding mechanism.

consumers, thereby reducing the long-haul traffic loads. That caching function may grow increasingly important as demand rises for full-motion video and other broadband applications.

9.3.3 Satellite Trends

In the near term, satellites will support development of a broadband telecommunications and information-processing infrastructure by extending the reach of terrestrial networks. Satellites complement wireline technologies like fiber optic cables, considered by most carriers and information service providers as the more basic and essential element in the development of broadband services. Satellites also can perform a gap-filler function by providing access to broadband telecommunications capacity in areas where fiber optic cables do not exist. In other remote areas, satellites may constitute the only available telecommunications resource, because even the twisted copper wire pair associated with conventional narrowband telephone service has not reached into many rural and remote locales throughout the world.

The satellite marketplace has matured and diversified in its 25 years of commercial service. The market continues to evolve from one dominated by government-owned carriers and cooperatives with an exclusive or oligopolistic share of national, regional, and international markets to one where private enterprise and competition dominate. Technological, business, and regulatory trends favor a more segmented, versatile, and competitive environment. The cutting edge trends in satellite development emphasize the commercial potential of a technology initially deployed for military, space exploration, and intelligence-gathering applications.

The old world satellite order favored government ownership and extensive regulation. Its advocates considered the market able to support only a few large operators, typically organized as cooperatives through agreements executed by governments. The new world satellite order combines deregulation, privatization, and entrepreneurialism to support competition and the view that private operators can operate efficiently and profitably.

Commercial ventures have begun to dominate the satellite marketplace. Governments have privatized the domestic or regional cooperative, and the INTELSAT global cooperative reshaped itself as a private commercial venture to compete effectively with increasing numbers of private market entrants like PanAmSat, AsiaSat, and GE Americom. Those commercial ventures seek to maximize revenues rather than average rates to promote widespread access to affordable satellite technology. They are quick to respond to individual user requirements with customized applications rather than a simple one-size-fits-all inventory.

Private satellite ventures have developed in many diverse markets:

- Direct-to-home television broadcasting;
- Transoceanic and transcontinental voice telecommunications and video program delivery;
- Enterprise networking, which integrates both voice and data requirements with access achieved via on site very small aperture terminal (VSAT) Earth station terminals;
- Mobile telecommunications accessible anytime, anywhere.

Those market niches reflect the increasing versatility of satellites. As the overall market matures and as more operators compete, individual carriers have begun to specialize.

9.3.4 A Changing Business Environment

The proliferation of satellite types and ownership patterns has created a changing and more competitive environment. Instead of a few operators with cradle-to-grave possession of satellites, a diverse and expanding set of operators will acquire title for a portion of a satellite's usable life. That creates the potential for the evolution of market segments to run the gamut from premium, noninterruptible service via state-of-the-art satellites in the best orbital slots to discounted service via aging satellites perhaps also under new ownership.

Users already have the option of securing lower cost service that lacks backup capacity and that may be preempted for use by customers taking a higher grade of service from the same carrier. In the future, different carriers will target different sectors of the market on the basis of such variables as price, backup, availability of in-orbit service restoration, reliability, age of the satellite, and orbital location.

The proliferation of carriers and market segments will create pressure on incumbent operators to respond to change. The international satellite marketplace has lacked significant competition in view of pervasive government involvement, limitations in the number of satellites that can operate in desired positions of the orbital arc, and concentrated market share held by cooperatives created by intergovernmental agreement. Early on, governments helped sustain the industry by executing treaty-level documents to establish a cooperative model for the deployment of international satellite capacity through INTELSAT and its maritime counterpart, Inmarsat. When satellite technology was making its initial crossover to commercial applications, the cooperative model helped spread risks, achieved scale economies and ensured that lesser developed nations could access cutting edge technology with limited investment and without financial handicaps in view of their low demand for service.

Now that the commercial satellite industry has matured, the managers at INTELSAT and Inmarsat had to restructure first replacing the cooperative status through corporatization and a private company spin-off, and later through outright privatization. Even as their mission remains the provision of ubiquitous access to satellite service on a global average-cost basis, they must increasingly respond to competition and marketplace imperatives. The quasi-diplomatic status of the INTELSAT and Inmarsat cooperatives accorded them privileges and immunities that have translated into financial and competitive advantages. That status also allowed the cooperatives to secure commitments from participating nations to avoid causing "significant" technical or economic harm when those nations authorized competing satellite systems.

Countries consider it increasingly possible to satisfy their commitment to promoting global satellite access while still authorizing some degree of competition. That can be seen in the following ventures: PanAmSat, Lockheed Martin, Loral, and GE Americom in the United States; AsiaSat and Asia Pacific Telecommunications in Hong Kong; an increasing variety of regional satellites like Indosat, Measat, Palapa, and Thaisat serving the ASEAN nations; Astra, British Sky Broadcasting, Eutelsat, Hispasat, and Telecom serving Europe and beyond; and systems in such diverse nations as Canada, Brazil, Mexico, Argentina, Saudi Arabia, Korea, Japan, Australia, Russia, Turkey, Israel, Iran, India, China, and Taiwan. The managers of INTELSAT and Inmarsat, recognizing the need to adapt to changed circumstances, agreed to relinquish treaty-level privileges and immunities in exchange for greater latitude to operate as commercial enterprises that can respond to competitive necessity with selective rate reductions.

9.3.5 Satellite Service Without Frontiers

Most geostationary orbiting satellite footprints traverse national boundaries, a technological feature that creates financial opportunities to aggregate traffic but also difficult political, intellectual property, and cultural challenges. Proliferating satellites may exacerbate concerns about so-called cultural imperialism, because more video program options may lead to audience migration from national programming to foreign programming. Despite the potential for satellites to operate without frontiers, national governments may attempt to impose border limitations by denying "landing rights" for particular satellites and by restricting the amount of foreign programming available to national cable television, broadcast, or direct broadcast satellite (DBS) operators.

Satellites are both a blessing and a curse in the broadband telecommunications and information-processing environment. On one hand they provide global access to news, information, entertainment, education, "edutainment," and "infotainment." The GII incorporates satellites to provide access in the vast regions of the world that do not qualify for broadband wireline deployment and that, in most instances, have not even secured narrowband access to POTS. On the other hand, many national governments are wary when they recognize that satellites make boundaries porous and the citizenry more vulnerable to outside influences.

The transborder nature of a satellite footprint all but eliminates mutually exclusive domestic and international markets. National governments have found it impossible to prohibit or regulate the use of satellite terminals and the extent of citizen access to the rest of the world. Miniaturization of satellite terminals and the increasingly integrated global economy make it both technologically unfeasible and commercially imprudent even to attempt to restrict foreign investment and involvement in telecommunications ventures.

9.3.6 Difficulties in Fostering and Sustaining a Level, Competitive Playing Field

Currently, the international telecommunications satellite marketplace is experiencing great stress and change. On one hand, the intergovernmental cooperative model largely has been dismantled with the privatizations of INTELSAT and Inmarsat. On the other hand, regulatory and marketplace uncertainty exists as to the viability of a robustly competitive satellite marketplace and what market niches satellites can fill. National governments and consumers have rejected the view that the marketplace cannot sustain viable competition while ensuring ubiquitous access to satellite service. Whether through internal desire to have greater flexibility to compete or through external prodding by governments, INTELSAT and Inmarsat no longer enjoy the quasi-diplomatic status that insulated the cooperatives from tax liability and the laws of host nations. Likewise, the organizations have privatized and lost the power to veto competition by claiming it would cause significant technical or economic harm. On the other hand, the large startup and operating costs for satellite service limit the number of competitors, and lately the number of operators has dropped through mergers and acquisitions.

The U.S. Congress recently enacted a law that would structure the transition to a competitive international satellite marketplace.[48] The transition requires the dismantling of an industrial structure in which Comsat operated as an intermediary between carriers and consumers down the service chain and INTELSAT and Inmarsat up the chain. A competitive marketplace does not require Comsat to serve as a carrier's carrier, with legal authority to process all satellite transponder orders and to foreclose direct access to INTELSAT and Inmarsat. Likewise, a competitive marketplace requires that former cooperatives do not leverage their previous quasi-diplomatic status and preferred access to orbital slots for an unfair competitive advantage.

Congress had to craft legislation that fairly balances the interests of all parties:

- The desire by the newly privatized INTELSAT and Inmarsat to compete without unnecessary limitations;
- The need to ensure that Comsat can survive in changed circumstances, including the absence of captive markets and easy markups;
- The legitimate claims of private ventures like PanAmSat that both Comsat and its former cooperative partners compete on a level, competitive playing field.

9.4 Service to a Mobile, Wireless, Networked Society

Satellites will perform an increasingly significant role in building a terrestrial wireless infrastructure to work in conjunction with wireline options. The

48. See Open-Market Reorganization for the Betterment of International Telecommunications Act, Public Law No. 106-180 (enacted Mar. 17, 2000). The law prohibits the FCC from issuing a license or construction permit to a privatized INTELSAT or Inmarsat unless the Commission determines that the nature of the privatization and granting a license would not harm competition in the U.S. telecommunications market. The law also sets a timetable and terms for privatization of INTELSAT and Inmarsat, codifies the right of direct access to their transmission capacity by all carriers and users, and prohibits the FCC from assigning by competitive bidding orbital locations or spectrum used for the provision of international or global satellite telecommunications services.

satellite component will fill the extensive gaps where terrestrial networks operate poorly or do not exist but where people require access to the rest of the world in the form of another person, corporate network, company database, or transaction system.

The unprecedented marketplace success of cellular radio and other mobile technologies confirm the demand for reliable, tetherless access to the rest of the world. Mobile telecommunications networks can enhance business productivity and efficiency. However, even when nationwide cellular roaming becomes possible, a variety of different operating standards will limit the prospect for a person to use the same transceiver when traveling abroad; in many nations, mobile radio service has only begun to extend widely beyond urban locales. LEO and MEO satellite projects have the potential to make personal communications global in scope. Those ventures include a constellation of nongeostationary orbiting satellites that provide an interoperating array of beams illuminating the entire globe. Individually and collectively, those systems can provide ubiquitous, wireless, digital coverage to pocket-sized telephones, albeit at costs well in excess of terrestrial wireline and even wireless rates. The cost of the systems, coupled with the ongoing buildout of terrestrial wireless options, currently challenge the commercial viability of most existing satellite ventures that offer only narrowband services. Broadband satellites should fare better in view of the more extensive options available and more robust demand.

Telecommunications planners have coined the terms *universal personal telecommunications* (UPT) and *global personal communication services* (global PCS) to identify communication options free of cords using available terrestrial radio options augmented by satellites. Numerous logistical and regulatory problems must be resolved to make that vision a reality, but the demand exists at least for terrestrial services. The United States, the United Kingdom, Germany, and other national governments have raised billions of dollars by auctioning off spectrum for use by cellular PCS networks.

The considerable optimism for terrestrial and satellite-delivered telecommunications stems in large part from the unprecedented increase in cellular radio service demand. In the 10 years after the mid-1980s, usage rose from near 0 to almost 30 million subscribers in the United States. Currently, there are approximately 100 million subscribers in the United States.[49] Yet

49. See Cellular Telecommunications Industry Association, Statistics and Surveys, Frequently Asked Questions & Fast Facts, available on-line at http://www.wow-com.com/consumer/faqs/faq_general.cfm.

even with such a steep and profitable rate of usage, cellular radio has achieved a market penetration of less than 50%. Mobile service entrepreneurs expect terrestrial options like PCS to achieve mass-market penetration. New digital mobile radio services like PCS have ample capacity, and providers typically offer free or subsidized handsets. Per-minute usage rates have dropped significantly below luxury levels and new rate plans offer baskets of large minutes of use available for local, roaming, and long-distance calling.

Mobile satellite service operators can perhaps ride the coattails of terrestrial mobile service market success by providing service to dual-mode transceivers, which cut over to the satellite option when terrestrial service becomes unavailable. If terrestrial systems can achieve profitability with a market penetration of less than 50%, it follows that global or regional satellite systems need only to acquire a small portion of the total wireless market to achieve success as well. Despite the relatively small number of subscribers needed, mobile satellite ventures nevertheless present substantial risk because of their cost (approximately $9 billion for Teledesic and $5 billion for Iridium) and high per-minute rates. LEO systems require extensive management information systems and network coordination to link as many as 228 fast-moving satellites. The Iridium satellite constellation communicated not only with ground stations but between satellites. Consumers have balked at a $3 per minute charge, but conditions already exist where access to the rest of the world comes at a price of $10 a minute or more (from hotel rooms, the high seas, and business communication centers in countries with unreliable conventional networks). However, the currently available LEO systems provide only voice and slow-speed data transmission without the capacity to meet broadband applications.

Several new satellite proposals provide broadband satellite options and perhaps will generate greater consumer acceptance. Hughes has proposed the Spaceway system, a constellation of geostationary orbiting satellites that will operate in the extremely high frequencies of 20 to 30 GHz. By becoming a first-time operator in that frequency band, Hughes will have ample spectrum available so it can provide more transponder capacity than currently available at lower frequencies. Teledesic proposes the commercial rollout of "Star Wars" technology, with 228 refrigerator-sized LEO satellites. Those satellites will operate as global web capable of providing the same kind of broadband functionality currently available only from terrestrial options.

9.4.1 The Global Phone Vision and Challenge

MSS ventures have achieved the promise of delivering anytime, anywhere telecommunications services. They can fill the geographic coverage gaps left by terrestrial wireline and wireless carriers and accordingly can help achieve perennially unmet universal service objectives. However, the future of existing MSS ventures to achieve marketplace success provides sobering insights into new technology risks in view of multibillion-dollar costs, use of various new orbital configurations, and the large number of satellites in the constellation needed for global coverage. But equally important are the Earth-bound regulatory and policy challenges that the developers of those systems encounter.

To deliver the much touted promise of the universal telephone, MSS carriers will need to accomplish the following tasks:

- Find ways to integrate a global or regional service into the mix of existing network services in such a manner as to provide seamless connectivity that augments rather than siphons off traffic and revenues;
- Resolve regulatory obstacles arising largely from the need to secure a global set of authorizations and licenses through many different types of administrative mechanisms, all of which serve primarily national or regional interests;
- Reassure incumbent telecommunications operators and their government regulators that MSS technologies will not jeopardize national sovereignty or security or evade national regulations and policies.

9.4.1.1 MSS Basics

Technological innovations in mobile communications have evolved to the point where we can envision ubiquitous communications network access[50] via handheld terminals:

> Wireless technology is evolving toward higher frequencies (inherently more information dense), with a range of clever compression algorithms to squeeze many conversations into a given frequency...[along with new satellite orbital options that expand the channel multiplication options]. This development, coupled with cheap electronics, permits mobility for

50. Already, satellites illuminate over 99% of the Earth's surface, and Inmarsat provides global access to land mobile services via terminals as light as 2 kg.

the user—in some situations, wireless has become an alternative to conventional wireline technology for basic services. [11]

Satellites constitute the driving technology for such service, because once launched they provide a readily available infrastructure that serves remote villagers and road warriors alike with footprints that illuminate an entire region of the world. Given such extensive geographic coverage, new MSS ventures need only secure single-digit market penetration to succeed. Regulatory mandates or enlightened self-interest (and the ability to carry additional off-peak traffic at insignificant incremental cost) presents the possibility that MSS networks also can contribute to achieving universal service objectives. Financial and regulatory issues continue to present major obstacles to the universal telephone vision.

9.4.1.2 Four Types of Satellite Ventures

The traditional model for telecommunications via satellite has involved space stations in geostationary orbit (GSO) 22,300 miles above Earth transmitting very large footprints that can illuminate more than one-third of the Earth's surface. Global cooperatives like INTELSAT and Inmarsat helped pool investment and share both financial and technological risks. Additionally, they enfranchise developing nations by offering small equity stakes and incorporating a universal service mission that requires cost averaging among dense and sparse routes.

Three new satellite system models have evolved:

- "Little" LEO satellites—small constellations of very small and relatively inexpensive satellites operating on VHF and UHF and providing nonvoice, data, and position-location services. Ventures in this model include Orbcomm and VitaSat.
- "Big" LEO, MEO, and ICO satellites—a constellation of more numerous, complex, and expensive satellites operating in the L-band (1.5–2.5 GHz) and providing voice, data, and position-location services to fixed and mobile users. Ventures in this model include Iridium,[51] Globalstar,[52] and ICO, Ltd.[53]

---

51. Iridium will operate a $5-billion global constellation of 66 LEO satellites about 400 miles above Earth that are able to provide voice, data facsimile, and position-determination services to handheld transceivers. Intersatellite links make it possible for the Iridium network to route calls between a number of satellites onward to a gateway

Earth station located near the call recipient. The other satellite systems lack that function and must rely on expensive switching and routing on Earth that may involve Earth stations far from the intended call recipient.

Iridium will provide spectrum conserving bidirectional transmissions (to and from the satellite constellation) on both a frequency and time division basis over the entire lower L-band (1610–1626.5 MHz). However, the FCC has announced its intention to require band segmentation to accommodate multiple applicants, thereby reducing the bandwidth available to Iridium in the United States. Likewise, coordination with the Russian Glonass navigation system, other LEO systems, and terrestrial operators may further reduce the bandwidth available.

Motorola created the Iridium concept and has spent almost five years and several hundred million dollars developing the technology and soliciting investors to form a global consortium. In 1993, an initial private placement of $700 million created a consortium comprising a geographically diverse set of investors representing different aspects of space, telecommunications, and venture capital industries: Khrunichev Enterprise, builder of the Russian Proton launch vehicle; China's Great Wall Industry Corp. operator of the Long March launch vehicle; Iridium Nippon, a Japanese investment group led by Daini Denden, a major cellular radio operator; Kyocera Corp., a diversified manufacturer; two venture capital groups, Mawarid Group of Saudi Arabia and Muidiri Investments BVI, Ltd. of Venezuela; incumbent carriers BCE, Inc. of Canada, Sprint of the United States, and STET of Italy; manufacturers of Iridium network equipment, Lockheed and Raytheon; and telecommunications enterprises in developing nations, including United Communications Industry of Thailand. In 1997, the venture raised $223 million through initial public offering of Iridium World Communications and sold south Pacific gateway rights for $100 million. Total assets and investment came to $2.9 billion.

52. Globalstar plans on operating a LEO constellation of 48 satellites in eight orbital planes. With fewer satellites operating in orbits about 800 miles above Earth, the $2.5-billion Globalstar network is projected to provide service at a cost of less than $0.50 a minute plus a monthly service charge of $60 to $70. However, less in-orbit resources means that the network will rely heavily on the widespread availability of gateway Earth stations to route calls. Globalstar initially was organized by Loral/Space Systems, a satellite manufacturer, and Qualcomm, Inc., an innovator in code division multiple access mobile radio technology and provider of vehicle location services via GSO satellites. In 1994, the venture announced that it had secured the infusion of several hundred million dollars from new investors, including Alcatel, a major French aerospace and telecommunications equipment manufacturer; France Telecom, the country's monopoly telecommunications carrier; Alenia, an Italian aerospace manufacturer; Deutsche Aerospace and DASA, aerospace manufacturers in Germany; Hundai, a major diversified Korean manufacturer; Dacom, Korea's second telecommunications carrier; AirTouch, the cellular radio spinoff of Pacific Telesis; and Vodafone, a British cellular radio operator.

53. ICO Global Communications was spun off from the Inmarsat global cooperative in January 1995. It briefly operated as a private company with plans to provide personal mobile global communications services. However, the venture filed for bankruptcy protection as a result of a general lack of investor and lender confidence in mobile satellite services. Cellular radio entrepreneur Craig McCaw has invested in a reorganized venture.

- Broadband overlay satellites, a constellation of LEO or GSO satellites operating in the Ka-band (20–30 GHz) to create a seamless web capable of providing wideband capacity primarily to fixed users. Ventures in this model include Spaceway/Expressway,[54] Teledesic,[55] Cyberstar,[56] Astrolink,[57] Celestri,[58] and Skybridge.[59]

---

54. Hughes Communications, Inc. submitted the Spaceway broadband Ka-band application to the FCC in December 1993. The company anticipated the first launch to culminate in an 11-satellite constellation. Initially, the Spaceway network will provide coverage to approximately 90% of the world's population segmented into four overlapping regions, each with two geostationary satellites. In 1997, Hughes proposed the $3.85-billion Expressway satellite network, the first commercial proposal to use the 40–50 GHz frequency band. The Expressway proposal calls for 10 orbital slots, with two satellites each. Hughes plans to launch the first Expressway satellite 50 months after FCC approval.

55. Rather than provide ubiquitous narrowband (less than 4,800 bps) capacity to mobile users, Teledesic will offer a global overlay of wideband functionality using Ka-band frequencies (20–30 GHz). The system will offer throughput rates in excess of 2 Mbps initially from 288 refrigerator-sized satellites, at a total cost of $9 billion. Like its visionary backers, who include Bill Gates and Craig McCaw, Teledesic pushes the envelop with an eye toward providing a ubiquitous, broadband GII. The system will commercialize a technology developed as part of the Strategic Defense Initiative that deploys observer (Brilliant Eyes) and interrupter (Brilliant Pebbles) satellites into a seamless, global array. In 1997, Boeing agreed to invest up to $100 million in the venture for a 10% share.

56. Loral Space and Communications Ltd. has proposed a comparatively modest three-satellite Ka-band system with an eye toward being the first to commence service in 1999. The company teamed up with Alcatel in 1997 to propose a $3.9-billion video and data venture called Sky Bridge, comprising 64 LEO satellites.

57. Lockheed Martin Telecommunications has proposed a network of nine satellites operating from five orbital locations and providing global coverage. The venture initially had plans to launch its first satellite in the first quarter of 2000, but has experienced delays in coordinating its spectrum use with terrestrial users and in securing all needed regulatory authorizations.

58. Motorola has proposed a $13-billion, Ka-band satellite network comprising both geostationary and LEO satellites. Unveiled in June 1997, the Celestri system will provide high-speed data and video transmissions to 99% of the globe beginning in 2002. The proposed system folds in parts of two previously proposed Motorola satellite projects, Millennium and M-Star. The initial backbone of the system will be one GSO and 63 LEO satellites. The LEO spacecraft will provide high-speed interactive communications around the globe, and the GEO component will broadcast to users in a send-only mode. The total Celestri system will have the capacity to transmit 80 Gbps.

Because an LEO satellite operates in an orbit much closer to Earth, it can communicate with handheld transceivers instead of the somewhat larger and heavier terminals needed to communicate with GSO space stations. The closer proximity to Earth, however, reduces the size of the satellite footprint and also eliminates synchronization with Earth, which means that the satellite speeds above a particular location on Earth from one horizon to the other in a matter of minutes.

LEO satellite networks require a constellation of space stations to provide global coverage. Because the satellites move quickly over any single point on Earth, the operator must find a way to hand off traffic from one satellite to the other or build such intelligence into the way Earth stations track and manage the flow of traffic. In either scenario, LEO satellites are moving targets, thereby requiring greater network management functionality than is needed for GSO satellite systems.

9.4.1.3 Regulatory Obstacles to Global Telephony

Notwithstanding the success achieved by global cooperatives like INTELSAT and Inmarsat, other MSS operators face daunting regulatory challenges to securing all the authorizations and licenses needed to provide global connectivity. Until they secure such permissions, a dichotomy exists between the promise of MSS footprint coverage and performance (i.e., the actual scope of permissible service).

Spectrum Allocations

Despite successfully securing a global consensus on spectrum allocations for many types of MSS,[60] operators must ensure that individual nations follow through with domestic allocations. Any nation's deviation from the global consensus presents the potential for incompatible frequency allocations,

59. Alcatel Alsthom of France and Loral Space & Communications Ltd have formed a strategic partnership to jointly develop, deploy, and operate a high-speed global multimedia satellite network through the coordination of Alcatel's LEO satellite-based SkyBridge project and Loral's GSO satellite-based CyberStar project. Services were planned to be introduced in the market through leased transponders in early 1998 and subsequently through dedicated geostationary satellites in 1999 and a constellation of LEO satellites in 2001.

60. See, for example, Robert M. Frieden, "WARC-92 and Low Earth Orbiting Satellites: A Case Study of the Process for Accommodating Spectrum Requirements of New Technologies," in J. Savage and D. Wedemeyer, eds., *Proceedings of the Fifteenth Annual Conference of the Pacific Telecommunications Council,* Vol. 1 (Honolulu: Pacific Telecommunications Council, 1993), pp. 271–287.

interference between service providers, and the need to segment the available spectrum into even smaller slivers allocated to particular services. The decision to operate in the congested L-band generates uncertainty over how much spectrum any venture can clear, because of the need to coordinate with existing or planned geostationary communications satellite systems, other satellite operators authorized to use the same spectrum, and other terrestrial users, including local multipoint video distribution services.

Licenses and Operational Authority

MSS operators must acquire national authorizations to operate in-country and for as many as four different spectrum use licenses for: (1) space segment, to transmit traffic to and from satellites as well as spectrum for intersatellite links; (2) network control operations to track satellites and send telemetry and perform network control functions; (3) gateway operations to transmit and receive traffic using feeder links; and (4) handset licensing and type approval authorizing the use of transceivers to access satellites.

Transborder Roaming

Subscribers to MSS will expect to use their transceivers across international borders. However, transborder cellular roaming has proved difficult and costly, with some nations balking at the prospect of allowing the use of transceivers licensed in another country. The ITU has organized conferences to consider ways for nations to recognize the licensing or authorization granted by other nations.[61] Absent such a mechanism for multilateral coordination, users may not be able to use their MSS transceivers in some countries, or they may experience inconvenience, customs, immigration problems, and delays in border crossings, which can reduce traffic volumes and the perceived value of MSS.

9.4.2 Integrating MSS into the Existing Mix of Delivery Options

MSS ventures must take every effort to integrate their networks into whatever mix of terrestrial wireline and wireless services exist. MSS service cannot

---

61. See International Telecommunication Union, *Informal Group—Memorandum of Understanding—GMPCS, Chairman's Report* (Feb. 18, 1997), available on-line at http://www.itu.int. pforum/gmpcs-mou/report-e.htm; Memorandum of Understanding to Facilitate Arrangements for Global Mobile Personal Communications by Satellite, Including Regional Systems (Feb. 18, 1997), available on-line at http://www.itu.int.pforum/gmpcs-mou/mou-e.htm.

simply overlay the globe as new standalone systems in a manner that cellular radio services have done in localities lacking reliable wireline facilities. MSS requires the use of gateway Earth stations and access to and from the PSTN so wireline users can originate calls to MSS subscribers and so MSS subscribers can terminate calls to wireline network subscribers. Likewise, MSS needs to be integrated with incumbent systems like international direct distance dialing and the global telephone numbering plan that make it possible for almost any telephone—wireline or wireless—to access any other telephone through a sequence of dialed digits.

9.4.2.1 MSS Operators and Accounting Rate Settlements

In view of regulator concern that MSS ventures not divert traffic or revenues from existing services and carriers, MSS operators may not be able to avoid having to divide toll revenues with incumbent wireline carriers under the current accounting rate regime. International carriers typically negotiate operating agreements with other carriers for the interconnection of lines and the onward delivery of traffic to intended call recipients. Such an arrangement provides for a correspondent relationship among carriers that includes the division of an accounting rate that ostensibly identifies the approximate per-minute costs the carriers would incur to link two international half-circuits[62] and for the two domestic "tail" circuits running to and from international gateway facilities. For routings involving two carriers,[63] the correspondents typically would agree to a 50-50 settlement rate, whereby the carriers would divide equally the accounting rate based on the view that they equally participated in routing the call.[64]

62. The half-circuit concept operates on the presumption that carrier correspondents achieve a whole circuit by linking two half-circuits at the theoretical midpoint of a submarine cable or at the satellite providing the transmission link. In the submarine cable scenario, each carrier has the responsibility to secure access to circuits linking transmission facilities on its territory to the location where the cable makes its landfall (referred to as the cablehead), possibly located in a different nation, and onward to the midpoint. For more background on international telecommunications operations and policy, see Rob Frieden, *International Telecommunications Handbook* (Norwood, MA: Artech, 1996).

63. One or more additional carriers participate in calls requiring transiting, the use of an intermediary carrier. For example, a carrier in the United Kingdom might use the transit facilities of a U.S. carrier to route calls destined for Hong Kong. For conventional IMTS traffic, a single transiting carrier would receive up to one-third of the applicable accounting rate.

64. The accounting rate system operates largely independent of the manner in which carriers invest in international satellites and submarine cable capacity. Carriers typically pool in-

It remains to be seen whether and how more than 150 national governments will exempt MSS operators from having to settle accounts according to the current system. If no exemption is received, accounting rates can add significant costs in addition to the satellite space segment even for traffic that traverses only a few hundred miles from the gateway Earth station to a recipient located across one or more international borders. Some nations look to accounting rate settlement surpluses as a major source of hard currency for infrastructure development. Absent a direct investment in an MSS venture, incumbent operators and their national regulators may consider any strategy to avoid the accounting rate as evidence that MSS operators do seek to bypass preexisting routing arrangements and divert incumbent carrier revenue streams.

Heretofore MSS operators have gone on record with summary conclusions that terrestrial network integration will present no problem.[65] However, they appear not to have addressed the matter vigorously in view of immediate concerns like satellite construction and finance. MSS operators appear to have assumed that a consensus will be reached under the auspices of the ITU[66] or that investors and gateway operators can negotiate routing

---

vestments, resulting in similar costs between nations using the same transmission facility, for example, direct-dialed IMTS between the United States and Sweden vis-à-vis calls between the United States and Finland. However, accounting rates may vary widely even for nearby or adjacent nations. In the above example, the United States–Sweden per conversation minute accounting rate was 0.12 Special Drawing Rights (SDRs) (US$0.16) in 1997, but the United States–Finland accounting rate was 0.35 SDRs (US$0.48). See U.S. FCC, International Bureau, Telecommunications Division, Policy and Facilities Branch, Consolidated Accounting Rates of the United States (May 1, 1997), available on-line at http://www.fcc.gov/ib/td/pf/consolar.xls (Excel file).

65. "Calls made via the Globalstar system are routed through the existing national infrastructure, ensuring that PTTs will receive their fair share of the revenue stream, and be able to exercise their customary authority. . . . The Globalstar system, for example, will be fully integrated with existing fixed and cellular networks, and does not by-pass the PSTN. It will therefore fully comply with national service preferences, and the system can deny service to unauthorized users." Globalstar World Telecommunications Policy Forum Presentation.

66. Key officers of the ITU have made consensus building for global mobile personal communication systems (GMPSCs) a priority. ITU forums have developed a set of principles and recommendations "reflect[ing] an emerging common understanding among policy makers and regulators and GMPCS operators about their common interest in achieving the early deployment of GMPCS systems, and realizing the full economic benefits of deployment, in a manner consistent with the goals of national policy in each ITU member country." ITU, *World Telecommunication Advisory Council (WTAC) Report to the*

agreements on favorable terms. ITU documents recognize the complexity in MSS-PSTN integration, but no concrete plan or consensus view has evolved.[67]

Unless special accommodations are made, MSS operators may have to interconnect lines and settle accounts in much the same way as incumbent international telephony providers. MSS operators must negotiate operating agreements that will have to contain many of the coordination, financial settlement, and database registration issues that initially confounded cellular radio operators. Despite daunting challenges in space and on Earth, MSS presents a new and potentially significant contribution to the quest for the universal telephone.

9.4.3 The Global-Phone Reality

The demise of the Iridium LEO satellite system offers plenty of lessons for this era of convergence, dotcoms, and fascination with technology. While we have the benefit of hindsight, at least some of the lessons appeared early in the project, for example, the challenge of securing operational authority from more than 150 nations and the unprecedented and daunting task of operating a global telephone system. The legacy of Iridium's $5-billion collapse may lie in providing weighty and valuable instructions in the challenges in offering the next best thing, the killer application in an environment not yet fully transitioned to respond quickly and responsively to change.

Even today the telecommunications marketplace moves glacially compared to Internet time. No matter how attractive the concept of anytime, anywhere telecommunications access, Iridium's developers had to mesh a global network with a marketplace only beginning to transition from

---

Secretary-General on GMPCS (Jan. 19, 1996), available on-line at http://www.itu.int/pforum/wtacrepe.htm. However, the most concrete consensus document addresses the transceiver portability and the willingness of nations to permit use of such devices even if they have been licensed by another national government.

67. "Rapid deployment of GMPCS on an economically viable basis will only be possible if GMPCS operators are able to interconnect with the preexisting public network, and if they can do so on technically and commercially reasonable terms. Further work on interconnection policy is desirable to clarify policy alternatives and assess the feasibility and desirability of achieving compatibility of national regulatory policies concerning GMPCS." ITU, *World Telecommunication Advisory Council (WTAC) Report to the Secretary-General on GMPCS* (Jan. 19, 1996), available on-line at http://www.itu.int/pforum/wtacrepe.htm.

government-centralized management and monopolies. Early on, Iridium's developers decided to play by the rules—no matter how restrictive—that required them to secure licenses, respect national sovereignty, and accommodate the expectations of incumbent carriers that Iridium generate new traffic and revenues. A global, borderless system operating in cyberspace had to subdivide itself into licensed operations in each and every national space, an obstacle that Internet ventures have only begun to confront.

Playing by the rules also meant that Iridium had to work within the unwieldy and time-consuming spectrum allocation process organized under the auspices of the ITU. The twin constraints of playing by the rules and a limited window of opportunity sank Iridium. A narrowband, voice-oriented global cellular radio system could work only as a predecessor to a second-generation broadband, datacentric network and only if a terrestrial option did not exist.

9.4.3.1 Losing the First-Mover Advantage

Few would dispute the view that first movers have to move fast or risk losing the advantage of being first to market. The international spectrum allocation system and the regulatory process combined to rob Iridium of that advantage. The ITU allocates and reallocates spectrum in blocks based on service definitions. Because Iridium and other systems proposed to offer a new type of mobile satellite service, the ITU had to refine its service definitions and reallocate spectrum to accommodate the new technology, a process that took more than four years to complete. Additionally, the licensing process in the United States and elsewhere typically creates filing windows that provide applicants, far behind in development and financing, to achieve parity in terms of regulatory agency consideration. Iridium may have offered commercial service two years in advance of other ventures like Globalstar, but the ITU process enabled terrestrial systems to extend widely into the hinterland in many places where Iridium might have been the only option. Likewise, the FCC licensing process compressed the market head-start. By the time Iridium and Globalstar got sufficient operational authority and launched enough satellites to commence commercial service, the world had changed and the market opportunity had narrowed.

9.4.3.2 A Less Captive Market

Global MSS ventures like Iridium relied on captive markets, that is, providing the only viable option for accessing the rest of the world. Captive customers recognize that their particular circumstances lend themselves to price gouging. But captive markets in telecommunications have narrowed to niches like hotel rooms, cruise ships, aircraft, and areas lacking cellular radio service. Captive markets also have a tendency to evaporate, because their large margins encourage arbitrageurs and other entrepreneurs to find ways to serve those markets legally or otherwise. In nations with monopoly telephone companies and high long-distance rates, callback and call-reorigination ventures offer dial tone from nations with lower rates, regardless of the legality of such service. Because Iridium committed early along to playing by the rules and to securing favor with incumbent, often monopoly, carriers, gray-market tactics were not possible.

9.4.3.3 Why Working Within the System Failed

The developers of Iridium recognized the potential threat their system presented to incumbent telephone companies. A global telephone company has the potential for bypassing the incumbent carrier's monopoly, bottleneck control of access to the rest of the world. Iridium could not have persuaded government delegations at the ITU to reallocate spectrum or government regulatory agencies to issue licenses without offering incumbents a share in call revenues even if the incumbent did nothing but agree to the availability of service in its country.

Quite simply, Iridium got the worst of both worlds. To work within the old world order dominated by monopoly, often state-owned enterprises, Iridium had to offer incumbent carriers a share of the revenue stream, and pay conventional international revenue sharing and local access charges. But in trying to abate the bypass concern, Iridium bound itself to pay significant additional charges for the use of incumbent carrier network facilities to originate and terminate calls. For most mobile calls, whether originated terrestrially or from a satellite or terminated over conventional wireline networks, ventures like Iridium had to pass onto consumers two rate components: (1) the satellite, space segment charge and (2) land-based charges to compensate the Earth station operator and the carrier that delivered a satellite call or originated one. Because the Iridium network design limited the number of ground stations, each facility covered a broad geographic span. The broad geographic exclusivity that design fostered also resulted in high land charges, because much of the traffic had to traverse long distances and cross international borders to reach intended call recipients.

Iridium's decision to play by the old rules juxtaposes with the decision by many new ventures keen on exploiting kinks in the old world armor, particularly the inability to prevent international traffic from leaking into domestic networks and vice versa. At a time when Iridium service providers had to tack on high land-line charges, other ventures offered drastically lower international long-distance calling rates. The $3 to $4 per minute that Iridium charged looked so much more out of line when increasingly visible options offered service at less than $1. While Iridium engineered a route-by-route revenue division system, based on the traditional international telecommunications settlements, other ventures had found ways to originate and terminate traffic exempt from such settlements, making it possible to make calls well below the prevailing, retail rate and at a fraction of the Iridium charge.

9.4.3.4 Inability to Dislodge Cellular Consumer Expectations

As Iridium was making its case for spectrum allocations and licenses, terrestrial alternatives proliferated and expanded. MSSs do not work until almost all the satellites have successfully reached orbit, while terrestrial systems can expand incrementally with the addition of single, new cell sites. Thus, while Iridium had yet to begin commercial service, consumer expectations about mobile telecommunications already had become more sophisticated and demanding. Cellular radio consumers initially expected to pay high rates and receive spotty, interruptible service, but improvements and marketing initiatives offered a better deal. Terrestrial consumers grew to expect a "free" handset offered in exchange for committing to a term of fairly reliable service accessible from indoors and many rural locales and only minor financial penalties for roaming (use of a phone outside its home calling area). Perhaps more important, the nature of mobile service changed as operators resorted to new strategies to expand market penetration. While average revenues per user declined, cellular operators increased their total revenues by offering large buckets of service with minutes of use charged at the same rate for local, roaming, and domestic long-distance use.

Satellite-based mobile service operators could not differentiate their service, no matter how more complex and difficult the feat of providing ubiquitous service. Consumer expectations matter more than legitimate explanations why mobile satellite handsets had to weigh and cost more and why they could not operate indoors. Additionally, cellular radio already had captured most of the "gee whiz" consumer experience of tetherless and remote access to the rest of the world. Consumers had grown blasé about the incongruity of being centrally located in the middle of nowhere yet still being able to stay in touch.

9.4.3.5 Relying on the Credo of "If We Build It, They Will Come"

Several years ago, in the movie *Field of Dreams*, Kevin Costner's character convinced us that it made perfect sense to build a baseball diamond in a corn field so that long dead players might once again enjoy the game. Movies work because we suspend disbelief. Telecommunications business plans work when a company offers something people will buy. Perhaps the Motorola engineers became intoxicated with the vision of a global cellular radio system. Certainly this author did as one of the first employees in Washington, D.C., assigned the task of evangelizing the LEO concept to secure FCC and other regulatory approvals. But for all its cutting edge technology and elegance, the Iridium network failed to find the 2 million to 3 million users out of the over 6 billion potential souls technically able to take service.

Iridium's backers belatedly found that they could not rely primarily on road warriors, for whom price matters little. In telecommunications, price does matter and demand appears elastic, as evidenced by substantial increases in the cellular radio and long-distance calling volumes that result when rates drop and by the fact that high rates stifle calling. Iridium could not erect a toll bridge or a bottleneck through which users would willingly pay $3 or more per minute. Apparently, price and ergonomic issues like ease of use and portability of handsets do matter. It did not help that Iridium's managers had to coordinate a commercialized United Nations, with a variety of investors having different social and commercial cultures.

9.4.3.6 The Difficulty in Vertical Integration and Ascending the Telecom Service Learning Curve

Iridium's major backer, Motorola, has achieved commercial success based primarily on its manufacturing prowess. Yet it was willing to risk several hundred million dollars on the assumption that it could leverage its expertise into the telecommunications services marketplace. Manufacturers considering moving up or down the value chain risk alienating preexisting customers by competing with them. For the most part, Motorola had been content to manufacture the handsets, chips, and other equipment for use by customers who would provide services, including mobile radio and cellular service, to end users. Diversifying into the services arena constituted a major departure, but perhaps Motorola had to take the initiative to spearhead the project. Over time its ownership interest in Iridium declined, and the satellite manufacturing and other operational contracts it received from Iridium offset a portion of its initial investment. But until it could entice other investors, Motorola took the lead.

Motorola and many of the later investors lacked experience in the satellite services marketplace and in the intricacies of international telecommunications policy making. Because few incumbent telecommunications carriers decided to invest in the venture, the Motorola-led collection of venture capitalists, equipment manufacturers, and newcomers had to confront the esoteric and Byzantine policy-making process. With so much at stake and a sense that policy makers failed to appreciate the time sensitivity of the venture, the Iridium backers may have come across as brash and unaware of the complexity in securing all the required operational authorizations.

9.4.3.7 The $5-Billion Contribution

Ironically, the Iridium venture has made satellite telecommunications a more legitimate factor in an increasingly Internet-centric environment. Never before has one project triggered the construction and launch of so many satellites. Prior to Iridium's one-a-month assembly line construction pace, one-of-a-kind designer satellites typically took 18 to 24 months to build. The smaller, lighter LEO satellites of Iridium supported a new orbital arc as an alternative to the congested geostationary one. Iridium's launch requirements supported the development of new options and proved the ability of a single rocket to insert five or more LEO satellites into proper orbit. Perhaps most important, Iridium proved the concept that a constellation of intersatellite links could support anytime, anywhere telecommunications access. Visions of broadband, Internet-in-the-sky satellite service will build on Iridium's contributions.

9.5 Fear and Loathing in Internet Access Pricing

The terms and conditions under which ISPs interconnect facilities provide a compelling case study of the similarities and differences between Internet and telecommunications networking. Even though both types of services use the same telecommunications transmission facilities, the enterprises that provide the networks have managers operating under different assumptions and cultures. Also, the manner in which the Internet became a "network of networks" [12] has favored a more generous cost-sharing arrangement compared to the pay-to-play, or pay-as-you-go system in telecommunications. Internet networking regularly requires the cooperation of several different ISPs that agree to carry each other's traffic. In telecommunications, the carriage of such "transit" traffic triggers a specific, agreed-to payment. Internet transiting often lacks the calibration and specificity in whether and how much one ISP owes others.

Until recently, ISPs interconnected facilities on a loose, rough-justice basis. ISPs sought to create the user impression that the Internet constituted a single integrated network whereby individual ISPs purposefully linked into a daisy chain of affiliated networks. As a result, ISPs have provided consumers with seamless access to most of the individual networks that make up what we call the Internet, often with a direct operating arrangement covering only the first and last of many network connections. The packet-switched, best-efforts routing nature of the Internet, coupled with switching and routing protocols, provides robust and diverse network access without each ISP having to negotiate interconnection terms with every other operator.[68]

The "network of networks" refers to the daisy-chaining of different networks into the appearance of one integrated and seamless network. By way of an analogy to aviation, consider the Internet a free fly zone, that is, an environment in which users can traverse all of cyberspace without significant restrictions. In the physical world of airspace, nations must secure "flights of freedom" to operate aircraft in the sovereign airspace of other nations, to land, to take on and deliver passengers, and to transport passengers from one location to another within a nation. In the virtual world of the Internet, users face little if any of those impediments, even though transiting another ISP's network imposes costs on the transit-providing ISP, while the airline flying over another country imposes few, if any, costs on that country. Better yet, for Internet travelers the transporter through foreign lands is none other than the networks owned and operated by foreign ISPs.

Internet travelers perceive that they need not secure and pay for transport beyond the circuit-switched public telephone network link to the ISP providing Internet access. Internet users might acknowledge that costs are incurred when they require the services of other ISPs to route traffic beyond the reach of their ISP, unlike an airline that flies over a nation. But such users might simply conclude that ISPs provide reciprocal and roughly equivalent switching and routing services for other ISPs, thereby obviating the need to meter traffic flows.

On the other hand, telecommunications carriers achieve similar connectivity with greater effort and specificity: the one-by-one accumulation of

68. See Joseph P. Baily, "The Economics of Internet Interconnection Agreements," in Lee W. McKnight and Joseph P. Bailey, eds., *Internet Economics* (Cambridge, MA: MIT Press, 1997), p. 161; R. A. Cawley, "Interconnection, Pricing, and Settlements: Some Healthy Jostling in the Growth of the Internet," in B. Kahin and J. H. Keller, eds., *Coordinating the Internet* (Cambridge, MA: MIT Press, 1997), pp. 346–376.

operating agreements.[69] While users expect their traffic to reach the final destination, regardless of the routing and networks traversed, each and every segment will have been previously secured through cash payments or facility swaps by the originating carrier

Because ISPs and their telecommunications counterparts have implemented different access pricing systems, what will become of each in an increasingly convergent, integrated marketplace? More specifically, if the Internet becomes the central infrastructure web for data and voice communications as well as telecommunications and information processing, then what pricing system will apply? Perhaps the answer lies in the changes occurring in access pricing in both telecommunications and information-processing markets. As telecommunications becomes more like a commodity, bit-transport business, carriers have begun to eliminate price differences based on type of service, for example, voice versus data, instead using a bandwidth or bulk minutes of use measure. As ISPs become more commercial and bottom-lined oriented, they have replaced rough justice, sender-keep-all (SKA) pricing, which typically involved no transfer of funds, to a better-calibrated system that imposes charges on ventures that cannot or will not take and deliver as much traffic as they generate. It appears that the access pricing and interconnection trends developing in both telecommunications and the Internet bring the now significantly different regimes closer in terms and conditions.

9.5.1 Internet Incubation Support of SKA Access Pricing

Technological ease in switching and routing Internet packet traffic, the interest in building out networks, and connecting ever increasing subscriber numbers, government involvement and the costs of metering traffic combined to distract ISPs from a profit-maximizing access pricing system. Unlike the metered, usage- and distance-sensitive telecommunications pricing regime, ISPs typically offered their consumers and other ISPs unlimited "all you can eat" (AYCE) access opportunities. Absent congestion and the need to satisfy ever increasing demand for bandwidth, which would subsequently occur, ISPs initially could accommodate everyone while accruing ever greater positive networking externalities. We can look back on those golden years as a time when Internet pioneers had a mission to build something special that

69. For a helpful primer on the Internet and the telecommunications policy implications of its development, see Kevin Werbach, "Digital Tornado: The Internet and Telecommunications Policy," OPP Working Paper No. 29 (Washington, D.C.: FCC, Mar. 1997).

could change the world. Concerns about whether everyone fully and fairly bore the financial burden appeared almost crass.

9.5.2 Traditional Internet Access Pricing Models

9.5.2.1 SKA Model

The SKA model allows ISPs to retain all subscriber payments without having to settle accounts with other ISPs that participate in routing and delivering traffic. The SKA model promotes the daisy chaining of unaffiliated networks and delivers global access to sources of information, commerce, and entertainment. This model has served as the primary template for Internet traffic routing, because of its administrative convenience and the willingness of ISPs to promote network connectivity regardless of whether traffic flows are symmetrical. SKA involves network interconnection without a metering mechanism, either because the parties do not care whether traffic symmetry exists, assume that such symmetry exists, or believe that metering and the settlement of financial accounts trigger more cost and inconvenience than a rough-justice agreement to accept and route onward each other's traffic.

The SKA model promotes positive network externalities and universal service, because smaller and rural ISPs typically enjoy opportunities to generate more outbound traffic for free carriage by other ISPs than they receive from other ISPs for carriage onward to another ISP or for terminating traffic. The opportunity to avoid paying a penalty for being comparatively less necessary and operationally more expensive than larger ISPs translates into an opportunity for users in rural and high-cost service areas to access the Internet on terms and conditions similar to what urban subscribers pay.

9.5.2.2 Peer-to-Peer Bilateral Model

The Internet-specific peer-to-peer bilateral model adopts SKA but with the expectation of traffic symmetry. Two unaffiliated ventures agree to use this model, which requires a direct and meterless connection, if and only if they have "the same, size, experience, technology and customer base" [13]. This model may eliminate the opportunity for Internet users in rural and high-cost areas to pay less than full-service costs if ISPs in such areas must resort to more expensive transit arrangements with bigger ISPs, resulting in a one-way transfer payment from the smaller ISP to larger ones.

9.5.2.3 Hierarchical Bilateral Model

Even before the threat of network balkanization, a hierarchy of ISPs had developed, based on geographic scope of service, available bandwidth, traffic volume, and subscribership. The hierarchical bilateral model applies when different types of ISPs agree to interconnect their networks. The terms and conditions of this two-party contract reflect unequal bargaining strength in the sense that a smaller ISP, denied SKA and other cost-free interconnection opportunities, now must persuade a larger ISP to handle its traffic. A negotiation in this context establishes a customer-provider, resale relationship rather than a carrier-to-carrier or ISP peering arrangement. Increasingly, this model predominates as most large Tier-1 ISPs treat smaller regional and local ISPs as "clients." Accordingly, the smaller regional and local ISP typically has to transfer funds to the Tier-1 ISP, because the larger operator has incurred a greater infrastructure investment burden and has the capacity and wherewithal to route the smaller ISP's traffic onward to another network or to the final destination.

While transfer payments occur in this model, it is important to note that no incentive exists for a likely transfer payment recipient, that is, the Tier-1 ISP, to discriminate or to deny interconnection with a smaller ISP that will pay for access. If this model continued to dominate the Internet, the free ridership and unintended subsidy problem could abate without any single, financially qualified network operator experiencing denied access to other networks. The smaller network operator would simply have to agree to make the necessary transfer payments and thereby have transit access to servers, e-mail recipients, and others via the switching and routing facilities of other, larger, and geographically diverse networks.

9.5.2.4 Third-Party Administrator Model

The third-party administrator model involves a neutral clearinghouse function managed by a paid administrator that might not even operate a network. Before relinquishing all Internet management responsibilities, the National Science Foundation operated network access points that served as public peering sites for the exchange of traffic between networks. Now commercial ventures (e.g., the Commercial Internet Exchange Association [CIX], Sprint, UUNET, and PSINet) perform the same function. Such businesses place greater emphasis on generating a profit from administrative fees and offer access to any ISP on a rational, traffic volume–based price structure.

The third-party administrator model works well when the administrator has the financial wherewithal to expand capacity and routing functions to meet demand and to maintain an adequate level of service. Third-party administrators must enforce requirements that ISPs maintain bandwidth and traffic-processing capabilities commensurate with vastly expanding traffic growth. Currently, some traffic exchange locations have become so bogged down with traffic that packets of information must be re-sent, or they are lost altogether.

ISPs, particularly ones with the largest traffic volumes and available transmission capacity, are off-loading some or all of their traffic onto private peering locations, because public peering points have become congested bottlenecks. That migration has the most adverse effect on smaller ISPs, which lack the facilities investment to interconnect individually with one or more of the Tier-1 ISPs at several locations, typically an individual ISP's point of presence. A small ISP could expect to provide its subscribers with access to just about any other ISP's network simply by interconnecting with the large number of peers at a third-party administrator's public peering site. When Tier-1 ISPs boycott such sites and agree to handle traffic of smaller ISPs only upon payment for capacity leases, the smaller ISPs must scramble to find substitute ways to access the major carrier's disparate networks, typically at several different locations and at higher costs.

9.5.3 A Maturing Internet: More Hierarchical, More Commercial

As the Internet grew and governments reduced or eliminated their financial underwriting, ISPs could not ignore access-pricing issues, particularly in light of increasingly more expensive bandwidth requirements and the proliferation of ISPs, including small ventures with limited budgets, subscriber bases, and network inventory. The largest, Tier-1 ISPs became the first to break rank and replace SKA with a demand for compensation from smaller ISPs. Even at the risk of balkanizing the Internet and imposing higher costs on smaller, often rural or non-North American ISPs, the Tier-1 ISPs could justify their new access-pricing regime. Paying for Tier-1 ISP services is rational and efficient and appropriately eliminates opportunities for free riders to transit the networks of other ventures without providing commensurate services themselves.

9.5.3.1 Internet Pricing Systems Now Looking More Like Their Telecommunications Counterparts

Tier-1 ISPs' demand for compensation from smaller ventures evidences a strain on the "all for one and one for all" service orientation that has made the Internet so accessible and integrated. Telecommunications networks have never had such porousness and ambiguity in routing arrangements, nor do telecommunications carriers expect traffic flows to even out. At least in theory, a telecommunications carrier knows the composition and routing options for all the traffic it carries. Telecommunications carrier correspondents know who originated the traffic and which carrier will bear financial responsibility for such carriage. The telecommunications pricing model works to eliminate free riders even though, economically speaking, the routing of either Internet or telecommunications traffic may trigger little if any incremental cost under circumstances where no congestion exists.

For Internet applications, near-zero carriage costs favored the absence of a pricing system, that is, a zero-cost arrangement for the termination of traffic at an intended destination or the intermediary transiting of traffic onward toward its final destination. For telecommunications applications, near-zero carriage costs favored creative pricing arrangements like usage-insensitive private lines and circuitous routing, which takes advantage of distance insensitivity, that is, the technical capability of routing traffic at the speed of light coupled with the financial capability of loading traffic at near-zero cost when time of day and other factors favor ample transmission capacity availability.

Even if a particular routing might trigger no incremental cost, telecommunications carriers have devised a pricing arrangement that attributes a cost for each call or at least for the assumed volume generated by a particular link. Such specificity means that telecommunications carriers bear the burden of acquiring or leasing all the transport facilities their customers need. Likewise, those carriers establish operating agreements with each other for direct service and for indirect, transiting service using the transmission facilities of another carrier. Accordingly, a telecommunications carrier identifies and pays for all facilities it owns and to which it has access.

ISPs historically have lacked such direct commercial linkage by contract. In exchange for a general offer of routing reciprocity, a single ISP had access to all other ISPs' network resources. ISP A, having an interconnection arrangement with ISP B, can tap into both ISP B's network and whatever interconnection and transiting opportunities ISP B has secured from ISP C and others. Such reciprocity works well and efficiently when all ISPs generally handle the same volume of traffic and operate the same type of transmission facilities. In this situation, metering would impose significant and unnecessary financial, administrative, and technical costs. Satisfaction with a rough-justice, unmetered exchange of traffic based on an assumption that traffic patterns were at parity even without the ability to measure traffic made it possible for ISPs to forego the expense and efficiency loss in having to erect systems for tracking traffic volumes. A larger percentage of each packet could handle actual traffic rather than contain the billing and other information needed if all ISPs had to meter traffic.

Such reciprocity does not work well when ISPs are not equals, an increasingly likely scenario as governments reduce or eliminate financial subsidies and as an increasing number of ISPs enter the marketplace. With the privatization of the Internet, a variety of commercial ventures have begun providing service with different subscriber numbers, inventory of transmission capacity, interconnection agreements with other ISPs, number of locations where the venture can exchange traffic, and locations where the venture does business.

Telecommunications carriers traditionally have established pricing arrangements that take into consideration the likelihood that the carriers will differ in one or more of the factors identified above. Telecommunications carriers agree to become correspondents for the purpose of securing traffic transiting and terminating service, but they do not expect each other to be true peers. In telecommunications, one carrier will operate in a country with more people, income, and competition, typically resulting in a comparatively higher volume of outbound calls than what a correspondent would generate in a smaller and poorer country. Under the anticipated and likely condition of asymmetric traffic patterns, telecommunications carriers have established a financial settlement arrangement whereby the carrier generating more traffic pays the terminating carrier for the comparatively greater use of that carrier's facilities. A similar arrangement might apply for Internet correspondents in the future.

In response to the combination of congestion and the potential for free-rider or subsidized ISPs, Tier-1 ISPs have established private peering points and have largely exited preexisting public peering points. Free access to those private peering points is limited to similarly situated, Tier-1 ISPs. Other ISPs can secure access to the new interconnection points for a price. The terms and conditions under which smaller ISPs secure access to Tier-1 ISP facilities have begun to parallel the contractual, private-correspondent telecommunications model: Smaller ISPs now must lease facilities from a Tier-1 ISP. Possibly in the future a Tier-1 ISP might implement a better-calibrated financial settlement.

The diversification and proliferation of ISPs occur at the same time as ISPs have grown more conscious of Internet traffic volumes and patterns. This subject has become more important to ISPs, because demand has grown so quickly and dramatically that operators have had difficulty in making timely system upgrades. Necessary transmission and switching capacity enhancements to keep up with demand force ISPs collectively to make substantial additional investments lest the Internet become bogged down and inoperable. ISPs that conscientiously make infrastructure upgrades rightly object to a cost-free interconnection arrangement available to ISPs that have failed to make commensurate investments.

But perhaps most important for our purposes, they demonstrate that the current Internet pricing system cannot prioritize traffic based on the willingness of the sender or the recipient to pay a premium price; reserve capacity; or offer more reliable service than best-efforts routing for a price higher than the zero-cost true peering or capacity leases of smaller ISPs.[70]

9.5.4 Traffic Patterns in Telecommunications and Internet Applications

Traffic patterns have always mattered in telecommunications, and carriers implemented a pricing mechanism at the outset to account for differences in flow. Traffic patterns similarly matter in the Internet, but an appropriately calibrated pricing mechanism did not exist at the outset. Accordingly, implementing one now will require time, difficult commercial negotiations, and perhaps a public policy and public relations campaign to convince the larger set of stakeholders that the overall health of the Internet depends, in large part, on the establishment of a new, fairer, and more rational pricing regime. Of course, beneficiaries of the status quo have every incentive to delay the date when they lose their significant financial windfalls.

A 1997 study conducted by the ITU succinctly stated that "[t]raffic patterns on the Internet are seriously distorted … [and not easily remedied because] there is considerable inertia" [14]. The study reported that in 1994 more than 80% of the total global Internet traffic traversed the United States largely due to the number of desirable World Wide Web sites located in that country. Under a telecommunications settlement scheme, service providers generating more outbound traffic than inbound would make transfer payments to the carriers transiting or terminating the comparatively more traffic.

Despite the importance of traffic patterns in both Internet-mediated and telecommunications environments, the Internet pricing system historically has operated from a rough-justice, value-of-service model. Because so many Internet subscribers seek access to U.S.-based Web sites and thereby derive comparatively greater utility and value from such access, the ISPs providing such access must largely bear the expense in engineering complete

70. Extensive efforts are under way to address these limitations. For example, a resource reservation protocol (RSVP) would provide a means to provide and charge a premium price for dedicated links.

links to U.S. sites. The balance of utility and value has skewed heavily to the United States, because of the attractiveness of what U.S. sites have to offer, comparatively cheaper international private-line access, and because the ISPs seeking access historically have had less to offer by way of both networking facilities and number of Web sites.[71]

Absent a change in the flow of traffic toward parity, U.S. ISPs can demand that interconnection take place in the United States as opposed to some equidistant location. The rationale for requiring offshore ISPs to self-provision the lines achieving such access all the way to a U.S. location lies in the assumption that such access enhances the value and commercial appeal of the offshore ISP. It can deliver to its subscribers U.S.-based content, access to the robust set of U.S. networks, and the use of U.S. networks for accessing the facilities and services of other networks beyond. Arguably, an offshore ISP, which does not establish peering arrangements throughout the world, could secure subsidized or underpriced access to U.S. ISP networking arrangements for global access.

Just as some ISPs and national governments object to self-provisioning in the face of migrating traffic patterns, U.S. ISPs object to what has been termed "hot-potato"[72] routing: the strategy of some ISPs to find a relatively nearby peering site for cheap off-loading of traffic onto other ISPs' networks that will provide transit services over the long haul to the final destination. In the worst-case scenario, an ISP far from the United States might secure access to the rest of the world courtesy of one or more U.S. ISPs and via one peering

---

71. "When foreign countries wish to connect to a U.S. backbone network they are obligated to cover the full costs of the international private line (both half circuits). There is some logic in this in that the countries connecting to the U.S. backbone network are not providing the U.S. backbone with much more than one network and a limited number of routes, all of which may be in just one country. The connecting network benefits greatly from connecting to the U.S. backbone; whereas, the U.S. backbone networks benefit much less from connecting with the outside network." ITU, *Challenges to the Network Telecoms and the Internet* (Geneva: ITU, 1997), p. 25.

72. "Because of 'shortest exit routing,' the standard means of exchanging traffic among networks, the data goes onto the receiver' s network at the earliest point. All networks have an interest in passing off the data like a hot potato, to the other network at the earliest point, since it saves the sender-ISP valuable bandwidth." Kenneth Neil Cukier, "Peering and Fearing: ISP Interconnection and Regulatory Issues," presented paper at the *Harvard Information Infrastructure Project Conference on the Impact of the Internet on Communication Policy*, Dec. 3–5, 1997. Conference Program is on-line at http://ksgwww. harvard.edu/iip/iicompol/agenda/html. The Cukier paper is on-line at http://ksgwww. harvard.edu/iip/iicompol/Papers/Cukier/html.

point the non-U.S. ISP accessed via self-provisioned international private lines. While seemingly burdensome to force self-provisioning, the ability to access the rest of the world via other ISP facilities presents quite a bargain, particularly in light of comparatively cheap private-line rates for access to the United States relative to other points.

Such transiting opportunities do exist, as evidenced by route tracing performed by the publication *TeleGeography*.[73] Intraregional traffic within Asia-Pacific and other regions of the world may transit the United States to take advantage of several combined factors: distance insensitivity in telecommunications routing, comparatively cheaper private-line rates to the United States than between nations in a region, and the robust network access opportunities accruing from having secured even just one network access point in the United States.

9.5.5 Pricing Responses to Changed Circumstances

Concerns about network congestion, burgeoning demand, diversification of Internet applications (including high-capacity video), free riders, and hot-potato routing collectively have caused ISPs to rethink the peering and SKA systems. Private peering has become the most recent interconnection model and the one most likely to involve some degree of discrimination, entrance requirements, or transfer payment. The private peering model involves the overlay of quasi-private Internets unavailable to every ISP or Internet user or available at a price. Private peering users purposefully deem their networks inaccessible to outsiders ostensibly to preserve network integrity and minimum quality of service levels. The migration to private peering, however, also results from the real or perceived need to safeguard a sizable and expanding investment from the congestive effects of free riders.

The onset of private peering does not eliminate public peering opportunities, nor does it necessarily reduce the Internet's robustness, seamlessness, and integration. It does, however, make the Internet more hierarchical in the sense that the Tier-1 ISPs have reduced the number of ISPs with which they peer on an SKA, cost-free basis.[74] It also makes the pricing structure arguably

---

73. Kenneth Neil Cukier, "The Global Internet: A Primer," Fig. 8, "Internet Traceroute Maps, July 1998, From Hong Kong to Japan via Silicon Valley, California," in Gregory C. Staple, ed., *TeleGeography 1999* (Washington, D.C.: TeleGeography, 1999), p. 121.

74. For an extensive collection of Internet access and pricing papers and articles, see Hal R. Varian, The Information Economy, The Economics of the Internet, Information Goods,

more rational and costly to smaller ISPs because they must pay more for the privilege of accessing and transiting Tier-1 carrier networks.

Currently smaller ISPs can maintain Tier-1 ISP access by leasing the Tier-1 ISP-specified number of private lines, priced on a flat, usage-insensitive basis. Economists assert that an even more efficient and equitable pricing system—particularly during network congestion—would involve settlements on a per-connection basis and perhaps even on a next-hop basis, that is, router-by-router settlement for all links between the content requester and the content provider, presumably for routes both to and from the server providing the content. Such precision in network management would require more extensive monitoring and perhaps more overhead and less payload content in packets. On the other hand, it would impose market-based discipline in the switching and routing of Internet traffic. ISPs would face appropriate incentives and burdens with fewer opportunities to hand off traffic without incurring attributable costs. Likewise, subscribers would face more exacting price signals and payment obligations, with the mixed result of perhaps having to pay for services previously available on a flat-rated, AYCE basis but also having the opportunity to pay for and receive premium (faster, better, more reliable) service.[75]

9.5.6 Emerging Internet Charging Issues

The diffuse and largely unregulated nature of the Internet will largely preclude any attempt to manage its evolution and to enforce specific policy outcomes. The migration from government sponsorship to private enterprise dilutes an already limited degree of control.

Notwithstanding those substantial limitations, governments do need to monitor changes in the Internet, particularly as they relate to public policy matters, for which government has a legitimate interest. Put another way, governments cannot shape the manner by which the Internet matures, nor can it mandate a particular outcome. But government should understand the nature and scope of the Internet's influence on markets and individuals.

Intellectual Property and Related Issues, available on-line at http://www.sims.berkeley.edu/resources/infoecon/ (viewed Feb. 9, 1999).

75. See David D. Clark, "Combining Sender and Receiver Payments in the Internet," in Gregory L. Rosston and David Waterman, eds., *Interconnection and the Internet* (Mahweh, NJ: Lawrence Erlbaum Associates, 1997), pp. 95–112.

Internet traffic patterns, coupled with emerging financial settlement mechanisms, will have a profound impact on who pays for needed infrastructure deployment and upgrades and how that payment is made. Stakeholders in the golden age of carefree, zero-cost settlements will balk at new pricing regimes that divest them of opportunities to ride the Internet without paying the full fare. Likewise, advocates for new pricing regimes may have efficiency arguments on their side, but they may fail to secure immediate relief, given the decentralized and largely ungoverned nature of the Internet. Relief or discomfort for either side probably will evolve on an incremental, often bilateral basis as ISPs negotiate and renegotiate the terms and conditions for interconnection.

9.5.7 Lessons from the Battles for Internet Access

Internet forecasting is a particularly risky undertaking. Nevertheless, one already can see that as the Internet matures and becomes more commercial, ISPs will have to address a different set of interconnection and pricing issues than what occurred during the developmental years. ISPs have turned their attention to issues like resolving problems pertaining to service quality, capacity upgrades, and cash flow. They worry less about network ubiquity and access and more about network reliability and ensuring that the expanding set of players fully and fairly pay its share for network access. In that environment, ISPs appear more businesslike and prudent in matters relating to interconnection and access. Like their telephone company counterparts, ISPs now need to care about both stimulating and accommodating demand. But unlike telecommunications carriers, ISPs face greater uncertainty about who generates demand and how to recoup costs incurred in satisfying such demand.

Telephone companies establish direct operating agreements with carrier correspondents, while ISPs are just beginning to realize the utility in such a model. In the not-too-distant future, most ISPs may apply an adaptation of the financial settlement methodology of telephone companies. The ISP charging model will provide greater structure and financial discipline to the Internet. The days are numbered for free riders and nonexplicit service subsidies.

References

[1] FCC, International, Common Carrier Bureau, *Accounting Rates and the Balance of Payments Deficit in Telecommunications Services*, Dec. 12, 1988, p. 29.

[2] Regulation of International Accounting Rates, CC Docket No. 90-337, Notice of Proposed Rulemaking, 5 FCC Rcd. 4948 (1990).

[3] Stanley, K., FCC, Common Carrier Bureau, Industry Analysis Div., *The Balance of Payments Deficit in International Telecommunications Services*, Dec. 12, 1988, p. 19.

[4] Implementation and Scope of the International Settlements Policy for Parallel International Communications Routes, CC Docket No. 85-204, Notice of Proposed Rulemaking, 50 Fed. Reg. 50 FR28,418 (1985).

[5] Report and Order, 59 Rad. Reg. 2d 982 (1986), *on recon.*, 2 FCC Rcd. 1118, *on further recon.*, 3 FCC Rcd. 1614 (1988).

[6] Regulatory Policies and International Telecommunications, CC Docket No. 86-494, Notice of Inquiry, 2 FCC Rcd. 1022 (1987), Report and Order and Supplemental Notice of Inquiry, 4 FCC Rcd. 7387 (1988), *order on recon.*, 4 FCC Rcd. 323 (1989).

[7] Regulation of International Accounting Rates, CC Docket No. 90-337, Phase I, 6 FCC Rcd. 3552 (1991).

[8] Accounting Rate Phase II First Report and Order, 7 FCC Rcd. at 560.

[9] 1998 Biennial Regulatory Review Reform of the International Settlements Policy and Associated Filing Requirements, IB Docket No. 98-148, Report and Order and Order on Reconsideration, 14 FCC Rcd. 7963 (1999).

[10] Cheong, K., and M. Mullins, "International Telephone Service Imbalances," *Telecommunications Policy*, Vol. 15, Apr. 1991, p. 117.

[11] Bond, J., "The Drivers of the Information Revolution—Cost, Computing Power, and Convergence," *Public Policy for the Private Sector*, Quarterly No. 11, Sept. 1997, pp. 227–228.

[12] Noam, E. M., "Beyond Liberalization—From the Network of Networks to the System of Systems," *Telecommunications Policy*, Vol. 18, No. 4, 1994, pp. 286–294.

[13] Baily, J. P., "The Economics of Internet Interconnection Agreements," in L. W. McKnight and J. P. Bailey, eds., *Internet Economics*, Cambridge, MA: MIT Press, 1997, 161.

[14] ITU, *Challenges to the Network Telecoms and the Internet*, Geneva: 1997, p. 25.

10

Privatization and Globalization in Telecommunications

The single most important change in international telecommunications occurred when governments, spanning the continuum of political philosophies, decided to reduce or eliminate their ownership interests in the incumbent national telephone company. Most governments, in developed and developing nations alike, no longer endorse the view that the telecommunications sector requires centralized government management, and increasing numbers now believe that competition accrues greater social benefits than government-sanctioned monopolies. Privatization in the telecommunications sector constitutes an essential catalyst for a host of additional significant changes, including liberalization of policies that affect the dominant service provider, deregulation of telecommunications in general, and globalization of perspectives and investments by many telecommunications and information-processing ventures.

10.1 The Rationale for Privatization

Privatization in telecommunications involves a change in the legal status of a nation's government-owned telecommunications carrier or equipment manufacturer to permit private investment. Privatized companies may retain a monopoly service franchise, but in many instances, nations may permit limited competition coupled with liberalization of regulatory requirements imposed on the incumbent carrier. At the same time that it privatizes the PTT, a government typically streamlines and reduces oversight to allow the privatized incumbent flexibility to respond to changing conditions, including the need to compete in some markets with one or more new facilities-based carriers and numerous resellers.

The onset of privatization represents an explicit recognition by governments that

> telecoms have become too important to be left to engineers exchanging incomprehensible acronyms. A rich country with outmoded laws that blocks innovation will set its companies at a huge disadvantage against foreign competitors; a poor country will find it immeasurably easier to attract foreign investors if it ensures that they have access to telephones and fax machines that work. So the thrust of the debate is everywhere the same: how to encourage the fullest possible modernisation of domestic networks?[1]

Simply put, "much of the recent enthusiasm for privatization is the rather simplistic notion that government ownership results in economic inefficiency" [1, p. 3]. Notwithstanding that view, some government-owned entities have performed quite well, either because they have achieved a sufficient degree of independence or because government has targeted telecommunications for subsidization from tax receipts. For example, well before their privatization, government-owned carriers in France, Germany, and Singapore successfully invested abroad and developed a global marketing presence through joint ventures even as they upgraded domestic facilities and operated state-of-the-art networks. France Telecom, Deutsche Telekom, and Sprint formed and then disbanded a joint venture aiming to provide global,

1. The following articles provide background on the subjects covered in this chapter. Mark Naftel, "Countdown to 1998: Status of Telecommunications Competition in Europe and Comparison With the United States," *Journal of Transnational Law and Policy*, Vol. 7, 1997, p. 7; Michel Kerf and Damien Geradin, "Controlling Market Power in Telecommunications: Antitrust vs. Sector-Specific Regulation. An Assessment of the United States, New Zealand and Australian Experiences," *Berkeley Technology Law Journal*, Vol. 14, Fall 1999, p. 919; David S. Barrett, "The European Commission's Antitrust Analysis of High-Technology Joint Ventures and Strategic Alliances: A Diverging Analysis?" *Transnational Law*, Vol. 12, Fall 1999, p. 475; Kristina Driscoll Carey, "Competition Law and the Privatization of Telecommunications Markets in the European Union," *Suffolk Transnational Law Review*, Vol. 22, Summer 1999, p. 747; Christopher J. Sozzi, "Project Finance and Facilitating Telecommunications Infrastructure Development in

one-stop shopping to large, multinational enterprises. France Telecom holds a 20.4% ownership share in the dominant Mexican telephone company, TelMex. Singapore Telecom has experienced difficulty in executing regional investment strategy (e.g., a rejected bid to acquire Hong Kong Telecom) but has developed a mobile telecommunications system with the London-based Virgin Group and also has invested in several U.S. and other foreign ventures.

Raymond Duch suggests the likelihood that a privatization initiative will occur depends on two critical dimensions [1, p. 6]:

- The ease with which challenges to established public policy can be mounted;
- The institutional resources available to threatened interests that resist any changes in the policy status quo.

Using those two variables, one can predict that pluralist nations like the United Kingdom and the United States will promote greater access to the political process and provide limited protection to incumbent institutions like telephone companies, despite their political clout and large base of employees. Nations slower to embrace privatization, like France, appear to emphasize stability and maintenance of political order and limited access to the political process. Nevertheless, even those nations cannot maintain the status quo in the face of change and political pressure from regional governing bodies (e.g., the EU, stakeholders, and the electorate). Corporatist nations, like Germany, concentrate on expanding the wealth and welfare of the populace and have done so by limiting access to the political process and by supporting successfully operating government-owned institutions. Such nations also are convinced that privatization enhances the national welfare.

Newly-Industrializing Countries," *Santa Clara Computer and High Technology Law Journal*, Vol. 12, Aug. 1996, p. 435; Rachelle B. Chong and Wendy Chow, "Financing Telecommunications Projects in Asia: A Promising Regulatory Perspective," *Federal Communications Law Journal*, Vol. 52, Dec. 1999, p. 1; Paula Barnes Sours, "The Impact of U.S. Regulatory Activity on Prospects for Implementation of the WTO Agreement on Basic Telecommunications," *North Carolina Journal of International Law and Commercial Regulation*, Vol. 23, Summer 1998, p. 645; Stefan M. Meisner, "Global Telecommunications Competition a Reality: United States Complies with WTO Pact," *American University International Law Review*, Vol. 13, 1998.

10.2 Why Government-Provided Telecommunications Services in the First Place

Ironically, economic analysis has been used to support both state ownership of telecommunications and privatization. The justification for government ownership lies in the view that telecommunications constitutes a natural monopoly. Under that paradigm, competition increases risk, results in wasteful duplication of facilities, prevents the monopoly from achieving economies of scale necessary to operate efficiently, and jeopardizes the prospects for achieving public-policy goals like universal service at affordable rates. A regulated monopoly can extend and subsidize services to accrue the social benefits like national cohesiveness achieved when most citizens have access to telephones. Regulation mandates or supports carrier-pricing decisions that set basic service rates at levels that would be unsustainable if business considerations (e.g., return on investment) were considered exclusively. The incumbent may claim that it alone can be trusted to serve all areas, lucrative or not. It also may raise national security concerns about the strategic nature of telecommunications, particularly in times of war or civil unrest.

France Telecom provides a case study for assessing how government ownership could deliver many of the benefits touted by advocates for privatization. France used to have one of Europe's worst telecommunications infrastructures. With legislated financial support, France Telecom accelerated system upgrades and speedily achieved near-ubiquitous digital service capability. It also distributed at no charge Minitel terminals for telephone subscribers to use in accessing databases, including on-line directories, and for communicating via electronic mail. At least in the case of France Telecom, "enlightened paternalism" achieved widespread deployment of cutting edge technologies throughout a nation, an outcome that might not result if a private company operated without close government scrutiny.

Proponents for privatization claim that monopolies seek to perpetuate the status quo, even in the face of technological innovations that reduce barriers to market entry and undermine justifications for a single service provider. Having invested in expensive technology, subject to regulatory limitations on how quickly the investment could be recouped, the incumbent may not innovate or make timely additional investments. But to foster goodwill among the electorate, the incumbent, perhaps in collaboration with elected officials, may pursue political and popular objectives, regardless of whether they might adversely affect economic efficiency. In application, that means that the incumbent may scrimp on infrastructure improvements while intentionally underpricing domestic POTS with subsidies from overpriced international and corporate services.[2]

Reformers claim that carriers and equipment manufacturers must respond to elasticities of supply and demand, that is, whether a particular class of consumers has alternative ways to communicate and their intensity of need for and sensitivity to changes in service and equipment prices, that is, demand elasticity.[3] Pricing too out of line with the available options encourages users to devise methods for bypassing overpriced services and, at the extreme, to relocate personnel and facilities to nations with better and cost-based telecommunications rates. Privatization or at least action that affords greater operational flexibility for the incumbent carrier results when government and the public believe that less governmental oversight and regulation can accomplish the following objectives:

- Better response to consumer demand;

---

2. "Often these policies confer benefits on a very diffuse electorate, the costs are borne by a small segment of society, and they negatively affect the efficiency of the public enterprise. One example is the subsidization of residential telephone rates with revenues from business and long-distance telephone tariffs. The costs of subsidizing a very diffuse segment are borne by a more concentrated constituency. Although such pricing policies tend to command widespread political support, they are inefficient from a strictly economic perspective because the tariffs have little relation to the cost of providing the services." R. Duch, *Privatizing the Economy—Telecommunications Policy in Comparative Perspective* (Ann Arbor, MI: University of Michigan Press, 1991), p. 30.

3. Adverse financial consequences to society, known as deadweight social losses, may increase when governments refuse to authorize competition or mandate privatization: "Small changes in the cost of supplying products and services can now have a very significant impact on the quantities of goods made available on the market. Similarly, consumers have become increasingly sensitive to changes in the pricing of telecommunications goods and services. As a result, the impact of subsidies on overall industry demand and growth have become increasingly magnified." R. Duch, *Privatizing the Economy—Telecommunications Policy in Comparative Perspective* (Ann Arbor, MI: University of Michigan Press, 1991), p. 102.

   Duch speculates that adversely affected consumers will become more inclined to seek remedies with the legislature and regulatory authorities. According to Duch, increased elasticities of supply and demand provide "a signal and an incentive for consumers and taxpayers to press for the reduction of entry barriers. Such a change in supply and demand elasticities has been occurring in the telecommunications industry over the last twenty years. As a result of dramatic technological advances in the telecommunications industry, capital investment costs have fallen, thereby increasing the elasticity of supply and dramatically reducing the cost of services. All of this has, in turn, increased the elasticity of demand."

- Faster incorporation of new technologies;
- Greater efficiencies than the government-owned incumbent carrier can achieve;
- Survival in an increasingly competitive and volatile environment intolerant of inefficiency and poor productivity;
- Compliance with preexisting public service mandates that include the duty to offer uneconomical but socially desirable services.

10.3 Privatization Models

While some nations may want to maintain restrictions on market entry and access by foreign carriers and equipment manufacturers, the philosophical and political justifications have lost much of their appeal. Nations have permitted foreign investment to augment what the government incumbent can generate internally or secure through loans. Typically, nations limit such investment to a minority position, thereby promoting the formation of strategic alliances that involve the following:

- Investment in a privatized incumbent (e.g., the privatizations in Mexico, New Zealand, Venezuela, Spain, Italy, and Portugal);[4]
- Creation of new ventures (or adding investments) in companies seeking to serve markets made accessible by deregulation and liberalization (e.g., a new carrier competitor to the incumbent, like Mercury in the United Kingdom);
- Teaming with an incumbent carrier to fund infrastructure development in a third country (e.g., Telefonica's investments in privatization and new carrier opportunities in Argentina, Puerto Rico, and Peru);
- Entering into strategic joint ventures with several incumbents and new carriers to develop a regional or global marketing capability (e.g., the Concert venture of AT&T and British Telecom).

4. Many privatizations have involved a series of public stock offerings that prevent a single company from acquiring an ownership majority. Shares in British Telecom, Cable and Wireless, Telefonica, Telecom New Zealand, Chile Telecom, and Telefonos de Mexico are traded on the New York Stock Exchange.

The pace of privatization opportunities has increased, particularly because many nations have made market access commitments, under the auspices of the WTO, that relax or eliminate foreign ownership and licensing restrictions in telecommunications. Incumbents facing competition at home, with the likely consequence of a decline in market share, aggressively scan foreign markets for investment opportunities that they believe will compensate for lost domestic revenues. Most major incumbent telecommunications ventures have made foreign investments with an eye toward securing closer proximity to users, new profit opportunities, and operating experience using foreign standards. Additionally, individual companies may perceive the need to develop business alliances with foreign companies instead of a unilateral attempt. A consortium or joint venture of multinational companies that have done business together may accrue better odds for winning the bidding sweepstakes for a privatizing incumbent carrier, or for a new service franchise.

10.3.1 Statistical Indices Showing Need for Larger Investment

Telecommunications is expected to remain a high-growth industry in many nations whose number of telephones per 100 inhabitants[5] ranges from the single digits to 25. Some developing nations have considered privatization as a way to accelerate improvements in telephone penetration and the overall telecommunications infrastructure. Privatization tenders may require bidders to agree that they will achieve certain benchmarks specifying minimum levels of investment commitments and network upgrades (e.g., number of new telephone lines that must be installed per year). The trend toward privatization of incumbent carriers has gathered momentum, presenting outsiders with possibly the most substantial investment opportunity: the opportunity to own or share ownership in a carrier with an exclusive or near-exclusive franchise in "reserved" services not subject to any facilities-based competition for some period of time.

The concept of privatizing incumbent carriers has become more attractive in large part due to deregulatory initiatives in the United Kingdom and

5. "Telephone penetration and related measures are important barometers ... [and] their utility with respect to international comparisons ... seems to center on their use as measures of comparative development." U.S. Department of Commerce, National Telecommunications and Information Administration, *Telecommunications in the Age of Information*, NTIA Special Publication 91-216 (Washington, D.C.: Government Printing Office, Oct. 1991), p. 193.

the United States that have been perceived as favorably stimulating the national economy and the telecommunications sector. At the time of such groundbreaking initiatives, however, most observers expressed concern and skepticism. In 1984, Britain's telecommunications monopoly became publicly traded and was forced to share a domestic duopoly with Mercury, a Cable & Wireless, Ltd. subsidiary. The boldness of Britain's initiative resulted in part from the shared philosophical views of then–Prime Minister Margaret Thatcher and then-President Ronald Reagan, as well as the belief that AT&T's divestiture and other U.S. deregulatory initiatives had stimulated innovations, efficiency, and beneficial competition. Notwithstanding actions taken in those two key nations, it would take several more years to achieve a growing consensus for reform.[6]

10.3.2 Public and Policy-Maker Support for Ending Status Quo

For privatization to occur, nations need to reject long-standing economic arguments that efficient telecommunications require a single service provider that can spread risk over a large captive-user population and achieve economies of scale and scope. National legislators and regulators require evidence that a more efficient telecommunications industry can evolve through privatization and market entry while not handicapping the ability of the incumbent carrier to recoup its investment in infrastructure development.

A marketplace orientation supports entry by new operators installing competing transmission facilities that are neither a waste of resources nor a threat to the incumbent's ability to provide essential services. In view of the expanded applications demanded by sophisticated users involved in key sectors of the international economy, many nations now consider the single incumbent unable or unwilling to provide all product and service solutions to user requirements.

---

6. In 1988, the European Commission (EC) issued a telecommunications directive ordering liberalizing the terminal equipment market, using its authority pursuant to Article 90 of the Treaty of Rome granting it power to ensure that "public undertakings and undertakings to which member states grant special or exclusive rights" do not thwart free competition, but operate in the general interest of the EC. The government of France challenged that action in the European Court. In mid-1989, the threat of another Article 90 directive on value-added network services prompted the EC ministers to order incremental deregulation from 1990 onward. "This prospect [of increasing deregulation in telecommunications] amounts to a minor revolution in all member states except Britain." N. Colchester and D. Buchan, *Europower—The Essential Guide to Europe's Economic Transformation in 1992* (London: Economist Books, 1990), p. 158.

For a marketplace philosophy to predominate, policy makers must decide that the incumbent carrier should no longer have a mandate to operate as one system with complete responsibility for all telecommunications requirements. Once the government decides to foster competition, the incumbent carrier must abandon its expectation that all revenue streams will remain free from encroachment by newcomers. In exchange for accepting the prospect of some competition, the incumbent typically receives significant relaxation of government oversight, including the migration from rate-of-return regulation, which prescribes the maximum level of profitability, to a price cap regime that creates incentives and financial rewards for efficiency.

A number of philosophical and pragmatic elements factor into the decision whether to privatize, the manner in which privatization occurs, and the corporate structure that results. The primary basis for acting falls into several different categories.

10.3.2.1 Real or Perceived Need to Stimulate Efficiency and Change

Some of the largest privatizations have occurred on the basis that a more commercial approach should replace a public utility service orientation and

> that transfer from the public sector to the private sector will conduce to greater efficiency and superior customer service by injecting entrepreneurial energy and freeing independent managers to formulate their strategies in accordance with more commercial motives. [2]

The privatizations in the United Kingdom, Japan, Australia, and New Zealand fit that model and place telecommunications in the context of a larger marketplace exit by government. U.S. companies that have participated in that type of privatization include BellSouth, with its participation in forming the second Australian carrier Optus, Verizon's major ownership interest in New Zealand Telecommunications, and SBC participation in the acquisition of Telekom, the South African incumbent carrier.

10.3.2.2 The Need or Desire to Secure Outside Capital

Many privatizations occur when national governments seek to encourage private financing of telecommunications ventures. This category includes privatizations of carriers with up-to-date telecommunications systems and high telephone line penetration rates, as well as those with obsolete equipment and low penetration rates. In the former case, adequate capital might have been available from government sources, albeit in competition with other public works. Privatizations in Portugal, Chile, and Malaysia provide examples here. In the latter case, capital is needed to expedite development of a functional infrastructure, even if all or portions of privatization franchise fees flow into the general treasury, as typically has been the case for incumbent carrier revenues in developing nations. Argentina, the Czech Republic, Hungary, and Mexico provide examples for that second category.

Differences in an incumbent carrier's conversion to partial or complete private ownership often depend on the scope of permissible foreign ownership and retained government control. Many nations, including the United States, have laws that limit the percentage of foreign investment. Often, governments retain a so-called golden share that represents a degree of ongoing oversight and perhaps veto power if they consider a business decision inconsistent with the national interest. Such limitations can adversely affect the number of bidders and the amount they offer.

10.3.2.3 A Change in Government Regime or Economic Policy

A broader government commitment to market-driven economies also promotes privatization. In the case of Eastern Europe and the Commonwealth of Independent States, telecommunications became one of numerous previously state-controlled industries targeted for privatization, often occurring in fits and starts. Germany also fits in that category. Once a stalwart for government control, Germany first divided the incumbent carrier into three quasi-private enterprises: banking, postal operations, and telecommunications. It later partially privatized Deutsche Telekom. Germany has impressed observers with its change in regulatory philosophy,[7] even as it confronts the expensive task of achieving resource and infrastructure parity between its two formerly separate nations.

10.3.2.4 Other Pragmatic Reasons

The newfound attractiveness of telecommunications privatization probably also results from recent imperatives and incentives. Telecommunications has substantially become less a public utility undertaking, like the provision of electricity or water, and more a competitive undertaking in which minutes of network use become indistinguishable (i.e., fungible) between different companies, and customers can migrate to carriers that provide the best deals.

---

7. See H. Riche, "Germany's TELEKOM: A New Way of Doing Business in a Liberalized Market," *Telecommunication Journal*, Vol. 58, X-1991, p. 711. Note, however, that "TELEKOM will remain the only provider of the telephone network in Germany" (p. 712).

Carriers have belatedly recognized consumer sovereignty and the need to satisfy user requirements or risk traffic and revenue losses as consumers migrate to other carriers and service providers. Carriers still match half-circuits, but they see real financial advantages in achieving market share outside their domestic markets. Accordingly, telecommunications service providers, particularly new carriers, no longer consider the customers of other carriers as being off-limits.

As never before, carriers vie for high-volume customers, particularly ones with multinational traffic streams. Very much like airline carriers, telecommunications administrations have to devise innovative ways to confer discounts to frequent travelers on their networks to retain loyalty, particularly where carriers can capture traffic volumes by erecting a hub for routing traffic throughout a region.

That newfound pragmatism encourages aggressiveness to the point of poaching the customers of other carriers. So far, the incentive to generate greater traffic volumes has resulted only in selective price cutting. Carriers have recognized the profit potential in providing greater customer service, network functionality, and general flexibility so that customers will design or reroute their networks to traverse the innovative carrier's regional hub. Customer networks increasingly serve specialized customer requirements, making functionality the key factor, with geography and political boundaries becoming less significant variables. Sophisticated users design intracorporate networks that can be reconstituted to route around outages and to exploit new arrangements with carriers.

Telecommunication carriers have to demonstrate greater flexibility to accommodate customer requirements, because the twin impact of technological innovation and policy liberalization all but mandates it. If somewhere within the region, resellers of leased lines and users have opportunities to exploit such innovations, then incumbents face the potential for lost traffic and revenues if they cannot or refuse to innovate.

The balance of power in service negotiations has tilted in favor of consumers. PBXs and other devices within the financial reach and operational expertise of end users make it easier to operate private networks that also can access the PSTN. Additionally, the permeability of regional private-line networks, transborder satellite footprints, resale of leased lines, accounting rate evasion, and a host of other factors favor customers. The prudent telecommunications administration heeds the call for one-stop shopping, heightened responsiveness to user requirements, and upgraded networks. A satisfied customer is less likely to exercise the freedom to lease or buy terminal equipment from new vendors, to shift traffic to a different carrier's facilities, or to relocate all or some operations to take advantage of upgraded networks elsewhere.

The smart telecommunications administration also will seek strategic alliances with other carriers and equipment manufacturers if the alliance consolidates needed skills and technologies, promotes competitiveness and enhances the possibility for replacing domestic or international traffic volumes lost to competitors. That newfound receptivity to collaboration with outsiders, who can inject new thinking, parallels quickening efforts by policy makers and regulators to spur innovation through privatization, deregulation, and a global or regional market orientation.

10.3.3 Handicaps and Advantages in the Privatization Sweepstakes

Privatization opportunities trigger significant interest and competitive bidding. In certain instances, some companies may face handicaps regardless of having readily available investment capital and expertise. Such handicaps include the following:

- Lack of a shared cultural heritage, like that benefiting Telefonica of Spain in its successful efforts to invest in Chilean, Argentinean, Peruvian, and Puerto Rican carriers;
- Alien ownership restrictions that limit foreign investment but that may have loopholes for investors in the region, for example, efforts to achieve more coordinated Europe-wide policies might result in broader use of a common currency and relaxation of foreign ownership restrictions on enterprises incorporated in an EU member nation;
- Primary use of equipment operating on standards different from what the privatizing carrier uses;
- Incomplete vertical integration (i.e., integration of telecommunications equipment manufacturing and telephony operating experience) that would enable the acquiring company to recoup its investment partially through an affiliate's equipment sales;
- Lack of government foreign aid in telecommunications that would enable the acquiring company to raise its bid on the expectation that it could secure some financial support from its government.

On the other hand, some prospective investors can achieve significant advantages in terms of access to commercial capital, operational and management skills, administrative expertise (including tariffing and working in both regulated and deregulated environments), and technological leadership.

10.4 Privatization Occurs in a Number of Ways

Nations increasingly recognize that privatization "makes it possible in practice to get rid of certain bureaucratic procedures and other inhibitions on the efficacy of the businesses concerned" [3]. Most countries readily accept that telecommunications has a significant impact on a nation's economic viability by providing an infrastructure for transacting commerce.

Nations pursuing a privatization campaign have a variety of options and models to consider. To assess fully the privatization opportunities available, a review of the types of privatizations may prove helpful. This section sets out an outline of several investment opportunities available to companies and individuals.

10.4.1 Public Stock Offerings

Many privatizations result in government issuance of stock that reflects ownership shares in all or a portion of the incumbent carrier. The sale may first concern one or more transactions involving a large portion of the available shares, a portion of which the purchaser may have to resell to meet foreign ownership limitations. Privatizations in Argentina, Australia, Brazil, the Czech Republic, Chile, Denmark, France, Germany, Greece, Hungary, Italy, Japan, Mexico, the Netherlands, New Zealand, Peru, the Philippines, Russia, Singapore, Spain, Switzerland, and the United Kingdom involved public stock offerings. Each case has its own particularities in terms of what percentage, if any, the government retained, the percentage of stock held by single investors, whether labor unions, pension funds, or subscribers received stock ownership opportunities and the extent to which government retained a golden share capable of vetoing any decision deemed contrary to the national interest.

10.4.2 Investment Opportunity in a New Facilities-Based Competitor

Some privatization observers are surprised when they learn that relatively few nations initially authorize facilities-based competition. Many nations first spin off the incumbent carrier into a private or quasi-private enterprise and only authorize niche market competition (e.g., cellular radio). Some nations that opened markets to facilities-based competition early on are Australia, Canada, Chile, Japan, Korea, New Zealand, Sweden, the United Kingdom, and the United States.

Opportunities to acquire the former incumbent carrier or to invest in a new facilities-based competitor present substantial financial rewards over time, but they do carry significant risk. Accordingly, the typical bidder in such sweepstakes is a major telecommunications company interested in developing a multinational presence.

10.4.3 Competition in Niche Markets

A number of nations, keen on accelerating facilities construction and attracting foreign investors to build networks, have funneled attention to niche markets outside the basic telephone service sector. The frequently used mechanism involves granting a franchise for niche markets like mobile services, for example, cellular radio, business teleports, which provide reliable international satellite access, and microwave voice and data networks that overlay the existing, often unreliable wireline plant. Franchise terms often include a timetable for building a particular network and may include a variety of benchmark service obligations. Nations can improve the attractiveness of a franchise by granting other financial incentives, for example, easier repatriation of funds and favorable tax treatment.

Three primary models are used for niche market franchise opportunities:

- *Build, own, and operate.* This model allows new domestic and foreign enterprises to invest, own, and operate a franchise. Examples include the variety of cellular radio franchises (e.g., BellSouth in Peru, Ameritech [now part of SBC] in Poland), private overlay microwave radio networks (e.g., Verizon/US West in the Czech and Slovak Republics), and digital data overlay networks (e.g., IBM's partnership with the Czech and Slovak incumbent carriers and Tesla, an equipment manufacturer).
- *Build, operate, and transfer.* The build-operate-transfer (BOT) model allows an outsider to generate profits from niche or even core, reserved services but only for a fixed time period, after which the franchise lapses and ownership of the plant transfers to the government franchiser. This option provides developing nations with the

opportunity to acquire cutting edge technology that they could not afford to buy, and enables them to take the title to the technology after granting a long-term franchise. Foreign enterprises participate in such tenders, because of the opportunities to establish a presence in new markets. The franchise terms dampen short-term returns, but participating foreign carriers and manufacturers consider longer-term opportunities, for example, establishing an early presence that may lead to other franchise awards and market share for equipment. Examples of BOT arrangements can be found in Thailand and Indonesia.[8]

- *Build, transfer, and operate.* This model allows an outsider to sell equipment and operational services (e.g., very small aperture terminal satellite networks) in nations desirous of foreign investment and expertise (e.g., China and Saudi Arabia), but which are leery of permitting foreigners to own and operate facilities. The foreign enterprise installs needed facilities, confers title to the government, and receives either installment payments or a portion of the operating revenues as compensation for both the equipment and facilities management.

10.4.4 Joint Ventures That Include Government Participants

Joint ventures are another vehicle for public and private enterprises to achieve a critical mass of finances, political pull, and operational skills. The structure also enables a government to participate in entrepreneurial telecommunications ventures without privatizing the incumbent carrier. EUTELSAT represents a cooperative approach that enables nations to sponsor satellite competition, including direct broadcast satellite television without expressly changing national broadcasting laws and regulations. Another example is the AsiaSat satellite venture with government owners, the China International Trade and Investment Company, and private investors (Cable & Wireless of Britain and Hutchison Whampoa of Hong Kong). Many Chinese ventures blend government and private investment.

---

8. See P. Smith and G. Staple, *Telecommunications Sector Reform in Asia—Toward a New Pragmatism*, World Bank Discussion Paper No. 232 (Washington, D.C.: World Bank, 1994), p. 55.

10.4.5 Limitations on Foreign Investment

Privatization and its foundation in law, policy, and deregulation favor foreign investment opportunities, as do regional and multinational trade agreements. However, privatization does not mean that governments that have closed markets to foreign investment or restricted procurements to domestic sources suddenly have become free marketers. Most nations impose a variety of limitations on foreign participation, including the outright prohibition or percentage limitation on foreign investment in telecommunications. Many nations limit competitive opportunities to niche markets, and even where competition is allowed, the incumbent facilities-based carrier retains plenty of opportunities to stifle competition, for example, providing inferior, more expensive, and inflexible interconnection to the PSTN, predatory pricing, unfair trade practices, and cross-subsidies.

Other factors may dampen the attractiveness of investment opportunities. Governments may place an unrealistically high price tag on the privatized carrier in view of political factors restricting the level of permissible profitability, as occurred in Puerto Rico. Governments also may impose burdensome benchmark service commitments that the winning bidder must make (e.g., a schedule for installing a minimum number of new telecommunications lines). In the best of circumstances, it is difficult to reduce to present value a revenue stream generated by a nation's telephone company, particularly where substantial up-front investment is needed to upgrade the infrastructure and the possibility that government might subsequently limit the scope of reserved services or allow additional market entry.

Some nations view privatization as a one-time opportunity to acquire hard currencies and to replace an obsolete physical plant. The daunting task of building largely from scratch a nation's telecommunications infrastructure should lower the franchise cost, but the promise of newly accessible growth markets abroad in conjunction with limited further growth in mature domestic markets often triggers a bidding war, particularly if the nation auctions off franchise opportunities. Carriers like AT&T, British Telecom, Cable & Wireless, Deutsche Telekom, Telefonica of Spain, Sprint, and France Telecom intend to develop a global marketing presence primarily to serve multinational telecommunications companies. Ownership of a geographically diverse set of telephone companies supports their participation in a global network. Such investments also reflect the need to find new markets, because domestic policies have opened previously captive markets to competition.

Part of the heightened interest in foreign telecommunications service opportunities may have resulted out of a defensive strategy. Some companies believe that market access restrictions will grow worse over time, particularly after the implementation of initiatives aiming primarily to improve intraregional trade and global competitiveness. That fortress view stimulates current investment on the grounds that an in-country operation will fare better in the face of ongoing or worsening discriminatory practices in equipment standardization, carrier procurements, testing and certification, and licensing of service providers.

10.5 The Next Step: Privatized Carriers Merge or Launch Strategic Alliances

The pace of technological innovation, privatization, experimentation with competition, and globalization provides unprecedented opportunities and risks. By far, strategic alliances have become the most common vehicles for such investments. Incumbents with declining control over domestic consumers find in alliances the opportunity to serve new markets as well as to acquire resources and expertise, perhaps unavailable within the organization. If an incumbent faces what it considers the worst possible policy initiative, loss of monopoly control, then an alliance with a large multinational enterprise is one, often successful way to bolster skills to retain a dominant market share or to acquire shares in new markets. For example, the incumbent Canadian carrier, Stentor, Inc., has developed a strategic alliance with MCI, in part to achieve access to MCI's cutting edge technological leadership in such services as virtual private lines and networks. Ironically, AT&T invested in a major long-distance service challenger.

Increasingly, incumbents cannot rely on cartel management of new technologies, facilities, and services. Previously, incumbent carriers held significant ownership interests and voting power in global satellite cooperatives like INTELSAT and submarine cable consortia. They were able to control the deployment of additional capacity by creating a quasi-official consultative process ostensibly to forecast demand, plan for new transmission facilities, and allocate investment shares. Incumbent carriers were a dominant force in regional policy making, primarily because operational and regulatory matters were controlled by single entities in most countries. Incumbent carrier coalitions, like the Conference of European Post and Telecommunication (CEPT), bolstered individual incumbent carrier powers by promulgating policy suggestions readily accepted by the few independent regulators.

In the last few years, a number of strategic ventures have responded to the deteriorating prospects for marketplace control by preexisting organizations. In the satellite marketplace, private systems, separate from the cooperative ownership structure, have been established to serve growing consumer demand for television without borders, for example, new networks for program distribution directly to home satellite antennas, satellite master antennas, and cable television systems. Such new enterprises include program packagers like Sky Television in Europe and Star TV in Asia and new carriers like Astra in Europe and Asia Pacific Telecommunications. Even incumbent carriers like Cable & Wireless have invested in separate satellite systems, particularly as such carriers acquire greater consumer acceptance and fewer service restrictions.

In the submarine cable marketplace, governments have accepted private ventures, separate from the conventional incumbent carrier–AT&T consortium approach. The consultative process cannot prevent market entry by operators of private submarine cables. The FCC has decided that private carrier capacity should not even be considered in the forecasting of demand, and the Commission has authorized new submarine cable projects on the basis of promoting competition at private investor risk, rather than on conventional grounds based primarily on traffic forecasts. Examples include PTAT-1 in the North Atlantic, a joint venture of Sprint and Cable & Wireless; the North Pacific Cable, a consortium that includes Cable & Wireless, Pacific Telecom of the United States, and International Digital Communications of Japan; the Fiber Optic Link Across the Globe first organized by NYNEX, now a part of Verizon Communications; and the numerous submarine cables of Global Crossing and 360 Networks.

10.5.1 Types of Alliances

With changing rules of the road, additional and more diverse opportunities exist for market entry by strategic alliances. Set out in this section is an outline of the various types of alliance opportunities.

10.5.1.1 New Facilities

The bilateral and multilateral nature of international transmission facilities promotes joint ventures. The PTAT-1 submarine cable in the Atlantic Ocean provides an example. Cable & Wireless envisioned the development of a global digital highway to link the world's trading centers and the various telephone companies it owned or managed. Cable & Wireless understood that most nations, including the United States, impose restrictions on "alien" ownership of transmission facilities. The Submarine Cable Landing Act of 1921 requires reciprocity in the landing of cables on U.S. soil. Accordingly,

C&W sought strategic partners in the United States to own and operate the western half of the PTAT-1 cable and the eastern half of the North Pacific Cable. Through investment in a Pacific and Atlantic submarine cable network, Cable & Wireless can provide a seamless link between Hong Kong, Japan, and other cities in the Asia-Pacific region and the United States and Europe.

10.5.1.2 Government-Created New Market Niches

Governments create investment opportunities when they negotiate bilateral agreements that authorize new types of services and carriers. An agreement between the United States and the United Kingdom legitimized the market for international value-added networks by distinguishing between networks that provide enhancements to leased lines and networks that simply resell basic services (e.g., long-distance telephone calling) via leased lines. The United States and Britain recognized the former as an exception to the general prohibition on leased-line resale. Other nations have joined that view, making the U.S.-U.K. bilateral agreement a basis for market growth, investment, and future revisions to ITU Recommendations.

10.5.1.3 New Carrier and Franchise Opportunities

The largest number of strategic opportunities results when governments license new carriers, franchises, and new technology trials. Those opportunities require less capital than incumbent-carrier privatization but still can involve quite large initial investments and provide lucrative opportunities.

10.5.1.4 Additional Local and Long-Distance Companies

A few nations, led by the United Kingdom, New Zealand, and the United States, have abandoned the natural monopoly view for core, basic services like local and long-distance telephone service. Other nations, including Australia, Canada, Korea, and Sweden, first created a duopoly (a monopoly shared by two competitors). For example, in the United Kingdom, a Cable & Wireless subsidiary, Mercury, generated the only alternative to the incumbent carrier, British Telecom. Cable television companies provided additional local exchange service competition, and in 1997 additional facilities-based carriers were authorized.

10.5.1.5 Special-Purpose Networks

Nations disinclined to authorize market entry for core local and interexchange services can still stimulate competition in niche markets. For example, AT&T, IBM, Westinghouse, Itohchu of Japan, and Deutsche Aerospace have teamed with the former Soviet Union's Aeronavigation Research and Development Institution to modernize the Commonwealth of Independent States' air traffic control system by 2005. Electronic Data Systems, Inc. (EDS) provides turnkey technology integration and systems support for Polish, Czech, and Slovak cellular systems.

Many nations, with developed or developing economies, have authorized two or more competitive mobile radio operators. Nations with obsolete, unreliable, and overtaxed wireline facilities have permitted foreign ventures to install microwave and fiber optic cable networks for business applications. Competition in this sector results, because decision makers believe the market can support multiple carriers or because the incumbent carrier has shown an unwillingness or inability to provide reliable service.

For example, Hong Kong has more than four facilities-based mobile radio telephone carriers, some of the lowest rates and one of the highest market penetrations for service in the world. Ameritech and France Telecom operate the Polish cellular radio systems. A joint venture between Sprint and RosTel, a new organization formed by the Central Telegraph of the Russian Ministry of Communications, provides new high-speed data networks.

10.5.1.6 Cable Television

Many nations award cable television franchises instead of allowing the incumbent telephone company to provide the service. In authorizing a separate cable television wire into the home, decision-makers have created the potential for facilities-based telecommunications service competition. Developments in digital, fiber optic communications present the potential for cable television operators to provide local telephone service in addition to video services. A variety of foreign enterprises have entered this market. For example, US WEST, Time Warner, Kinnevik of Sweden, Tele-Communications, Inc. (now part of AT&T), and United International Holdings are partners in cable television ventures in the United Kingdom, Sweden, Norway, and Hungary.

10.5.1.7 Manufacturer-Carrier

Manufacturers pursue joint ventures with carriers to develop markets and to test equipment. For example, Motorola teamed with Cable & Wireless to develop personal communications networks in the United Kingdom. Andrew Corp., a U.S. manufacturer of microwave and satellite and LAN equipment, has teamed with the Moscow metro subway to develop a high-speed fiber optic network.

10.5.1.8 Multiple-Carrier Alliances

Perhaps belatedly, the world's telecommunications carriers have acknowledged consumer sovereignty and the need to become more responsive. Carriers have expanded the array of available service options and have convinced increasing numbers of users to trade internal telecommunications management for outsourcing: relying on the expertise of carriers and systems integrators to provide and manage all necessary design, negotiation, procurement, coordination, and project management functions. Such enterprises deliver a turnkey network, ready for use.

The combination of new multiple-carrier global alliances and success in persuading users to rely on outsiders for telecommunications management presents numerous investment opportunities for enterprises. In fact, the prevailing view seems to be that if a carrier does not become a member of one or more alliances, it will lose market share, perhaps much more.[9] Many global alliances manage international private networks used by multinational businesses or provide some sort of quasi-public networking option, commonly known as virtual private networks or software-defined networks.[10]

10.5.2 Why Alliances Fail

While press releases herald the onset of yet another strategic alliance, little is reported when parties abandon a venture. Failed alliances demonstrate that not all investment opportunities result in profits and successful blending of corporate cultures and objectives.

Telecommunications alliances generally result when two or more enterprises perceive a strategic opportunity for higher revenues and profits from

9. "Speaking in London at the Third Economist Conference on Telecommunications ... , DBT's Director General of International Affairs, Klaus Grewlich, stated boldly that 'to survive, operators must become global players,' adding that those who do not master globalisation 'may become candidates for not surviving, even in their own markets.'" "Diversification: Safeguard or Suicide?" *Public Network Europe*, Vol. 1, No. 10, Oct. 1991, p. 60.

10. See J. Williamson, "Navigating Through the Global Managed Network Maze—Outsourcing: Hit or Myth?" *Telephony*, May 4, 1992, p. 9. A European publication reported that by 1994 this market would have a value of $6 billion. "BT's Global Push for Outsourcing," *Public Network Europe*, Vol. 1, No. 10, Oct. 1991, p. 13.

collaborating rather than individually marketing. Ventures fail when they do not stimulate demand, capture additional market share, or result only in customer migration from one set of services to new ones. If the venture has to price its new services at a discount to rates otherwise available, the venture's success cannibalizes participating carriers' revenue streams. Some of the global one-stop shopping ventures and virtual private networks may generate user churn (approximately equal numbers of new customers and existing customers canceling service) and migration between services instead of growth, but users increasingly demand such discounted service packages.

Joint ventures and alliances in telecommunications require extensive coordination. While incumbent carriers used to share similar objectives and philosophies, varying degrees of privatization, liberalization, market entrants, and public policy initiatives have created significant differences in approach that may not be fully understood by joint-venturing parties. The alliance may stimulate vastly different strategies among investors, and the operating climate and strategy for success may vary by country, particularly because regulatory policies vary.

In Europe, for example, the United Kingdom stands at the vanguard for promoting marketplace competition and deregulation. Countries like Belgium, Germany, the Netherlands, Sweden, and Switzerland stand midway, having retained complete or partial government ownership for a period of time after initial privatization but also permitting limited competition by resellers. France, Greece, and Italy have lagged behind, primarily on political grounds, including a lack of consensus on the merits of privatization.

Blending the national telecommunications policy-making posture and a company's individual business plan presents an even greater diversity of cultures and objectives. For example, some telecommunications companies have learned how difficult it is to develop a successful venture that combines expertise in telecommunication equipment and services. AT&T's joint venture with the Italian manufacturer Olivetti, British Telecom's acquisition of the Canadian telecommunications equipment manufacturer Mitel, and IBM's attempt to vertically integrate with investments in MCI and Rolm failed to achieve desirable synergies. Other negative impacts on the likelihood for alliance success result from changing corporate strategies or the need for restructuring due to poor financial performance. An unexpected change in regulatory policy also can have an adverse impact, for example, permitting more quickly installed master antenna television systems to compete with cable television in Hong Kong.

10.6 Globalization Strategies

The most successful alliances occur when each participating enterprise can make a contribution and concludes that the investment of time, money, and effort fits with overall corporate objectives. The parties must understand the markets they will attempt to serve, including the climate and policies of the nations where they will operate. Similarly, the parties must achieve a synergy that creates concrete disincentives for any participant to exit on the assumption that it could perform all roles and therefore should capture all financial benefits. If the venture seeks to establish a global presence, it should have a geographically diverse composition, including representatives operating in all key markets.

The parties must commit to the long-term, something that many corporations and business environments cannot support. Likewise, they must conduct preliminary research to determine whether corporate cultures, regulatory climates, and trade factors promote a positive collaboration. A successful alliance also may result from less obvious and external factors than expertise, financial resources, access to markets, proximity to customers, and reputation. Likewise, the opportunity to erect an alliance also may depend on personal relationships, a company's desire to evidence its global orientations, labor and tax considerations, and the desire to operate in an environment using different technical standards.

The ability to contribute something meaningful to the venture also may depend on whether a country's technology transfer policies permit sale of computers, fiber optic cables, and other high-technology innovations, despite national security concerns. For example, before the relaxation of technology transfer restrictions, U.S. manufacturers complained that they could not propose state-of-the-art fiber optic cable transmission capabilities for projects in the Commonwealth of Independent States and China. Concerns about technology transfer of strategic or military importance and below-cost pricing to attract business have restricted the use of foreign launch facilities by U.S. firms or firms launching U.S.-manufactured satellites. Additionally, a venture may have to satisfy public policy goals of regulatory administrations in the host nation (e.g., contributing skills training, committing to local employment, and providing venture capital).

Over the last 10 to 15 years, the global economy has grown more integrated and interdependent. For example, consumers have grown to expect that merchants everywhere will accept credit cards. A complex global network must exist to process credit card verification orders across borders, time zones, bank systems, and different carrier networks in a few seconds. Newspapers like the *Wall Street Journal*, the *New York Times*, the *Washington Post* (through a joint venture that publishes the *International Herald Tribune*), and *USA Today* seek a global presence to acquire and disseminate news. "Cross-border business has been driven forward by three main things: falling regulatory barriers to overseas investment; tumbling telecommunications and transport costs; and freer domestic and international capital markets in which companies can be bought, and currency and other risks can be controlled" [4]. Additionally, foreign investment no longer automatically triggers fears of lost sovereignty[11] and control in most nations.

A key technological characteristic of telecommunications is its ability to make the data flows and broadcasts of one nation accessible in other nations. Some nations, primarily developing ones, still inveigh against transborder data flows and cultural imperialism. However, most nations acknowledge that transborder operations expand the utility derived from a telecommunications investment, help spread costs over a larger set of users, and enable enterprises to achieve scale economies.

Globalization exploits fundamental technological characteristics of telecommunications. For example, satellite footprints traverse national boundaries; three geostationary orbiting satellites can provide service to all populated regions of the world. Globalization results when corporate enterprises perceive strategic, tactical, or defensive reasons for investing in a foreign country. It can be more easily achieved when foreign governments treat firms on a relatively equal basis regardless of their nationality and where goods, services, transactions, capital, and labor can move freely.

Foreign investment also occurs simply because of the need for closer proximity to customers. Companies that think globally and act locally, can reduce costs (e.g., transportation) and risk (e.g., currency volatility). Additionally, they can minimize the potential for harm from discriminatory policies on imports, through duties, taxes, and subsidies. Telecommunications enterprises also have globalized in part simply to keep pace with their multinational customers.

Telecommunications equipment and service providers have to make strategic decisions on how to maximize profits and market share in a global economy. They can choose from the following plans of action:

- Compete from afar by not investing to create a foreign presence;

11. "[I]n the political domain the power of satellite telecommunications to transmit text and pictures cheaply has done more to weaken the grip of governments than any scheming capitalist giant." "Multinationals Back in Fashion," in "A Survey of Multinationals," *The Economist,* Vol. 326, No. 7804, Mar. 27, 1993, pp. 6–7.

- Compete in-country, where allowed;
- Generate market share in-country through acquisitions;
- Cooperate through joint ventures.

Until quite recently, historical, political, regulatory, and economic forces made competition unlikely. Try as they might, equipment manufacturers succeeded, if at all, in capturing only a limited market share in a region or a cluster of nations outside the home market. Manufacturers might compete in open procurements, but the combination of political and financial pacts, foreign aid tied to procurement of the donor nation's equipment, cultural affinity, and industrial policy tended to favor a national hero.

In the services arena, carriers matched half-circuits and entered into correspondent relationships that respected national sovereignty and positioned carriers in a cooperative posture. Carriers still match half-circuits, invest collectively in consortia and cooperatives, agree on toll division arrangements, and split responsibility for the maintenance of transmission facilities.

Governments and users, not telecommunications incumbents, have stimulated the urge to compete. By eliminating absolute monopolies through authorized market entry by resellers and even facilities-based carriers, governments have forced incumbent carriers and manufacturers to consider how to make up for lost market share. Users have grown more sophisticated in articulating their service requirements and more adept at pursuing routing opportunities that reduce cost and best satisfy their needs. Often the potential for demand stimulation at home, even if the incumbent becomes a more efficient operator, pales in comparison to the opportunities abroad.

The greatest opportunity for a company to globalize may lie in the pursuit of infrastructure improvements in nations that lack a reliable network and have telephone line penetration rate below 25 per 100 inhabitants, particularly if per-capita income will so increase that the majority of the population can afford what heretofore constituted a luxury. Many of those nations recognize the need to allow more qualified outsiders to assume the logistical and financial responsibility for speedy development of an advanced telecommunication infrastructure. Newly independent nations, for example, the split-up Soviet Union, and nations no longer subject to central planning in a political bloc, for example, Eastern Europe, now can pursue long-overdue infrastructure improvements without first receiving outside authorization.

The next best opportunities lie where national trade policies favor some degree of market access and foreign investment. Several nations with vast populations and single-digit line penetration per 100 inhabitants have accelerated efforts to remove telecommunication bottlenecks and outages, recognizing the drag on the economy. Such nations include China, India, Malaysia, and Indonesia.

Telecommunications enterprises can acquire foreign market share overnight by investing in incumbent wireline and wireless operators (e.g., Deutsche Telekom's investment in Sprint and Voicestream, a cellular radio company). Many incumbents have vast cash reserves available, the product of a captive customer base, the absence of a requirement to flow through revenues to the national treasury, and, perhaps, relatively ineffectual rate regulation. Foreign investment occurs when an enterprise thinks it can achieve operational synergy with the acquired company while building on the company's existing market share.

Many globalizing telecommunications enterprises prefer to pursue joint ventures and secure global or regional alliances with counterparts of equal capitalization, for example, AT&T's Concert venture with British Telecom. Limitations on foreign ownership support that alternative, as does the recognition that ventures marketing global services need geographically diverse participants.

Many incumbent players face new domestic competition and substantial threats to the status quo. Global and regional alliances represent a recognition of an individual firm's limitations and the real or perceived benefits accruing from an affiliation with similarly situated enterprises. A company with a long operational history in a particular region is better equipped to know the particular features, limitations, customer requirements, and other features specific to the area.

Telecommunications ventures typically involve substantial investments, which a number of players are better able to share. Similarly, the risk attendant to any particular venture can be spread across a number of investors. If a venture proposes to operate in a large geographic area and to assume end-to-end responsibilities, a strategic or tactical alliance of geographically diffused players is essential. It remains to be seen whether such ventures will have staying power and constitute the basis for a new industrial structure or simply are short-term efforts to pool investment and reduce risk.

10.6.1 Types of Global Alliances

This section presents a list of ways enterprises agree to structure a global alliance.

- *Cooperative equity arrangements.* Carriers pool investments in costly satellites and submarine cables through global cooperatives like INTELSAT, regional cooperatives like EUTELSAT, and submarine cable consortia.
- *Cooperative bilateral agreements.* Carriers negotiate accounting rates and other contractual terms and conditions for the acceptance and routing of traffic and the division of toll revenues.
- *Memoranda of understanding.* Companies agree to exchange personnel and cross-license technology. For example, MCI's investment in the dominant Canadian carrier Stentor included an agreement to cross-license proprietary virtual private networking technology that uses software for speedy creation and disassembly of private networks.
- *Joint ventures.* Companies negotiate short- or long-term contracts to bid on projects involving, for example, the installation and maintenance of equipment and services on a regional or global basis by the AT&T-British Telecom Concert venture.
- *Consortia of foreign telephone companies and local partners.* Major new foreign-license opportunities for a second facilities-based carrier (e.g., Australia's second carrier Optus, a consortium that includes Britain's Cable & Wireless; BellSouth of the United States; and Australian partners AMP, an insurance company, and Mayne Nickless, a transport company) or the privatization of the single incumbent carrier favor the creation of a consortium that comprises foreign players with telecommunications expertise and local partners with money to invest, national citizenship, and political ties.
- *Consortia of foreign telephone companies, foreign equipment manufacturers, and local partners.* An extension of the preceding model adds a manufacturer that provides expertise in ventures requiring new types of equipment. Numerous cellular radio and personal communication network ventures provide an example.
- *Consortia that include large-user groups.* Sophisticated, high-volume telecommunications users join in projects that should better serve their requirements. Examples include the consortia of Cable & Wireless, Itochu (a Japanese trading company), Toyota, Pacific Telesis (now part of SBC), Merrill Lynch, Singapore Telecom, and US Sprint, which developed the ASPAC transpacific fiber optic, submarine cable.

- *Joint ventures targeting niche markets.* A number of ventures target profitable niche markets rather than go after the telephone company franchise. Examples include value-added networks, outsourcing, cellular radio, satellite teleport, and microwave networks that overlay the obsolete wireline network.

10.6.2 Challenges Faced by Alliances

Alliances require partners to share objectives and to identify likely long-term benefits. The partners must devise a structure that maximizes the contributions each brings to the venture and allows the venture to evolve as conditions change. The parties must share a vision for the venture and agree to make it financially difficult for anyone to exit.

Telecommunications alliances face industry-specific challenges. The alliance may have investors subject to different degrees of government oversight and domestic competition. The venture may have difficulty defining a single, consistent corporate culture, because its investors may have inconsistent perspectives based on the extent to which they have privatized and abandoned prosocial obligations. Complex regulatory duties, like universal domestic service, may challenge the attention and financial wherewithal of investors to concentrate on making the alliance work.

The nature of the venture's mission vis-à-vis other activities of the investors presents an additional challenge. Strategic alliances may involve strange bedfellows who compete vigorously for business in other markets. For the alliance to work, the parties must carve out markets where they agree to collaborate rather than compete, no matter how attractive an option to go it alone is. The benefit of synergy must persist. Otherwise, individual investors will perceive a financial advantage in exiting the alliance.

National carriers and equipment manufacturers have rushed to join alliances, often out of concern that the failure to do so will result in lost market opportunities, particularly in the face of trade actions that do more to bolster a region's marketing prowess than to make the region accessible to outsiders. The proliferation of regional trading pacts (e.g., the Single Europe Act and NAFTA) creates concern that the marketplace will concentrate into an oligopoly of successful alliances that have a robust marketing presence in each of the three key markets of the world: Europe, the Americas, and Asia/Pacific.

It remains to be seen whether alliances are a productive response by corporate managers to their individual firm's limitations and the inability to operate efficiently everywhere, or a passing fad abandoned when investors become disillusioned by mounting financial losses and covenants not to make individual market entry. For the time being, the following justifications for tactical alliances seem strategic enough:

- Spreading risks;
- Sharing in high-cost ventures;
- Improving the information flow between equipment manufacturer and service provider;
- Promoting economies of scale and scope;
- Reducing the time from innovation to market rollout;
- Sharing research and development costs;
- Finding an expedient way to access an otherwise closed or partially closed market.

References

[1] Duch, R., *Privatizing the Economy—Telecommunications Policy in Comparative Perspective*, Ann Arbor, MI: University of Michigan Press, 1991, p. 3.

[2] Oxman, S., "Privatization: A Strategic and Financial Perspective," *Record of Proceedings, 6th World Telecommunication Forum, Part 3 Regulatory Symposium: Competition and Cooperation in a Changing Environment*, Geneva: ITU, Oct. 9–11, 1991, p. 211.

[3] Carlsberg, B., "Regulator's Perspective in the Role and Purpose of Regulation," *Record of Proceedings, 6th World Telecommunication Forum*, Part 3, 1991, pp. 7, 10.

[4] "Multinationals Back in Fashion," in "A Survey of Multinationals," *The Economist*, Vol. 326, No. 7804, Mar. 27, 1993, p. 6.

11

Developing Trends in International Telecommunications

Forecasting developing trends in telecommunications has become more difficult as an environment increasingly driven by the Internet and e-commerce generates greater volatility and faster changes. However, some conclusions about the consequences of recent changes can be determined, along with an assessment of how those changes may trigger even more significant future developments. While the Internet does not change everything about the structure, function, and culture of telecommunications, it has become clear that the characteristics of data processing and communications markets increasingly dominate older characterizations of a cozy, clubby, and predictable public utility environment.

11.1 An Internet-Driven Economy

The World Wide Web site of Business 2.0, an Internet commerce publication, suggests 10 driving principles of the new economy: matter, space, time, people, growth, value, efficiency, markets, transactions, and impulse.[1]

1. See *Business 2.0,* "The 10 Driving Principles of the New Economy," March 1, 2000, available on-line at http://www.business2.com/research/principles/2000/03/01/14842.

11.1.1 Matter

The new economy provides alternatives to many transactions that heretofore involved the traditional physical delivery of goods. Consumers still require plenty of physical materials and will continue to rely primarily on bricks-and-mortar institutions and facilities. Nevertheless, "[p]rocessing information is dramatically more powerful and cost-effective than moving physical products" [1]. Companies like Yahoo! have achieved multibillion-dollar valuations despite having few physical assets: The value accrues from serving as an information broker that can provide consumers with user-friendly access to desirable content. Yahoo! also operates as an audience broker, providing advertisers with access to a demographically desirable audience. Amazon.com offers the largest inventory of books, but much of what it offers depends on close management information system integration with book publishers and quick access to their inventories.

Telecommunications and the Internet facilitate the transmission and processing of often valuable but intangible assets, for example, information, news, intellectual property, electronic funds transfer, and financial services. Internet mediation will not completely replace bricks-and-mortar industries, but it increasingly accrues a comparative advantage for commerce involving intangibles in all or part of the transaction. While there may be a bricks component to the transaction (e.g., a warehouse that holds the book, compact disc, grocery, and other physical product), point-and-click buying can provide a faster, better, smarter, cheaper, or more convenient option in the sale of the physical asset, in addition to the management of features and functions needed to effectuate the sale. Other Internet-mediated options involve services that do not require a physical product (e.g., paperless airline tickets and hotel reservations secured from home rather than a physical visit to a travel agent).

See also, Aileen A. Pisciotta, "Regulation of International Communications in the Age of the Internet: Lagging Behind the Future," *International Lawyer*, Vol. 33, Summer 1999, p. 367; Henry H. Perritt, Jr., "The Internet as a Threat to Sovereignty? Thoughts on the Internet's Role in Strengthening National and Global Governance," *Indiana Journal of Global Legal Studies*, Vol. 5, Spring 1998, p. 423; Marcus Maher, "An Analysis of Internet Standardization," *Virginia Journal of Law & Technology*, Vol. 3, Spring 1998, p. 5; Steven R. Salbu, "Who Should Govern the Internet? Monitoring and Supporting a New Frontier," *Harvard Journal of Law and Technology*, Vol. 11, Winter 1998, p. 429; Neil Weinstock Netanel, "Cyberspace Self-Governance: A Skeptical View From Liberal Democratic Theory," *California Law Review*, Vol. 88, March 2000, p. 395.

11.1.2 Space

Internet users do not consider distance and space any sort of impediment. In cyberspace, the Internet constitutes a single cloud from which any user can access any other user and any Web site regardless of location. That distance insensitivity results from the manner in which individual Internet network operators agree to interconnect their facilities and the access pricing policies generally in force that support connectivity for a fixed price, which previously may have been set at zero.

The distance-insensitive pricing structure of the Internet results from a conscious pricing decision made by operators. Contrary to the observations of some, it does not result from the complete "death of distance" as a factor in the cost and charges for service. The laws of physics still apply in the sense that it costs more to route traffic halfway around the globe than across town. However, that does not mean that the charge for such routing traffic is commensurate with the cost. Like the Internet, telecommunications charges can be averaged and "postalized" with one fixed price regardless of the distance covered. "Snail-mail" postal services typically offer fixed rates, and increasingly long-distance telephone companies do so as well, at least for in-country services (e.g., $.10 a minute for calls anywhere in the United States).

Profound competitive consequences result from distance insensitivity in terms of the cost and the reach of access via the Internet and telecommunications. If distance has become equalized, then a distant supplier of goods or a service can become a competitor to a locally based supplier. Put another way, the potential for robust competition increases when a medium like the Internet makes it possible for anyone, anywhere to compete for business generated by anyone, anywhere. Amazon.com has sold books to 1.5 million people in 160 countries. Those consumers use Amazon because it provides a faster, better, smarter, cheaper, or more convenient alternative to local options.

Internet mediation does not signal the demise of local merchants any more than the arrival of large, national retailers like Wal-Mart and Target results in the elimination of a downtown central business district filled with local merchants. However, it does mean that some local merchants will fail unless they can adapt and provide local consumers with such better service and other desirable features that the consumers forego Internet-mediated options. The local store that provides better customer service and has a desirable item in stock may get the sale over an Internet vendor that can offer only next-day or second-day delivery. A local service provider may generate the trust and confidence of consumers uncomfortable with buying services such as stocks, insurance, mortgages, prescription drugs, or air travel tickets via the Internet. On the other hand, once a consumer accepts mediation, whether via the Internet or telecommunications, a vendor, customer service representative, and other elements of a commercial transaction can be located anywhere.

11.1.3 Time

The concept of time changes in the new world information order. It becomes insignificant in terms of operating hours, because the Internet offers 24/7 service. It collapses in the sense that consumers expect fast, convenient, and immediate access. It offers less continuity in the sense that circumstances and conditions change more quickly and with less trajectory from prior events.

Time compression means that the life cycle of a product or service shortens, as does the time between manufacture and delivery of a good and between order and delivery of a service. Operating on Internet time means that even physical products, like an Internet router, may become obsolete in months rather than years. Internet time also means that vendors have to operate more quickly and more efficiently yet also provide customized solutions. For example, Dell Computer offers consumers the ability to specify and customize the features of a personal computer while guaranteeing a delivery time of a few days. Dell largely eliminates the concept of the one-size-fits-all assembly line by according consumers the opportunity to select from a far larger set of options than previously available. Data mining and information processing also provide Internet ventures with the ability to customize a marketing pitch to individuals and companies soon after learning of their potential interest in a product or service.

11.1.4 People

For all the mediation by computers and telecommunications networks, human resources matter more so than ever before. People make the Internet work. They generate, process, and manipulate data, and they create the technology to do so. In the new world information order, brain power can accrue dividends well in excess of physical labor, particularly because knowledge generates power. Great value accrues from leveraging and creating a business plan to exploit ideas. While many Internet and dotcom stocks may fail, many others will represent the accrual of great value from a core idea and the execution of that idea largely through brain power.

11.1.5 Growth

Internet-mediated communications and commerce have the potential to generate faster and larger growth, however that growth is measured. Just as money in an economy has a multiplier effect given its turnover through numerous transactions, the Internet can stimulate an economy through a similar multiplier effect. Positive networking externalities accruing from additional points of communications and more Internet subscribers provide an example that the whole becomes more valuable with additional parts. Other types of growth measures include increases in the valuation of Internet companies as a result of subscriber growth and the expectation that increased revenues and profits follow as well.

Telecommunications and information-processing networks appear to evidence what stock market analysts call "hockey-stick" growth. After relatively flat revenues and other static financial measures, once they reach a critical mass of use and consumer acceptance, growth skyrockets. The stock prices of many telecommunications and information-processing companies experienced triple-digit growth in a matter of weeks once investors identified those enterprises as market leaders.

11.1.6 Value

Telecommunications networks and the Internet expand value both in terms of an individual's perception of utility from use and the enterprise value and other financial measures for the company involved. Consumers make investments in telecommunications goods and services with the expectation that they can work more efficiently and effectively. Positive networking externalities accrue when the utility accruing to any one user grows with the ability to access more users. For example, a cellular telephone becomes a more valuable personal communication device when it can be used in more locations, at home and abroad, with the installation of additional networks, and when the device can access more users. Cellular radio operators realize that principle when they trade off profit margins in exchange for more users and higher usage. Average revenue per user actually increases when carriers offer larger baskets of air time available for local, long-distance, and roaming use. For the Internet, becoming the preferred application, software, Web site, portal, and so on, translates into higher sales, greater revenues, higher name recognition, and the potential to benefit from having established the de facto standard and becoming a market leader.

11.1.7 Efficiency

Telecommunications and Internet mediation offer consumers and businesses opportunities to achieve the benefit of greater efficiency whether measured by savings in time, money, or effort. Efficiency gains can occur when intermediaries are eliminated but just as easily when new businesses offer simplicity, one-stop packaging, convenience, and intelligence. An example of the former would be the ability to book a flight directly with the airline providing the service. But an example of the latter would be the use of new travel Web sites that offer air travel ticketing coupled with business and leisure travel advice and assistance, lowest-fare searches, and a variety of add-on features that offer consumers a sense of community and the opportunity to plan and prepare for travel.

11.1.8 Markets

Telecommunications traditionally has brought businesses and consumers together in a way that positively extends the reach of both consumers and suppliers. For example, toll-free WATS numbers provide consumers with ready access to distant suppliers without their having to decide initially whether making a preliminary inquiry is worth the up-front expense. The Internet uses the market extension capabilities of telecommunications and does so in an accessible, user-friendly manner, one that removes the barriers of distance, language, and time of day. Point-and-click access, for both business and consumer markets, reduces transaction costs, another example of efficiency gain, while promoting market competitiveness. When competitors are just a click away, suppliers have fewer opportunities to price gouge or scrimp on customer service. Telecommunications and Internet mediation make markets work better, an outcome that generally favors consumers and efficient operators.

11.1.9 Transactions

Telecommunications and Internet mediation make transactions more efficient and global. But they also offer the potential for personalizing marketing, products, and services. Radio and television broadcasting provide mass marketing opportunities, the ability to advertise general-consumption items to a large population at a low cost per viewer. While the Internet extends the reach of mass marketing to a global potential (i.e., anyone with Internet access), it also can use information processing and data mining to know the wants, needs, and desires of individual consumers. The ability to reach an individual personal computer or information appliance means that Internet commerce can be mass-customized, that is, structuring a marketing pitch and offering tailored goods and services.

11.1.10 Impulse

Internet time operates at a faster pace than what operators of other media have experienced. Broadcast messages could trigger desired outcomes only after consumers received the message and subsequently acted on it (e.g., by buying the product when seeing it displayed at a store). The Internet provides opportunities for immediate gratification of a consumer want, need, or desire and the opportunity for advertisers to compress the time between a sales pitch and the ability of receivers to respond to the pitch. The Internet blends an attractive multimedia, graphic sensory experience with point-and-click technical capability to maneuver from marketing pitch to completed purchase in short order. Consumers can buy and consume spontaneously simply by responding to on-site marketing pitches integrated with desired content as a banner or other type of link to more information and consumption opportunities.

11.2 Migration from a Voicecentric to a Datacentric Infrastructure

If the 1980s and 1990s confirmed the importance in the migration from analog to digital networking, the new millennium and the new world information order depend on the development of an Internet-centric infrastructure. In the not too distant future—if it has not already happened—data communications will generate more revenues and require more bandwidth than voice communications. For several years now, voice traffic has generated single-digit growth in developed nations, while data communications accrue triple-digit increases. Internet traffic doubles in a matter of months for some nations. Some forecasters expect voice to ride on the data links at little if any additional cost, given the predominance of data networking, the bandwidth allocated, and the ability of data-oriented networks, including the Internet and private intranets, to switch and route all kinds of digital media.

A datacentric telecommunications infrastructure confirms marketplace and technological convergence, but the more critical issues involve what stakeholders will do in terms of packaging and pricing services as well as how they cooperate in the joint provision of services. Formatting data in a digital mode and greater reliance on a one-size-fits-all Internet may appear to facilitate access and connectivity. Yet we can predict that bottlenecks will persist and individual enterprises will continue to have every incentive to exploit them to the detriment of competitors. Put another way, just because technology facilitates and improves connectivity does not mean that the end results at the consumer level will show greater convenience and enhanced welfare. A data-oriented medium lends itself to expanded applications like that required to make e-commerce a reality. Whether the network achieves the even greater versatility to support a broader array of services will depend in large part on resolution of the same types of issues that challenged everyone in the prior voicecentric environment. Those issues transcend the medium or the predominant format: access terms and conditions, affordability, network connectivity, customer service, and network user friendliness (i.e., the ease with which consumers can access services).

11.2.1 Technological Innovations Promoting Internet Diversification

Although the typical Nethead would vigorously deny it, the Internet will have to become more like a telecommunications network—with all its Phonehead-driven architecture and mind-set—to solve pressing concerns about quality of service, billing, reliability, and security. Many consumers will not tolerate the insecure, best-efforts, shared-routing topology of the current Internet. Those users will pay a premium for secure, reliable, and predictable traffic routing of a sort readily available from telecommunications networks.

In fact, the kind of virtual private-line architecture designed to address the Internet's weaknesses has a direct service counterpart in telecommunications networks. Perhaps it makes sense to think of the Internet as becoming more like its telecommunications backbone even as telecommunications networks adapt many of the pricing, architecture, and network design features of the Internet. Virtual private networks (VPNs) stand at the midway point for both systems. In telecommunications, VPNs offer a cheaper alternative to consumers who can tolerate the slight diminution in service quality, reliability, and security when the carrier substitutes a dedicated link with a software-defined one that may be partitioned from the public network or shared among several users on a demand-assigned basis. For Internet users, VPNs offer a more expensive alternative to best-efforts routing, in addition to an increase in service quality, reliability, and security.

Internet VPNs employ technological innovations that allow the operator to "tunnel" a complete link using previously acquired or leased telecommunication lines. From the user's perspective, the Internet operator has engineered an apparently dedicated line, thereby ensuring a reliable and secure link. Rather than physically set up a dedicated pathway, the Internet operator adds routing and service priority information to the headers of each packet to be transmitted. Routers and other types of switching equipment can read this header information and route the traffic on a priority basis via a preferred and previously established inventory of transmission facilities. Typically, the routing replaces line interconnection at broadly available network access and public peering points with private, less congested private peering points. Virtual private networking allows traffic to route through what appears to be the Internet, but which in effect is a partitioned part of it available only to a predesignated set of users, who benefit by having more direct and more reliable links, with less potential for congestion, dropped packets, and queuing at bottleneck points along the way.

VPNs provide users with the best of both worlds: ease of use and functionality that appears embedded within the Internet and technical and operational features akin to preexisting dedicated and private telecommunications networks. VPNs may have the effect of balkanizing the Internet in the sense that not all of what we consider part of the Internet will be available to all users. VPN users will benefit by having access to and from the "public" Internet, and special and more expensive Internet-plus features. In application, VPNs help legitimize the centrality of the Internet for both telecommunications and information processing, but they also incorporate some of the pricing, quality of service, and interconnection standards and expectations of preexisting telecommunications networks.

11.2.2 A Maturing, More Hierarchical Internet

As Internet industry segments mature, many governments wind down and eventually terminate their role as incubator and anchor tenant. In many nations, including the United States, government helped promote Internet use and proliferation of the Internet infrastructure. Having concluded that the Internet has reached a critical mass, most governments now endorse the view that a largely commercial and private environment will best serve the national interest. Most nations favor a privatized Internet but not an environment that could be deemed laissez-faire or completely unregulated.

The retained regulatory mechanism in most nations is ill-equipped to address interconnection, access pricing, and universal service issues. As the Internet becomes a major medium for a variety of private and commercial transactions, activities previously considered illegal or warranting government oversight similarly will trigger such government involvement when the Internet provides the medium or conduit. It remains to be seen whether and how governments, particularly ones predisposed not to regulate or interfere with the Internet, will respond to market dominance by major ISPs and other developments that may limit competition and possibly reduce innovation.

The privatized, commercial Internet has evidenced economic characteristics similar to those of telecommunications networks.[2] Operators have accrued favorable economies of scale and scope through growth. ISPs accrue economies of scale by expanding both customer base and the inventory of bandwidth, thereby achieving operating efficiencies and lower per-unit costs. ISPs accrue economies of scope by augmenting monthly access subscriptions with other Internet-mediated services, including electronic commerce and advertiser-supported access to desirable content.

To accrue positive economies of scale and scope, telecommunications and Internet operators alike have worked to expand their customer base, available bandwidth, number of interconnection sites, and the content that they host as opposed to providing access to content housed elsewhere. Massive, multibillion-dollar mergers and acquisitions evidence the desire to achieve scale and scope economies in a speedy fashion: acquiring the market share of a competitor rather than migrating customers and revenues from competitors.

The quest to accrue scale and scope economies constitutes one of the major reasons the Internet has become more hierarchical,[3] with a small set of major carriers operating the key backbone routes and capturing a large market share however measured (by bandwidth, number of subscribers, minutes of use, revenues, number of discrete "hits" to internal web sites, number of discrete domain numbering system sites internal to ["hosted" by] the

2. For a helpful background on the nature of telecommunications regulation in the context of competition policy, see Michel Kerf and Damien Geradin, "Controlling Market Power in Telecommunications: Antitrust vs. Sector-Specific Regulation. An Assessment of the United States, New Zealand and Australian Experiences," *Berkeley Technology Law Journal*, Vol. 14, Fall 1999, p. 919.

3. For additional background on the impact of a hierarchical Internet industry structure on universal service policy objectives, see Rob Frieden, "Last Days of the Free Ride? The Consequences of Settlement-Based Interconnection for the Internet," *Info*, Vol. 1, No. 3, June 1999, pp. 225–238; Rob Frieden, "Without Public Peer: The Potential Regulatory and Universal Service Consequences of Internet Balkanization," *Virginia Journal of Law & Technology*, Vol. 3, Fall 1998, p. 8, available on-line at http://vjolt.net/.

network, etc.). The small number of major backbone ISPs, coupled with an increasingly commercial orientation, has made it possible for Tier-1 ISPs to demand and secure payments from smaller ISPs for access to their networks and the content they host.

Without judging the merits of whether a more hierarchical Internet achieves economies of scale and scope, the reduction in number and concentration of market share, however measured, has made it possible for Tier-1 ISPs to secure a superior bargaining position vis-à-vis smaller ISPs:

> As the cooperative, nonprofit ethos of the Internet began to fade, some providers began to have second thoughts about connecting directly to one another (through open peering). Today, large backbone providers such as AT&T, Cable & Wireless, GTE, PSINet, Sprint, Qwest Communications and UUNET consider one another peers and don't hesitate to connect to each other. However, they often spurn smaller... ISPs." [2, pp. 42, 44]

While technical and operational factors do affect the Tier-1 ISPs' interconnection decision-making process, the "main reason for not peering, however, is economic" [2, p. 46]. Through the use of traditional economic and antitrust/competition policy analysis,[4] the potential exists for a small number of Tier-1 ISPs to shape the typical terms and conditions for interconnection and coordination among most Internet operators.

11.2.3 Effects of a Hierarchical Internet

A more hierarchical Internet with a small number of backbone Tier-1 ISPs puts such operators in a position to set access terms and conditions. Later in this chapter, we consider whether Tier-1 ISPs might embrace traditional

---

4. See, generally, W. Kip Viscusi, et al., *Economics of Regulation and Antitrust*, 2d ed. (Cambridge, MA: MIT Press, 1995), p. 377. For background on United States antitrust law and policy, see Richard A. Posner, *Antitrust Law: An Economic Perspective* (Chicago: The University of Chicago Press, 1976), p. 8; Herbert Hovenkamp, *Federal Antitrust Policy: The Law of Competition and Its Practice* (Stanford, CT: West/Wadsworth, 1994); Georges J. Alexander, "Antitrust and the Telephone Industry After the Telecommunications Act of 1996," *Santa Clara Computer and High Technology Law Journal*, Vol. 12, 1996, p. 227. For background on antitrust and telecommunications-specific rules in the European telecommunications markets, see Paul Nihoul, "Convergence in European Telecommunications—A Case Study on the Relationship Between Regulation and Competition Law," *International Journal of Communications and Law and Policy*, Vol. 2, 1998/99, p. 1.

telecommunications interconnection and access pricing models. For purposes of this section, it is important to note that already Tier-1 ISPs typically charge smaller ISPs for access to backbone networks. That charging arrangement largely eliminates the SKA model that helped make the Internet so accessible, integrated, and interconnected. Now, only Tier-1 ISPs agree among themselves to the SKA, cost-free peering model. All other ISPs have become, in effect, customers of the larger backbone ISPs. Unless a smaller ISPs has ready access to several Tier-1 ISPs, the potential exists for bottlenecks to arise, coupled with the ability of the single Tier-1 ISP controlling that bottleneck to engage in a price squeeze (raising the costs of smaller ISPs for a service element needed by all competitors and supplied by one or few operators).

At the very least, a hierarchical Internet industrial structure results in different types of ISPs incurring different costs for access to both content and carriage. Such price differentials may simply reflect the costs incurred by larger ISPs and heretofore avoided, at least partially, by smaller ISPs, which have been able to hand off traffic for the long haul to the larger ISPs without full payment for such transit services. But the existence of unequal bargaining power may also lead to price gouging and discrimination, both of which may adversely affect the breadth and the price of Internet access.

11.3 Effects of Market Convergence

The prediction of converging telecommunications and Internet pricing models may come across as controversial and counterintuitive. Certainly, the two markets currently have substantially different pricing policies. While local telecommunications services may be offered on a flat-rate basis, most long-distance services are usage sensitive, meaning that consumers pay as a function of usage. Domestic telecommunications charges may be postalized and averaged (i.e., charged on a flat per-minute charge for any call), but international calling still has a particular rate per country, reflecting some correlation with distance, call volume, and supply-of-transmission capacity. The nature of Internet access, from a consumer (end-user) point of view, seamlessly blends access to content and the telecommunications transport needed to acquire and deliver the content. Users expect their ISPs to deliver content quickly and effectively regardless of where the content is physically hosted. Most consumers have grown accustomed to unlimited, AYCE access pricing or at least a large initial number of usage hours before additional metered charges apply.

If telecommunications usage triggers distance and usage-sensitive charges, and Internet usage typically is flat-rated, then how and why would those different pricing models converge? The answer lies with the change in motivation and strategy among most ISPs from one of promotion to one of cost recovery and profit seeking. Particularly because ISPs have to acquire massive additional amounts of bandwidth to accommodate burgeoning demand for broadband applications like real-time delivery of video, there is little difference between the transmission facility investment and expense profile of a telecommunications carrier and of a major ISP. ISPs cannot hope to recoup their facilities construction and lease expenses just from their end users, particularly when the nature of Internet connectivity means that unaffiliated ISPs frequently use each other's transmission services, and there typically is no symmetry, as there is in telecommunications, between sender and receiver. Internet communications often involves a short, narrowband request for content from a user, followed by a much longer-in-time and higher-in-throughput response from a server.

Tier-1 ISPs already have eliminated SKA opportunities for all but an equally large ISP to prevent smaller ISPs from tapping into an unaffiliated ISP's network without charge. Tier-1 ISPs cannot tolerate free riders, even though that was the likely outcome of a pricing model that ignored differences between ISPs in terms of customer base, inventory of bandwidth, number of interconnection points accessed and number of interconnection agreements with other ISPs. Rough-justice agreements worked only when traffic volumes and other measures were generally equivalent or when promotional strategies favored a temporary obliviousness to such differences.

Because Tier-1 ISPs have to recoup their infrastructure investments and cannot saddle their own end users exclusively, the burden has to be shared by advertisers and other ISPs. ISPs that provide backbone and transit services now impose access charges based on traditional telecommunications criteria such as bandwidth, mileage, and usage. However, the charging ISPs cannot replicate the telecommunications settlement procedures, because of the difficulty and expense in metering all traffic and attributing a charge to the potentially massive number of ISPs that might transit a single ISP's network. While economists and large ISPs might want to see a specific settlement rate per minute of use or kilopacket transited, a less exact system has evolved. Tier-1 ISPs operate like a carrier's carrier in that they lease bandwidth and lines to other ISPs. In effect, the smaller ISPs becomes a client or reseller of the larger ISPs network. This access charge regime closely parallels how facilities-based telecommunications carriers sell or lease capacity to other carriers and resellers.

11.4 Interconnection and Access Pricing

In a competitive marketplace, governments typically want to have only limited, if any, involvement in an ISP's decision whether to interconnect with another ISP and on what terms and conditions. If the market can ensure fairness in interconnection decisions and access pricing based on costs, then governments should step aside and refrain from imposing "legacy," old world order, and common carrier interconnection responsibilities, and from assessing the reasonableness of access charges. In the Internet's case, governments already largely have stepped aside, or they never got involved in the first place. Accordingly, should interconnection and access problems arise—and they have—governments will have to decide whether to get involved.

Whether governments should get involved largely depends on an assessment of market conditions, the definitions used, and the assumptions made. Traditional economics and antitrust analysis consider the functional equivalency or substitutability of a product or service in determining the relevant product or service market. Markets can be defined as including all goods and services considered by consumers to constitute an alternative to the others. Economists measure the substitutability of products and services in terms of cross-elasticities.

A dichotomy in terms of market entry costs and opportunities exists between local and regional ISPs, on the one hand, and long-haul, national and international ISPs, on the other hand. The commercial aviation marketplace provides a helpful example of a similarly dichotomous market. Few financial or other barriers exist to preclude the creation of a new airline. With a handful of airplanes leased by a fully leveraged venture, a new airline enters the marketplace. Absent barriers to accessing airport terminal and landing space, the airline can serve a few routes and provide significant competition to incumbents. However, no one would mistake such small and incremental competition as coming close to fostering robust and full competition to the service a major incumbent carrier provides. A nation like the United States might have hundreds of national airlines; nevertheless, six major carriers control more than 70% of the total market as measured by industry-appropriate criteria ("revenue miles" and "seat miles").

Depending on our perspective and market definitions, the commercial aviation marketplace in the United States can be characterized as either robustly competitive or oligopolistic, notwithstanding low barriers to market entry and a general decline in fares since deregulation stimulated market entry. Despite the absence of bottlenecks in terms of access to capital, airport terminal space, and runway landing slots, few airlines compete for long-haul traffic or offer a thoroughly national and international route system.

We would have a harder time justifying the view that a few ventures dominate the market for Internet access and Internet services if those markets were defined in the context of the total number of ventures pursuing some aspect or element of the multifaceted Internet marketplace. Conversely, a narrower definition of the Internet marketplace, emphasizing the market share held by Tier-1 ISPs, could support the view that those operators share market power and the ability to extract high rates and impose "unfair" terms ("monopoly rents" in economics).

The Internet marketplace does appear to evidence parallels to commercial aviation. While a nation might have hundreds, it not thousands, of ISPs, the overall market segments into a large percentage of the total ISP number serving single localities or regions, with a limited number of ISPs operating the major long-haul backbone networks needed for national and international services. The startup costs for a local ISP are evidence of the limited barriers to market entry. A new ISP can enter the marketplace simply by leasing a few local trunks from the LEC to provider subscribers with access to a modem bank for access to and from the Internet secured by the interconnection of those local lines with a few interexchange carrier lines that access the transit services of a larger ISP up the hierarchy of ISPs in terms of size and reach.

On the other hand, a major backbone Internet operator does not appear overnight. The Tier-1 operators must have the financial and operational wherewithal to construct or lease and manage a nationwide network of high-capacity lines. Few enterprises can amass the needed investment and coordinate a national or global network. Accordingly, it should come as no surprise that most of the Internet Tier-1 ISPs are subsidiaries or affiliates of major telecommunication carriers.

The potential for anticompetitive practices and leveraging bottlenecks exists in both industries. In aviation, absent government ownership or effective regulation, the airport operator could discriminate in favor of one particular airline in the manner in which it assigns (or denies) access to space in the airport terminal and opportunities to take off or land aircraft. In the Internet, access to the local loop and the backbone networks of Tier-1 carriers may be viewed as constituting essential facilities, whose access terms and conditions could choke off or stimulate competition.

11.4.1 Tier-1 ISPs and Limited Regulatory Burdens

The analogy between ISPs and commercial airlines breaks down somewhat on the matter of government regulation. ISPs incur substantially less government oversight than their airline counterparts, for four primary reasons:

- Governments regulate the telecommunications transport function performed by the carriers who lease facilities to Tier-1 ISPs;[5]
- Notwithstanding its growing importance, the Internet has not approached the status of public utility or functional equivalent to telecommunications;[6]
- Most governments have purposefully embraced a hands-off strategy with an eye toward promoting entrepreneurialism and private initiatives;
- Until recently, ISPs themselves have emphasized connectivity and global reach even if the network interconnection, access, and pricing policies employed to reach that goal (e.g., open peering and SKA) reduced profitability and resulted in the possibility that some ISPs would bear disproportionately greater financial burdens to build out the network infrastructure.

---

5. For example, in the European Union, Council Directive No. 90/387/EEC, art. 3, O.J. L 192/1, 1990, p. 2, establishes baseline principles that facilities-based telecommunications carriers must apply when leasing lines and interconnecting with enterprises providing value-added services. While the carriers do negotiate terms and conditions in a commercial, arm's-length atmosphere, Open Network Provision principles direct the providers of the underlying transmission capacity to offer access on terms and conditions based on objective criteria that must be transparent and published in an appropriate manner and that guarantee equal and nondiscriminatory access in accordance with community law. See Gunter Knieps, "Interconnection and Network Access," *Fordham International Law Journal*, Vol. 23, 2000, p. 90.

6. Regulatory asymmetry can work when the products or services involved do not constitute functional equivalents. However, proliferating and developing Internet services have begun to include features that consumers may consider as unregulated substitutes for regulated telecommunications services, for example, Internet telephony. "In general terms symmetric regulation means providing all suppliers, incumbents and new entrants alike, a level playing field on which to compete: the same price signals, the same restrictions, and the same obligations.... But all forms of asymmetric regulation contain an intrinsic bias toward some firms or technologies." Mark Schankerman, "Symmetric Regulation for Competitive Telecommunications," *Information Economics and Policy*, Vol. 8, 1996, p. 55.

Simply put, governments have not installed a regulatory regime for the Internet, because they believe one is not needed. Advocates for regulatory involvement in interconnection and access pricing issues will dispute that by emphasizing the need for closer antitrust and competition policy scrutiny using a better-calibrated market definition. Advocates for regulatory relief allege that the consolidation in the long-haul market segment accords the Tier-1 ISPs the power to distort the bargaining process and to extract supra-competitive, overly generous compensation for access to and transit through their networks. Those advocates believe the competitive playing field has tilted in favor of the Tier-1 ISPs, which can exploit the inelastic demand for their transport service and the content they have available. Some regulatory advocates would characterize the Tier-1 ISP networks as "essential facilities"[7] and bottlenecks, because all long-haul Internet traffic must traverse those facilities in much the same way, because that traffic might have only one local-loop-routing option provided by a regulated, common carrier incumbent LEC. If government adopted the view that Tier-1 ISP networks constitute essential facilities or bottlenecks, then that government has an economic and legal rationale for applying regulatory instruments aimed at improving the terms and conditions for access, including the interconnection and access charges Tier-1 ISPs impose on smaller ISPs.

Notwithstanding market consolidation by the Tier-1 ISPs, governments have yet to adopt the view that the long-haul Internet access marketplace is oligopolistic and incontestable. First, most small ISPs continue to have access to several Tier-1 ISPs, despite mergers among the telecommunications carriers that lease lines to ISPs. While the small ISPs object to having to pay for access to Tier-1 ISP facilities, such access remains available from

7. The "essential facility" doctrine in antitrust/competition policy supports government intervention to mandate access by competitors to a facility or service provided by one competitor based on the following assumptions: (1) the competitor has the ability to exert monopoly power over the essential facility, that is, to deny access or provide discriminatory access, including the imposition of higher access rates on competitors and thereby leading to a price squeeze; and (2) competitors cannot practically or reasonably duplicate the facility. See Daniel Glasl, "Essential Facilities Doctrine in EC Antitrust Law: A Contribution to the Current Debate," *European Competition Law Review*, Vol. 6, 1994, p. 306; William B. Tye, "Competitive Access: A Comparative Industry Approach to the Essential Facility Doctrine," *Energy Law Journal*, Vol. 8, 1987, pp. 337, 346. But compare with Phillip Areeda, "Essential Facilities: An Epithet in Need of Limiting Principles," *Antitrust Law Journal*, Vol. 58, 1989, p. 841; Allen Kezsbom and Alan Goldman, "No Shortcut to Antitrust Analysis: The Twisted Journey of the 'Essential Facilities' Doctrine," *Columbia Business Law Review*, No. 1, 1996, p. 145.

multiple Tier-1 ISPs. Second, no evidence exists to support the view that the Tier-1 ISPs have conspired or coordinated efforts to fetter smaller ISPs with discriminatory interconnection terms and conditions. Tier-1 ISPs operating in the United States now require access and transit payments from smaller ISPs, regardless of their location. However, the imposition of higher, distance-sensitive charges on ISPs places a comparatively greater financial burden on ISPs in remote locales, particularly ones far from major peering points and server farms containing the most desirable content. Third, there has been no indication that Tier-1 ISPs have engaged in a strategy to raise smaller ISPs' costs of doing business with an eye toward driving them out of the market. Tier-1 ISPs have not engaged in predatory pricing of access. On the contrary, every ISP wishes it did not have to charge AYCE rates to end users.

The strongest case for government intervention lies where self-correcting marketplace outcomes cannot be relied on to remedy short-term problems:

> A cautious approach would be to reject any possibility of mandatory access except where it is "essential" to the existence of competition. If applicants for access can plausibly invent around the network monopoly, establish their own competitive networks, or join other networks that may not be equivalent but are acceptable alternatives to the dominant network, that arguably might eliminate any consideration of court-ordered access. [3]

Advocates for a hands-off approach to interconnection and access-pricing issues emphasize the suitability of marketplace remedies, that is, discriminatory or unfair access terms and conditions should generate incentives for smaller ISPs to set up their own competitive networks or collectively join with other smaller ISPs to create a rival long-haul network. Likewise, they consider the profits accruing to Tier-1 ISPs appropriate rewards for risk taking and achieving marketplace success. Expropriating some or all of the monetary fruits of Tier-1 ISPs' labors simply rewards free riders and risk-averse players. A hands-off approach also absolves government officials of having to resolve equity and operational issues for which they may have no particular skill or impartial template.[8]

8. See Leonard W. H. Ng, "Access and Interconnection Issues in the Move Towards the Full Liberalization of European Telecommunications," *North Carolina Journal of International Law and Commercial Regulation,* Vol. 23, Fall 1997, p. 1.

11.5 The Quest for Economies of Scale and New Profit Centers

Technological convergence and marketplace convergence offer telecommunications and information service companies a blessing and a curse. As never before, technological innovations make it possible for a single company to provide a wide array of products and services. Likewise, markets no longer operate mutually exclusively from others. Technology makes it possible for single pieces of equipment and individual service providers to offer voice, data, audio, graphics, and video applications. Digital markets, where "a bit is a bit is a bit," do not divide into separate and discrete submarkets for voice, data, and so on. Incumbents risk losing market share to new competitors even as they have new market access opportunities.

The quest for market share and economies of scale has prompted incumbents and newcomers alike to embark on risky and costly undertakings. Internally and through mergers and acquisitions, companies strive to become one-stop shops for users' complete telecommunications and information-processing requirements. Ideally, a firm would like to integrate vertically and horizontally so it can proactively acquire new profit centers and reactively shore up revenues lost when competition erodes market share for existing products and services.

11.5.1 Vertical Integration

Vertical integration occurs when a single firm operates through part or all of a product or service chain (e.g., from manufacture, through distribution channels, down to serving the end user). Broadcast television networks have vertically integrated by producing, syndicating, distributing, and broadcasting content. The predivestiture Bell System likewise vertically integrated by having a corporate presence in the manufacture of telecommunications equipment and the delivery of local and long-distance services to consumers. However, in both telecommunications and information processing, many firms have chosen not to integrate vertically. Such reticence stems from the recognition that they should concentrate on what they do best, for example, manufacture equipment rather than operate it, and the sense that purchasers of goods and service do not appreciate having to compete with suppliers. For example, Cisco manufactures equipment that makes the Internet operate while refraining from using the equipment it manufactures to provide Internet services.

In a time when markets are increasingly contestable and competitive, incumbents cannot expect to retain dominant market share without innovating and competing vigorously. They must compensate for revenue and market share losses and also pursue new market opportunities. A telecommunications incumbent facing competition in local and long-distance telephone services at home has good reason to acquire a manufacturing presence, the capability of providing each and every tier or aspect of a service and new services. For example, technological innovations make it possible for incumbent telephone companies to retrofit the telephony plant to provide a variety of services that other ventures have offered, possibly through the lease of the telephony plant. An incumbent telephone company can become a transporter of video bitstreams (e.g., cable television), and having decided to enter the video business, it might vertically integrate into the creation of video content.

11.5.2 Horizontal Integration

Horizontal integration refers to the extension of a firm into adjacent markets and technologies that compete with its core activities. For example, a wireline telephone company has the skills and resources to become a successful wireless radiotelephone company. Telecommunications and information-processing companies have aggressively entered adjacent markets, because technological innovations make it cost effective to do so and real or perceived new profit centers await. Broadcast television networks, facing declining audiences due to competition from cable television and the Internet, realized the need to horizontally integrate into each of the adjacent markets. For example, NBC operates several cable television networks, including CNBC, as well as several Internet sites, including the Web information and entertainment portal MSNBC.

Successful horizontal integration occurs when a firm can leverage and extend its skills and competency into related markets that are nevertheless new to the company. Akin to accruing economies of scope, successful horizontal integration evidences the ability of a single firm to operate in two or more market segments that overlap to some extent. If NBC has achieved a competitive advantage in the creation of attractive content, it can exploit that skill in several media that to some extent compete for consumers' eyes, ears, hearts, minds, and income.

11.5.3 One-Stop Shopping

Vertical and horizontal integration are technical terms for describing the gold rush to become the preferred single source for just about every telecommunications and information-processing need, want, and desire. Largely through acquisitions, AT&T has expanded its array of products and services well beyond its predivestiture reach. For U.S. consumers, AT&T wants to be the preferred local and long-distance telephone company, cellular radio carrier, Internet access provider, portal to content available via the Internet, cable television company, and supplier of all the equipment needed for those services.

In pursuit of a one-stop-shopping marketing campaign, companies risk losing market share in core areas for lack of attention and effective business play execution even as they seek to acquire new and possibly large shares in new markets. AT&T's voluntary divestiture into four separate companies corroborates this point. Vertical and horizontal integrations are occurring at such a quick pace that many companies see the benefits in acquiring market share through mergers and acquisitions of ventures that already have succeeded rather than risk the time and uncertainty of internally developing the skills to excel in new markets.

11.6 Real or Perceived Unquenchable Demand for Bandwidth

The mergers, acquisitions, and market diversification efforts of various companies also occur in anticipation of growing consumer demand for the products and services that make the telecommunications and information processing marketplace. In particular, demand for more service and an expanded set of services should stimulate the need for much more bandwidth, that is, more telecommunications transmission capacity and more private and virtual private lines that provide the links for the Internet.

11.7 Conclusion

This is an exciting and confusing time in telecommunications as more challenges and opportunities arise. With the ascendancy of the Internet, it may seem that the locus of control shifts from telecommunications carriers to information processors. Still, telecommunications provides the essential transport function, and as such both incumbents and newcomers have plenty of opportunities to thrive in the information age.

With the Internet's growing importance, convergence of technologies and markets has become real. So too must practitioners and students realize the need to understand the culture of Phoneheads and Netheads and to listen and speak in those sometimes separate languages.

References

[1] *Business 2.0*, "The 10 Driving Principles of the New Economy," available on-line at http://www.business2.com/research/principles/2000/03/01/14842.

[2] Angel, J., "Toll Lanes on the Information Superhighway," *Network Magazine*, Vol. 15, No. 2, Feb. 2000, pp. 42, 44.

[3] Pitofsky, R., "Antitrust Analysis in High-Tech Industries: A 19th Century Discipline Addresses 21st Century Problems," *Texas Review of Law & Politics*, Vol. 4, Fall 1999, pp. 129, 138; see also Teece, D. J., and M. Coleman, "The Meaning of Monopoly: Antitrust Analysis in High-Technology Industries," *Antitrust Bulletin*, Vol. 43, 1998, p. 801.

Appendix A
Telecommunications @ the Millennium†

The Telecom Act Turns Four

Office of Plans and Policy*
Federal Communications Commission
William E. Kennard, Chairman
February 8, 2000

Four years after the 1996 Telecommunications Act became law, the vision of robust competition laid out in the Act is beginning to unfold. This report examines developments in the industry and their impact on consumers and the economy from just before the Act to the present – a survey of the telecommunications marketplace at the millennium.

- **The telecommunications industry has grown dramatically since 1996 and is increasingly a foundation of economic growth for the whole economy.** Revenues in communications services have grown 17 percent since 1996 while the broader

† This document, prepared by the U.S. Federal Communications Commission, provides a timely and insightful commentary on current regulatory and markeplace conditions.

* This staff report does not necessarily reflect the views of the Federal Communications Commisssion or any of its Commissioners.

information technology sector (which includes related industries that rely heavily on communications infrastructure) has contributed more than one-third of the growth of real output in the economy between 1995 and 1998.

- **Prices have continued to fall even as demand has surged.** As the price of communicating over the nation's networks (wireline and wireless alike) has fallen, use has increased. The fall in prices, improved quality, and increased availability has been a direct benefit to consumers, who now pay less for more. It has also lowered firms' costs of doing business, releasing resources for other investments and helping to keep all prices down – which ultimately benefits consumers.
- **Competition among networks and technologies underlies these benefits.** Private companies have responded to increased competition and changes in technology and consumer demand by investing in their systems, by deploying new services, and by lowering their costs.

The Telecommunications Industry is Booming

The telecommunications industry has grown since 1996, creating 230,000 new jobs and generating $57 billion[1] more revenues.

Revenues in communications services, which include all telephone services, radio, cable and broadcast television, and certain other services, have grown from $331 billion in 1996 to $388 billion in 1998, a growth of 17 percent in real terms (Figure 1). That figure does not include the rapid growth in sales of communications equipment—telephone handsets, central office switching equipment, etc.—where revenues have grown $26 billion, 24 percent, between 1997 and 1999. With the growth in output, employment in the communications equipment and services industries has grown from 1.6 million in 1996 to 1.8 million in 1999 (Figure 2).

Growth and service improvements in the telecommunications industry also have important secondary impacts on the rest of the economy. While the above figures present an industry growing at a healthy pace, they understate

1. All figures are 1998 dollars unless otherwise noted.

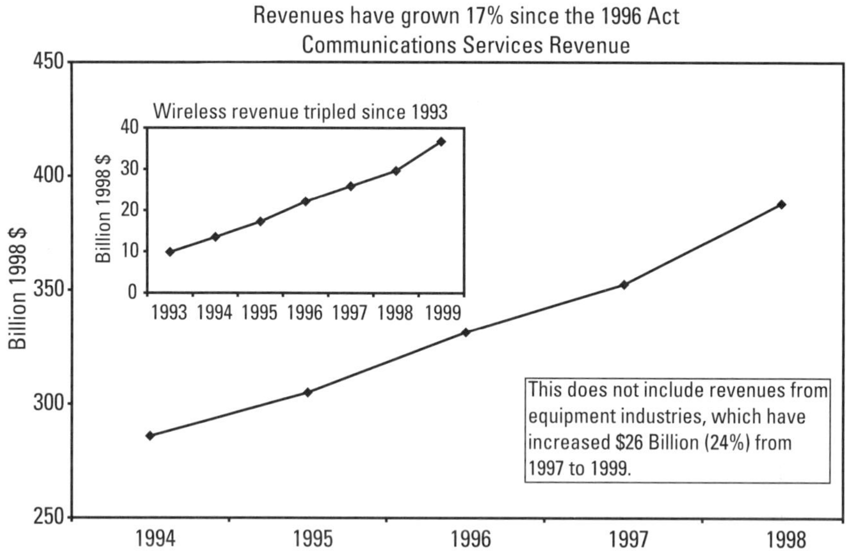

Sources: U.S. Census Bureau , Multimedia Telecommunications Association, CTIA

Figure 1

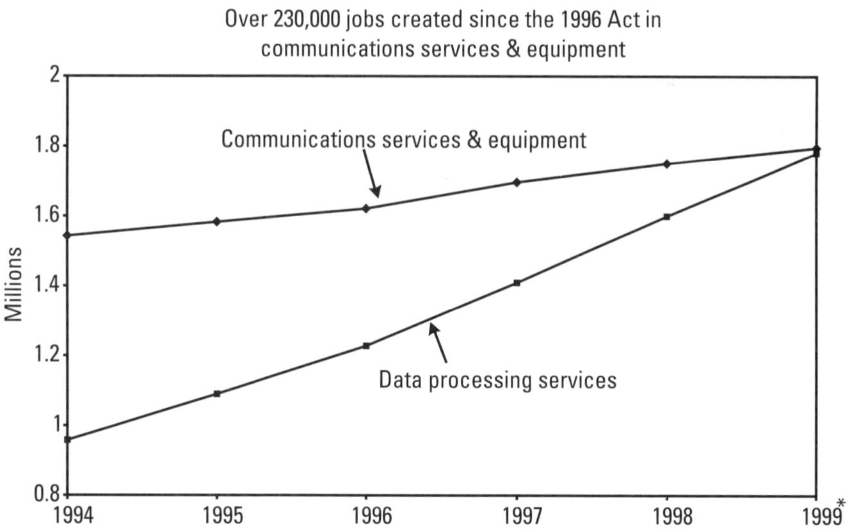

From Feb. 1996 to Dec. 1999. * preliminary

Source: U.S. Department of Labor (Bureau of Labor Statistics)

Figure 2

the full impact improvements in the telecommunications sector have had on the economy. This is because communications are end products as well as inputs to the production of other goods and services.

By providing cheap, efficient underlying infrastructure and services, the communications industry has played a pivotal role in the development of a host of data and communications related industries that could not exist without it. For example, the Internet, and electronic commerce generally, is built on the communications infrastructure.

The "digital economy" is changing the way we do business, carry out our day to day lives, and interact with each other. New virtual markets are being created to more efficiently trade goods and services. Companies like General Motors and Ford are now developing Internet trading networks to link their thousands of suppliers. These systems could electronically process upwards of $300 billion in transactions every year. We are only beginning to understand the profound positive effects these changes will have on the economy.

These secondary linkages are, of course, difficult to measure. One approach is to look at the growth in industries that rely heavily on communications infrastructure and services. For example, employment in the data services industry has grown by 666,000 since February 1996 (Figure 2). **In one study, the Department of Commerce estimated that information technology producing industries (the producers of computer hardware and software, data services, and communications equipment and services) directly contributed, on average, over one-third of the real economic growth between 1995 and 1998.**[2] That same study found that, because of the declining prices of their products and services, these industries reduced overall inflation by 0.7 percentage points in 1996 and 1997.

The benefits of the digital age are being extended to all Americans. The FCC has implemented rules pursuant to the Act that require all telecommunications products and services to be made accessible to people with disabilities. The Act also provides the FCC with important tools to update universal service policies to ensure that Americans in remote and rural areas receive service.

By 2006, it is estimated that half of the U.S. workforce will be employed in industries closely linked to information technologies. It is important that all Americans be able to participate in the emerging digital economy. Therefore, the 1996 Act established the Schools and Libraries

---

2. U.S. Department of Commerce (1999). *The Emerging Digital Economy II.*

Universal Service program (E-Rate) to provide affordable telecommunications services for schools and libraries, especially those in rural and economically disadvantaged areas.

Nearly three times as many public schools were connected to the Internet in 1998 (89 percent) as were connected in 1994 (35 percent), while schools in low income areas went from 19 to 80 percent connected quadrupled (Figure 3). However, at the end of 1998, only half of all classrooms were connected to the Internet. Some data suggest that one million classrooms were connected to modern networks in 1998 and 1999. This program is on the road to ensuring that all American children will have the technology tools to thrive in the global information age.

Prices are Falling as Demand is Increasing

In general, prices are falling.

Prices for wireless and long distance telephone service have fallen dramatically over the past years. Long distance prices (international and domestic), as approximated by average revenue per minute, have fallen by 34 percent since 1993 (Figure 4). For residential consumers, domestic long distance

More than 90% of public schools are now connected to the Internet
Percent of public schools connected to the Internet

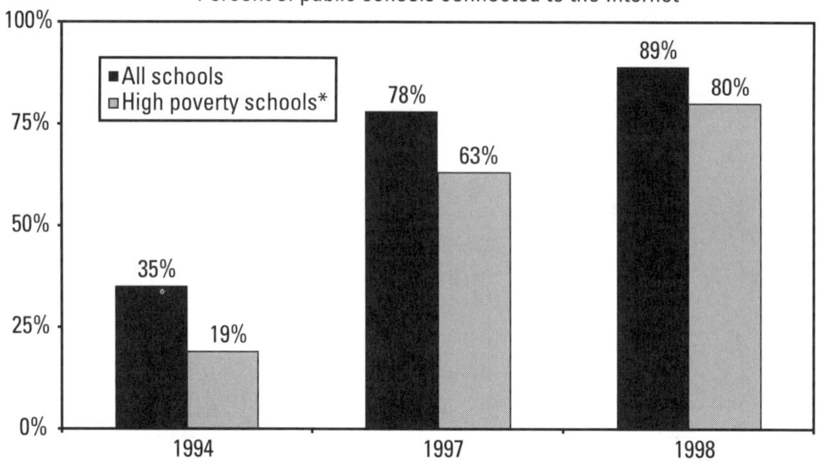

Source: National Center for Education Statistics

* High poverty schools defined as those with 71% or more of students eligible for free or reduced-price school lunches.

Figure 3

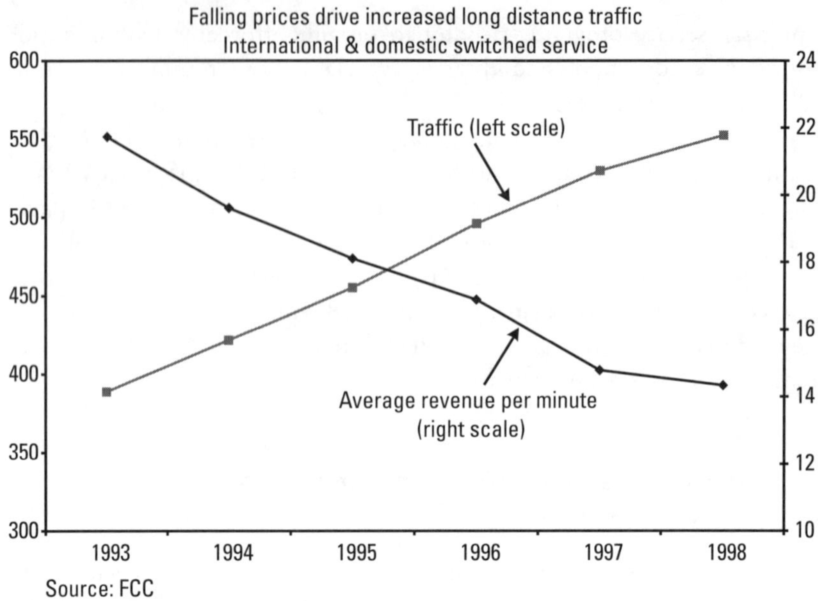

Figure 4

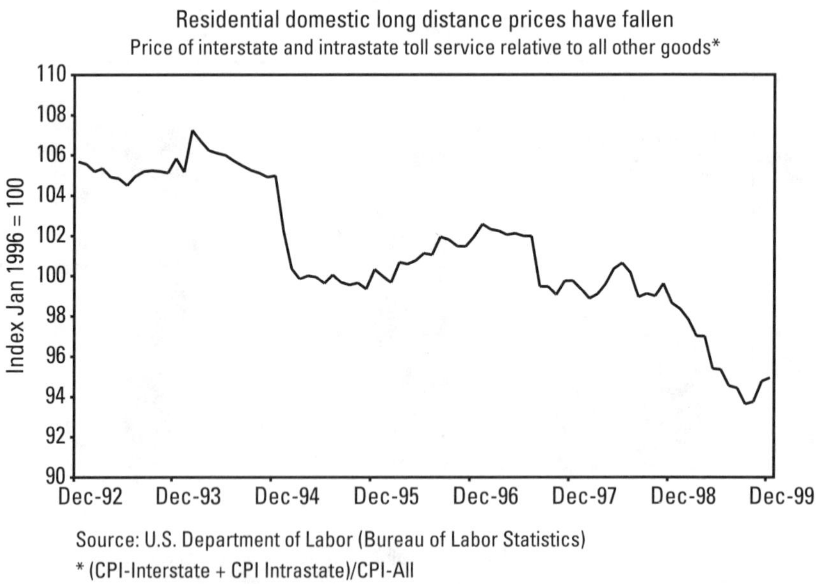

Figure 5

prices relative to other goods and services have fallen by 10 percent since 1993 (Figure 5). New calling plans now offer consumers long distance rates of only 5 cents per minute. International long distance prices have fallen by more than half since 1993. Mobile prices have fallen by 35 percent since 1993, and average monthly local bills have dropped by $30 – from $70 in 1993 to $40 in 1999 (Figure 6).

The only areas where prices have not fallen significantly since the 1996 Act is in local telephone and cable television services. Local rates have declined only 2 percent relative to all other goods since the Act. Although cable rates have decreased in terms of the price per channel received, overall cable bills have risen by 1.8 percent between 1998 and 1999 in real terms.

Usage is increasing.

With the drop in prices, Americans are making more long distance calls and mobile phones are becoming commonplace. Wireless subscriptions in the

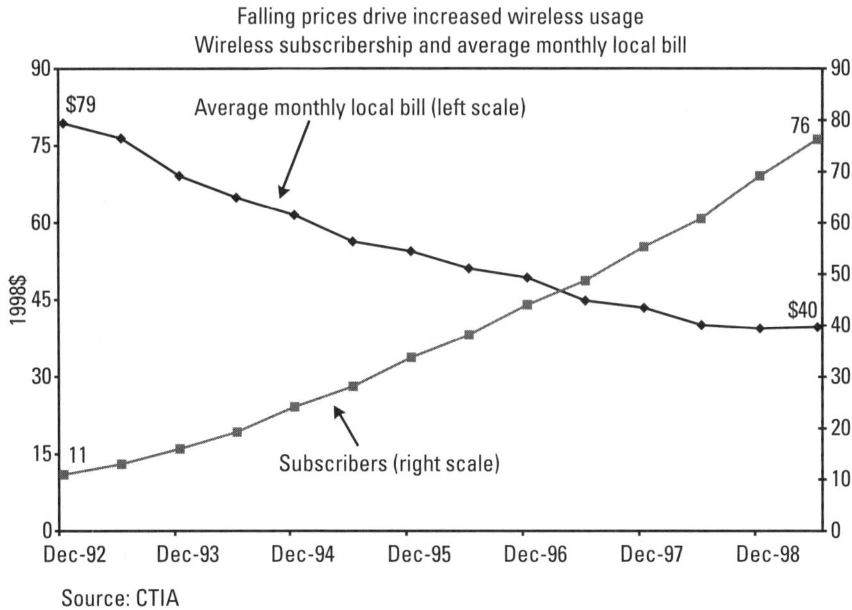

Figure 6

U.S. grew from 13 million in 1993 to over 76 million in 1999 (Figure 6), and are projected to increase to 149 million by 2003. In other words, the penetration rate of mobile service is expected to go from one in ten Americans in 1993 to six in ten in 2003. Similarly, since 1993, long distance traffic has increased 42 percent (Figure 4).

The Internet is driving growth in data traffic.

The Internet was in its infancy in 1996 when there were only 14.3 million host computers (computers with an IP address through which users can "log-in") connected to it in the world and only 27 million Americans using it (Figure 7). Today, there are more than 44 million Internet hosts (Figure 8), and nearly 80 million users in the U.S. Every day more people use the Internet in their business and personal lives. Families communicate through electronic mail and businesses are using the Internet to link their ordering systems to manage their inventories more efficiently. Electronic commerce has grown from next to nothing in the early 1990's to around $70 billion in 1999, and is projected to exceed $1 trillion in the next few

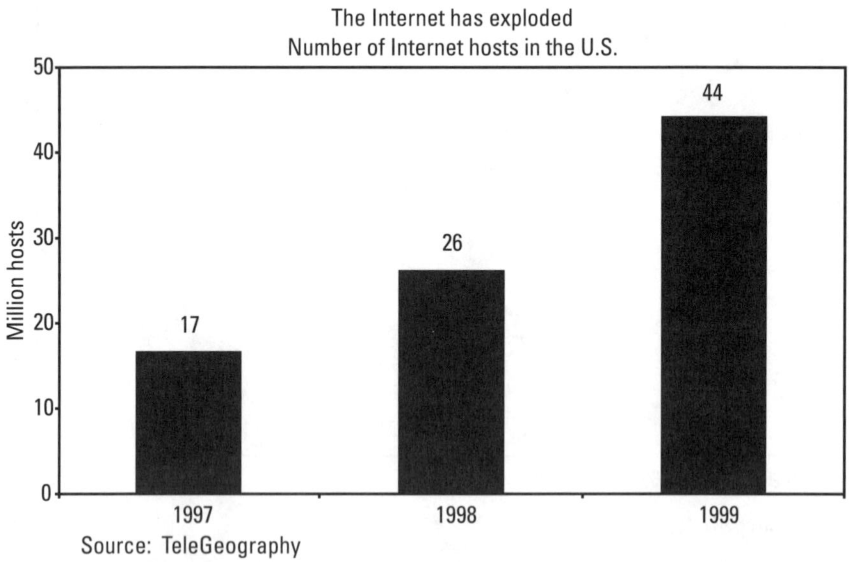

Figure 7

Internet users grew threefold from 1996 to 1999
U.S. Internet population*

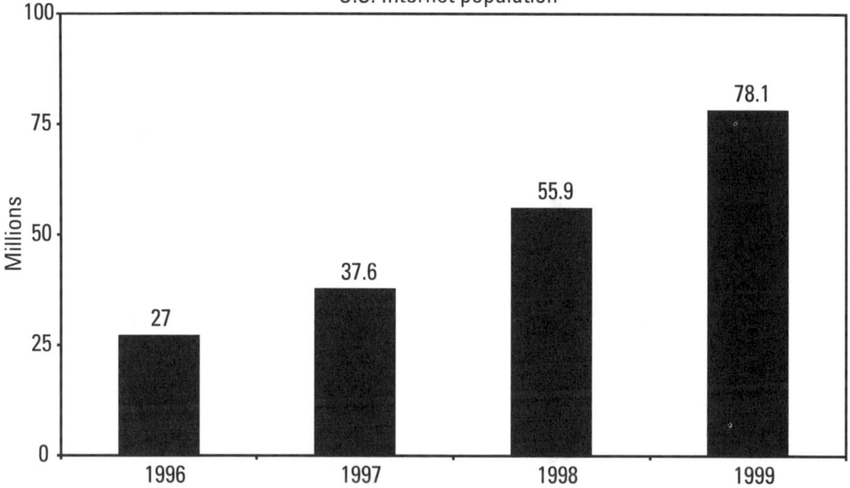

* Defined as those over 18 years old having accessed the Internet in the last 30 days.
Source: Mediamark Research

Figure 8

Residential broadband is born.

- When the 1996 Act was written, only a handful of households in a few select test markets were being offered broadband access to the Internet (Figure 9). Today, 1.7 million customers connect at speeds at least 25 times faster than a 28.8 Kbps modem. Cable modem service is now offered to nearly 30 million homes.
- Cable companies' service offerings have spurred telephone companies (both incumbents and their competitors), to deploy Digital Subscriber Line (DSL) technologies to deliver broadband access over telephone lines. This burgeoning competition is fueling deployment of broadband access and helping to keep prices low while improving service.
- Broadband Internet access is not just about faster net-surfing. It enables a fundamentally different Internet experience. Though the narrowband Internet is capable of transferring video and audio files, only a few would think of it as a substitute for

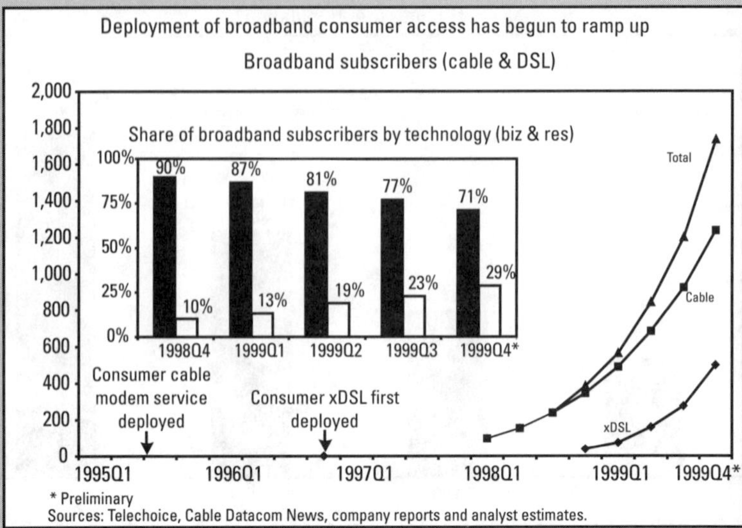

Figure 9

television or radio service. But a high-speed broadband link lets one download high quality video in real-time and has little problem playing audio that is "streamed" from net-radio stations on the Internet.

years. This growth is one of the drivers causing data traffic to double every 100 days.

Industry is Investing and Innovating

Companies are responding to demand, competitive pressures and new opportunities by investing heavily in their networks.

Competitors and incumbents invested roughly $25 billion more in 1999 than they did in 1995 (Figure 10).[3] These investments are driven by increasing demands on the nation's networks for voice and data traffic, to build new

3. Since comprehensive estimates of investment in different sectors based on similar methodologies and sources are not available, the amounts in Figure 10 should not be directly compared. For example, some investments by cable and wireless companies appear in both those categories and the CLEC category.

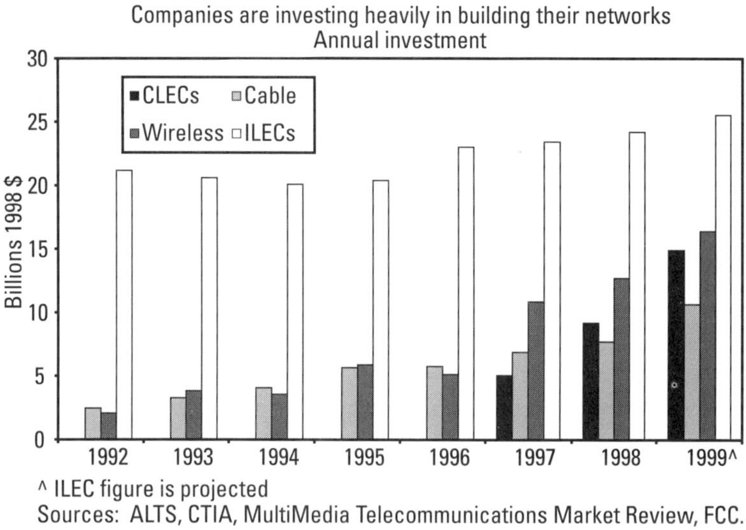

Figure 10

networks to compete with incumbents, and to offer better services to consumers in response to competitive pressure. In order to accommodate the rising traffic, companies have been investing heavily to expand their high-speed networks, and have nearly doubled the extent of the nation's fiber system (Figure 11). As demand for global communications has increased, the industry has responded by dramatically expanding capacity. By 2001 there is expected to be roughly 140 times as much trans-oceanic capacity as was available in 1996.

New entrants are building new networks.

Since the 1996 Act opened markets to competition, new entrants have invested ever larger amounts in building out their facilities and developing new services to compete with incumbents. They are laying fiber across the country, and installing switching equipment to link customers on their networks. Competitive Local Exchange Carriers (CLECs) have installed over 800 voice and 1,400 data switches, and laid 162,000 route miles of fiber through the end of 1999.

Similarly, wireless providers have, in the matter of a few years, built out entirely new networks across the country, and have been able to keep pace

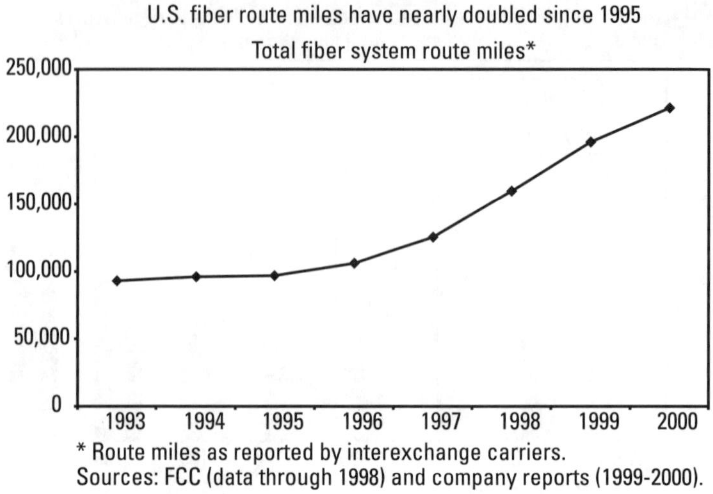

Figure 11

with increasing demand as more and more people discover the advantages of going mobile.

Not only have these new wireless operators brought competition to virtually every major market in the U.S. (Figure 12), but they have also facilitated the creation of truly nationwide wireless networks. For example, a business person could take a trip visiting clients in Boston, St. Louis, and Los Angeles and never leave the national networks of operators like AT&T and Sprint.

... and incumbents must upgrade to compete and grow.

Faced with competition in their traditional markets, incumbents have renewed their investments. Cable companies invested nearly twice as much in 1999 as they did in 1995. They are upgrading their cable systems to advanced two-way systems that can deliver not only more video channels, but high-speed Internet access and telephone service as well. Similarly, investment by incumbent local exchange carriers (ILECs), while flat in the early 1990's, grew by 19% between 1995 and 1998.

In the wireless industry, incumbent cellular operators have converted large sections of their networks from analog to digital technology in order to compete with the fully digital networks of broadband PCS operators.

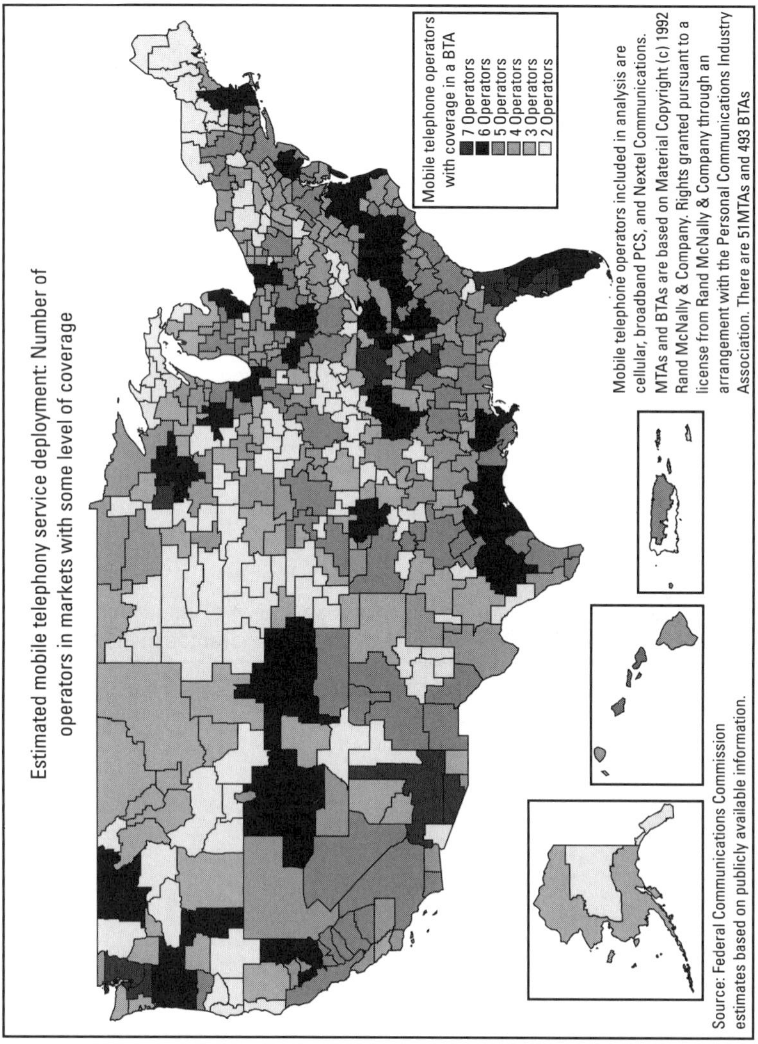
Estimated mobile telephony service deployment: Number of operators in markets with some level of coverage
Mobile telephone operators with coverage in a BTA
7 Operators
6 Operators
5 Operators
4 Operators
3 Operators
2 Operators
Mobile telephone operators included in analysis are cellular, broadband PCS, and Nextel Communications.
MTAs and BTAs are based on Material Copyright (c) 1992 Rand McNally & Company. Rights granted pursuant to a license from Rand McNally & Company through an arrangement with the Personal Communications Industry Association. There are 51MTAs and 493 BTAs
Source: Federal Communications Commission estimates based on publicly available information.

Figure 12

The industry is becoming more innovative.

With competition, firms have been under increasing pressure to cut costs, introduce new products, and improve service. As one measure of this increased innovation, the number of telecommunications patents granted annually by the U.S. Patent Office grew from 3,744 in 1994 to 7,674 in 1998. These patents cover a wide range of innovations such as the digital coding and decoding of conversations on wireless networks that have improved the sound quality of mobile telephones and allowed many more conversations to be carried over the same amount of spectrum. Overall patenting activity has been growing throughout the U.S. economy, but the telecommunications sector has grown more than three times as fast as the overall rate from 1994 to 1998. Telecom's share of all patents has grown from less than 1.5 percent in the early 1990's to 3 percent during the first half of 1999 (Figure 13).

Competition is the Driving Force

The 1996 Act provided a blueprint for competition among a variety of suppliers using different technologies to compete on a level playing field to deliver services to consumers. The Act mandated the elimination of barriers

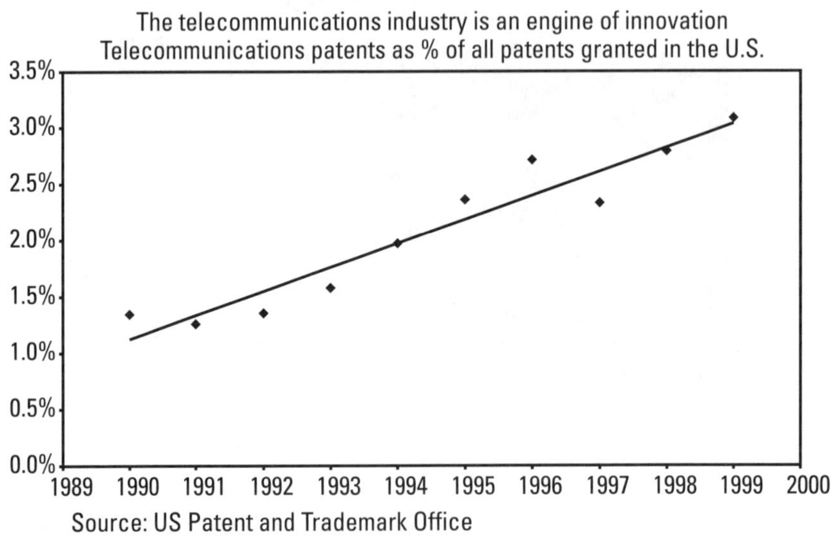

Figure 13

preventing companies from competing against each other—letting cable, wireless, long distance and other companies into local telephony and establishing the conditions by which the Bell companies could enter the long distance market. On the fourth anniversary of the Act, we see competition in the local market beginning to take hold and are beginning to see competition across multiple platforms to deliver services to consumers.

Competitors in the local telephone market have invested billions of dollars to build networks, and are successfully competing with incumbent carriers. According to some estimates, new competitors in local markets (both independent CLECs and the local arms of large long distance companies) have been adding over 1 million lines per quarter. Their share of the local market has increased to 4 percent of lines served and over 6 percent of local service revenue (Figure 14).

Local competitors have been particularly successful in the business market, where competitors have added 65 percent of all new lines deployed in the third quarter of 1999 (Figure 15).

Local competition is developing at a rate comparable to the development of competition in the long distance market.

In the long distance market, competitors that didn't exist before the 1970's now have over half of the long distance market as measured by revenue. However, in 1979, long distance competitors only had a one percent share of the long distance market. It took those competitors more than four years to take six percent of the market from incumbents. Similarly, in 1996, when the Act was passed, competitors had a one percent share of the local market. But it has taken them only two and one half years (through the second quarter of 1999) to reach the six percent mark (Figure 16).

Competition from satellite companies has eroded cable television's market share, forcing them to upgrade their cable systems.

Cable television companies have seen their market share fall by over ten percentage points in the 1990's (Figure 17). Between 1998 and 1999, competitors (mainly DBS operators such as DirecTV and Echostar) took 2 out of every 3 new multi-channel video subscribers (Figure 18). This competitive pressure has forced cable companies to upgrade their cable systems. Upgraded cable systems offer consumers more video choices, but also allow them to offer broadband Internet access and even telephone service. As Figure 17 shows, more people have more channels to choose from. Over 30

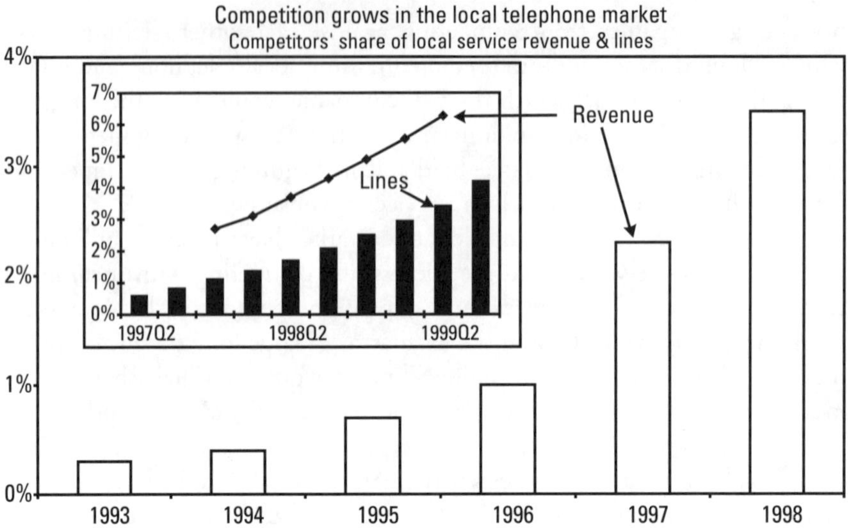

Sources: FCC, Merrill Lynch, Deutsche Banc Alex. Brown

Figure 14

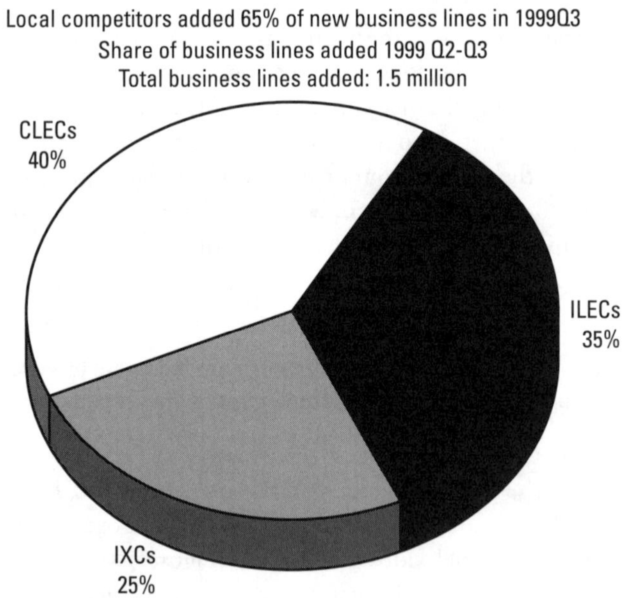

Source: Deutsche Banc Alex. Brown

Figure 15

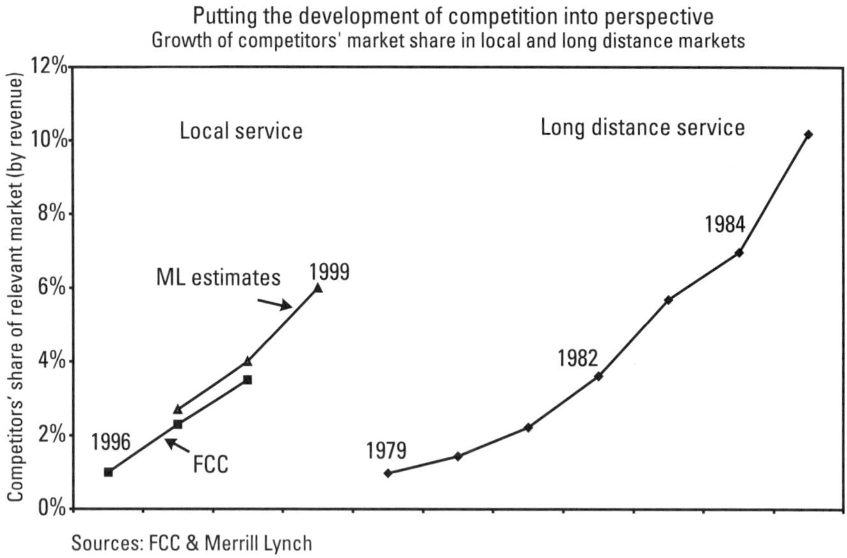

Figure 16

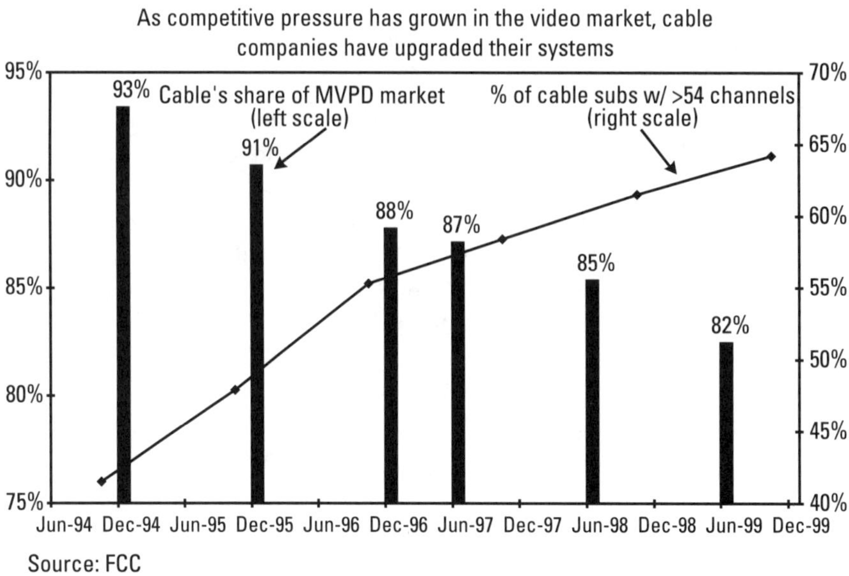

Figure 17

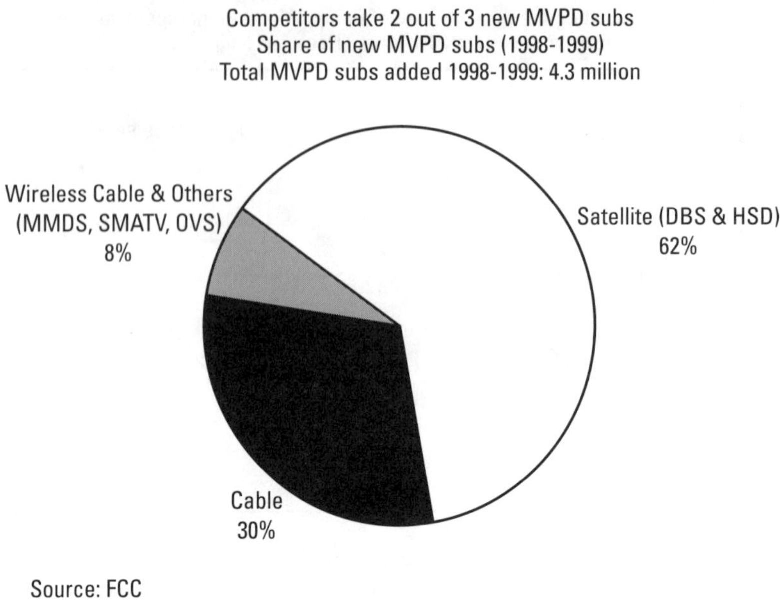

Figure 18

New York State is the Bellwether of Competition to Come

- New York was one of the first states in the country to require their incumbent telephone networks to interconnect with new entrants. On December 22, 1999, based on the extent of local competition, the FCC approved Bell Atlantic's application to provide long distance service in New York. Bell Atlantic, the New York State Public Utilities Commission, and local competitors in New York, have all worked hard to bring competition to the local market in New York state.

- In New York, competitors began acquiring the numbering codes that are necessary to route calls over their networks in the second quarter of 1994 (Figure 19). The number of competitors operating in New York has grown quickly since then. Competitors began entering other states in earnest only after the 1996 Act was passed.

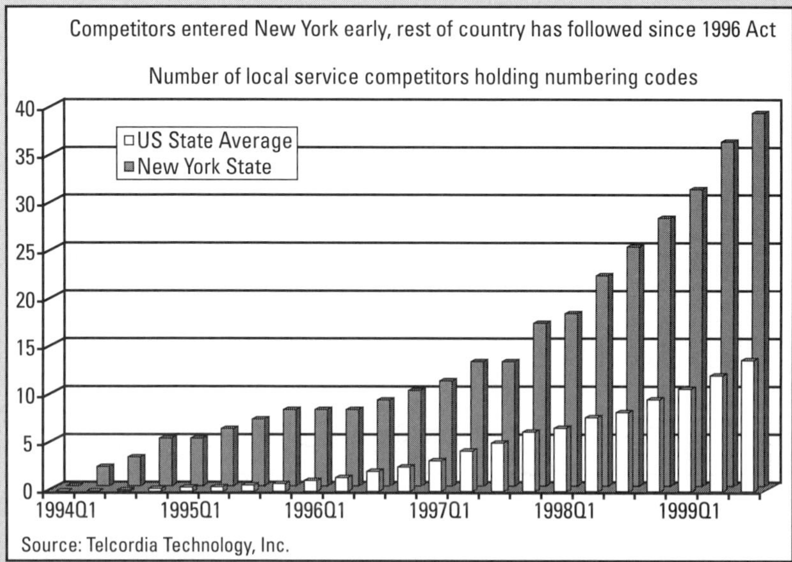

Figure 19

percent of the homes passed by cable TV can receive broadband Internet access today. About 130,000 customers already get telephone service from their cable company, and nearly 50 percent of all households are projected to be able to do so by 2005.

The wireless explosion.

Since the FCC first auctioned spectrum licenses in 1994, over 8,000 licenses have been awarded and new entrants have quickly developed wireless networks throughout the country. The 1996 Act further facilitated the development of competition in wireless markets by requiring local telephone companies to interconnect with new entrants and by establishing the ground rules for those interconnection agreements. These have made more competition possible in both mobile and fixed wireless networks.

In 1994, most Americans could choose between, at most, two competitors for their wireless service. Today, 94 percent of Americans have three or more providers in their home market, and over 75 percent have at least five operators competing for their business (Figure 20). With dramatic price

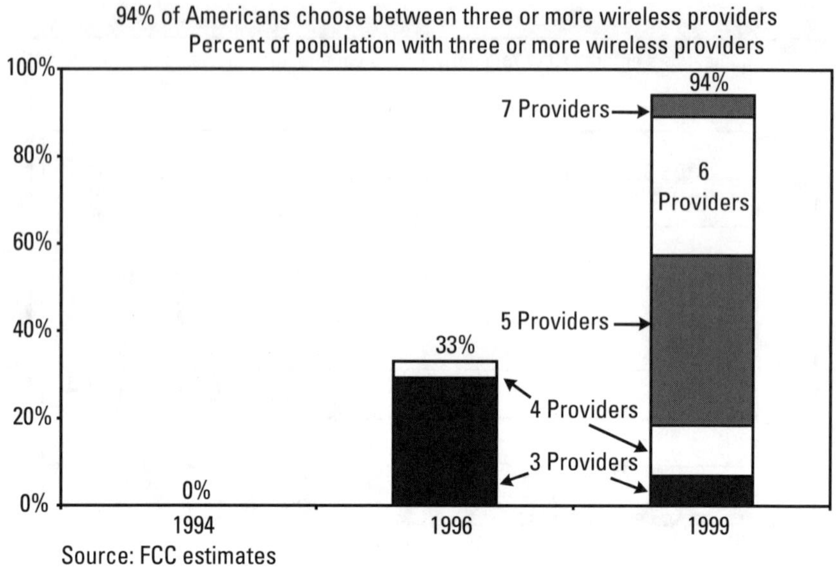

Figure 20

decreases and quality improvements in mobile telephone services, more people are beginning to see wireless telephones as substitutes for their wireline services. Along with the growth in the number of mobile telephone users, the average number of minutes each customer talks has more than doubled from approximately 100 minutes per month to between 200 and 300 minutes per month. This usage growth has helped increase the wireless industry's share of the overall voice market. In 1995, wireless voice minutes represented less than two percent of all voice traffic. By 1999, this had increased to more than seven percent, with the expectation that wireless will account for over 10 percent of all voice traffic during 2000.

The Act has changed the way people look at spectrum.

The Act, by opening the local market to competition, has turned spectrum into a third way to get into homes and businesses for video, voice, and data. Two CLECs, WinStar and Teligent, are already providing a full range of voice and data service to small and medium sized businesses using their spectrum in most of the largest markets. A third CLEC, Nextlink, is in the process of launching similar technologies and services in 25 markets during

The Bottom Line

In 1996, Congress and the Administration envisioned that the Telecommunications Act would be a sweeping update of our communications laws that would usher in a new competitive era in telecommunications. That era has indeed arrived. Consumers are beginning to reap significant benefits unleashed by the Act by enjoying an array of new services at lower prices.

2000. Others operators (such as Sprint, and MCI/Worldcom) are converting wireless one-way video systems to two-way networks to provide voice and high speed Internet access services to the residential market.

Glossary

accounting rate A unit of currency negotiated by international carriers as the amount of compensation to cover the cost of completing an international call. This amount includes all satellite or submarine cable transmission costs and domestic tail circuits used to link an international gateway with the call originator and recipient. Carriers typically divide the accounting rate in half when settling accounts.

Additional Plenipotentiary (APP) An extraordinary Plenipotentiary Conference of the International Telecommunication Union held in 1992 to consider recommendations on how to streamline operations. The APP significantly revised the Constitution and Convention of the ITU.

Administrative Procedure Act An Act of the United States Congress setting forth required procedures that administrative and regulatory agencies must use to provide for public participation and due process in the setting of rules and policies.

ADR Alternative Dispute Resolution. The use of conflict resolution alternatives to litigation and conventional administrative channels, including negotiations managed by a private facilitator or mediator.

ad valorem tariff A tariff calculated as a percentage of the value of foreign goods clearing customs.

advance publication The process by which nations inform other nations of future satellite orbital arc requirements through the submission of information about the satellite and the desired orbital location. The Radio Regulations Board, formerly known as the International Frequency Registration Board, of the International Telecommunication Union provides the forum for dissemination of orbital arc requirements and avoidance of resolution of actual or potential interference between satellites.

algorithm The application of a computing principle that sets the foundation for processing or transmitting data.

analog Information transmitted by varying modulation over a radio carrier wave. Voice and music originate in analog form and are converted for transmission over digital facilities like fiber optic cables.

ANI Automatic Number Identification. A feature in advanced telecommunications routing that provides the calling party's telephone number as the call is set up. Telephone companies have packaged this feature as a way to screen calls in residences and as a marketing and billing tool in commercial applications.

ANSI The American National Standards Institute. A nonprofit organization addressing standard setting, primarily by certifying other expert bodies to formulate standards in a narrow area of expertise.

APEC Asia Pacific Economic Cooperation group. An association of nations in the Asia/Pacific region that addresses trade and development work, including telecommunications and electronic commerce issues.

arbitrage A brokering function where a business acquires bulk capacity and resells it to individual users who singularly could not qualify for the discounts accruing to large volume customers.

Arianespace The world's first commercial space transport company established in 1980 by 36 European aerospace and electronics companies, 13 major banks, and the French space agency. The company launches satellites from a facility in Kourou, French Guyana.

ASEAN The Association of Southeast Asian Nations. A forum of Southeast Asian nations for addressing issues of mutual economic, technical, and trade issues. Many of the ASEAN nations have leased satellite capacity on the Palapa satellite system operated by the Indonesian government.

Asia-Pacific The region encompassing the Pacific Rim nations and countries bordering the Indian Ocean. This region has experienced the greatest economic growth that in turn has stimulated explosive development of telecommunication infrastructures.

Assembly of Parties The forum where representatives of nations, which have associated with the INTELSAT Agreement and Inmarsat Convention, meet to address major policy issues involving the cooperatives.

asynchronous transfer mode An advanced form of data transport whereby users can dynamically select the amount of bandwidth and transmission speed required and change the assignment as conditions warrant.

Atlantic Ocean Region (AOR) One of three major geographical regions, identified by a prominent body of water, that outline the largest possible satellite footprint. A satellite hovering above the equator midway between North America and Europe will have a footprint illuminating all of the Atlantic Ocean and most of North, Central, and South America; Europe; and Africa.

autonomy The assertion of independence and sovereignty which, if recognized by other nations, confers the right to participate in the International Telecommunication Union, INTELSAT, and other forums where policy and investment decisions are made.

balanced loading A regulatory policy requiring international carriers to activate satellite and submarine cables on a prescribed ratio or even basis with an eye toward bolstering use of the more expensive or developing transmission technology.

band A term that refers to a specific range of frequency spectrum. For example, the C-band refers to frequencies used to uplink and download traffic to satellites: around 6 GHz uplink and 4 GHz downlink.

bandwidth The total range of frequencies necessary to accommodate a spectrum using signal without distortion. Typically, bandwidth requirements grow as the amount of information to be transmitted increases. For example, an AM radio signal contains 10 kHz while an FM channel contains 200 kHz and a broadcast television signal contains 6 MHz, or 6,000 kHz.

bandwidth on demand The ability of users to adjust up or down the amount of capacity leased from a carrier, primarily through new technologies that make it easier to aggregate bitstreams.

barriers to market entry Structural, regulatory, and trade limitations to competition by domestic or foreign enterprises. Such barriers include prohibitions or limitations on the amount of foreign investment in a telecommunications service provider, reserving service monopolies to the incumbent carrier, imposing duties and other financial impediments on the equipment or services, and subjecting foreign carriers and manufacturers to more burdensome certification, testing, and licensing burdens.

basic rate interface The least common denominator of capacity in Integrated Services Digital Networks comprised of two 64 Kbps bearer circuits and one 16 Kbps data channel.

basic services The switching, routing, and transmission of voice or data provided by facilities-based carriers traditionally subject to common carrier regulation. Enhanced service providers use basic services as "plain vanilla" building blocks over which additional services and features are added.

bearer circuit A transmission pathway that can be subdivided into channels of less bandwidth or data throughput through multiplexing. For example, a bearer circuit with a total capacity of 64 Kbps can be subdivided into four channels with the capacity to transmit one voice grade channel.

beggar-thy-neighbor policy Attempts by one country to reduce unemployment and to increase domestic output by raising tariffs and erecting nontariff barriers to reduce imports. This strategy has proven risky because it can provoke retaliation.

Bell Operating Company (BOC) The telephone service operating companies of the regional holding companies divested from AT&T. For example, the regional holding company Verizon operates several BOCs serving customers from New England to Virginia.

bilateral trade agreement A formal or informal agreement between two nations.

bit "Binary digit." The smallest unit of measurement in the transmission of data having only two values called 0 or 1.

bit error rate The extent to which a digital transmission network generates an error. The rate can be reduced through error detection and correction.

bit rate Also known as "throughput," bit rate is the speed at which bits are transmitted, usually expressed in terms of bits per second. For example, a conventional dial-up modem can handle data at the rate of up to 56,000 bps.

BOO Build, Own, Operate. A method for infrastructure development where a foreign enterprise agrees to build and operate a facility. This arrangement creates a service franchise and the incentive for the operator to upgrade and maintain facilities without fear of nationalization.

boomerang box A device that provides dial tone to callers in another country thereby providing them a virtual presence in a location with lower outbound international long distance telephone rates.

boresight The center point on Earth where a satellite's radiated transmissions are strongest.

BOT Build, Operate, Transfer. A method for infrastructure development where a foreign enterprise agrees to build a facility, operate it for a specified time period, and then transfer title to the national government or carrier who will then take over operations after having had time to develop operational and management expertise.

bottleneck A facility or portion of a route where traffic aggregates, often allowing the operator or service provider the opportunity to charge monopoly rates and to engage in anticompetitive practices.

boycott A refusal to deal commercially or otherwise with a person, firm, or country.

broadband A medium offering high capacity, high speed services achieved by using large wide bandwidth and digital transmission.

browser Software that provides a convenient and easy-to-use interface for accessing content on the Internet.

BSA Basic Serving Arrangement. A term used by the Federal Communications Commission in its *Third Computer Inquiry* to identify the components in generic connections between Enhanced Service Providers and facilities-based carriers providing basic services. BSAs consist of the access links between facilities as well as the transport, routing, and functions.

BSE Basic Serving Elements. A term used by the Federal Communications Commission in its *Third Computer Inquiry* to identify optional network features, available on an unbundled, à la carte basis such as ANI.

BT An acronym referring to British Telecom, the dominant local and international carrier in the United Kingdom.

BTO Build, Transfer, Operate. A method for infrastructure development where a foreign enterprise agrees to build a facility, transfer title to the national government or carrier, and operate the facility at a profit for a specified time period after which a domestic carrier may take over operations. This arrangement enables the government to maintain a greater degree of facilities control than if a private foreign entity held title.

bundled The process by which two or more, possibly segregated, features are offered jointly.

bypass The use of alternative facilities or services to those of incumbent carriers (e.g., the use of cable television systems or wireless technologies to access interexchange carrier facilities rather than the incumbent telephone company's wireline network).

byte An intermediate unit of data transmission capacity comprised of 8 bits.

C3I Command, Control, Communications, and Intelligence. Tactical requirements of the military.

cablehead The oceanfront location where a submarine cable makes landfall and where power, amplification, and multiplexing functions take place.

caching Temporary storage of Internet content (e.g., frequently viewed World Wide Web pages at servers closer to a particular user).

callback services Providing outbound international long distance calling capabilities to customers in another country through the use of a device that processes a request for dial tone. Callback service can be initiated by a conventional inbound international long-distance call that is terminated before triggering a toll charge—the process of importing outbound telephone calling capability into a location having high domestic and international calling rates.

call-reorigination The process of reversing the routing of an international telephone call for purposes of providing consumers cheaper outbound calling opportunities.

C-band The portion of the radio spectrum (Earth-to-space at 6 GHz, space-to-Earth at 4 GHz) allocated for satellites providing service between fixed points or broadcast services.

CCIR The International Radio Consultative Committee of the International Telecommunication Union, reformulated in 1993 with portions of its portfolio assigned to the newly formed Radiocommunications and Telecommunication Standards sectors.

CCITT The International Telegraph and Telephone Consultative Committee of the International Telecommunication Union, reformulated in 1993 with portions of its portfolio assigned to the newly formed Radiocommunications and Telecommunication Standards sectors.

CDMA Code Division Multiple Access. A process for deriving more throughput and accommodating increasing demand for service by subdividing spectrum into various code sequences.

CEI Comparatively Equivalent Interconnection. A precondition imposed by the FCC before the BOCs could provide basic and enhanced services without structural separation.

cellular radio A terrestrial radio service designed primarily for mobile applications using microwave transmitters whose low powered operations enable frequency reuse and integrated service throughout a region.

CEN European Committee for Standardization. A regional standard-setting body in all fields except electrical and electronic matters.

CENELEC European Committee for Electrotechnical Standardization. A regional standard-setting body in electrical and electronic matters.

central office A telephone company facility that provides centralized management of switching, routing, and line transport functions that may traverse other facilities closer to the end user.

CEPT The Conference of European Post and Telecommunication Administrations. An assembly of European postal and telecommunications regulatory authorities aiming to harmonize standards, policies, and initiatives for the ITU and other global forums. Diversifying constituencies and establishment of a separate standard setting body (the European Telecommunication Standards Institute) challenge the ability of CEPT to establish significant and uniform policies.

circuit switching Physically linking two or more points via dedicated circuits.

CIS The Commonwealth of Independent States. An affiliation of sovereign states that formerly comprised the Soviet Union.

CITEL Inter-American Telecommunications Conference. An assembly of nations that participate in the Organization of American States with a mission to address radio and telecommunications issues, including ones to be addressed by upcoming conferences of the ITU.

collection rate The end user charge for service, typically set out in a tariff.

colocation The physical interconnection of lines and equipment owned and operated by different carriers typically on the premises of the major incumbent carrier.

commercialization of space Eliminating government subsidies and management so that private ventures can compete fully and fairly in space-related ventures.

common carrier A legal and regulatory classification that requires a telecommunications facility or service provider to serve any user within a certificated geographical region, and to provide service in a nondiscriminatory manner typically through public tariffs.

Communications Act of 1934 The primary United States law establishing the Federal Communications Commission and the general scope of broadcast and common carrier regulation.

Communications Satellite Act of 1962 The United States law that created Comsat Corporation as the sole signatory to INTELSAT and established general satellite policy. Congress amended this Act to permit the acquisition of Comsat by Lockheed Martin Corp., the privatization of INTELSAT, and more robust international satellite service competition.

comparably efficient interconnection The requirement of the Federal Communications Commission that the Bell Operating Companies provide a plan demonstrating that they will not discriminate against unaffiliated enhanced services providers when the BOC decides to provide similar services. While the *Second Computer Inquiry* required the BOCs to provide enhanced services through a separate subsidiary, the *Third Computer Inquiry* eliminated structural separations provided the BOCs filed a CEI plan for each enhanced service.

comparative advantage A fundamental international trade concept that views nations or regions having the ability to produce certain goods or services as being more efficient than other nations. By emphasizing production where a comparative advantage exists, nations can efficiently use available natural and human resources, and trade for goods and services where the nation has a comparative disadvantage.

compatibility The ability of users and carriers to interconnect equipment, lines, and facilities while maintaining services with a reasonable degree of reliability and quality.

complementary products Products that add to the value and utility of a product (e.g., sugar with coffee and modems with personal computers).

compression The application of techniques for reducing the amount of frequency spectrum or channel capacity needed to derive a circuit (e.g., 4 to 1 compression provides the means for deriving four channels where previously only one channel was available).

Computer Inquiries A set of proceedings of the Federal Communications Commission beginning in the 1970s designed to erect a regulatory system that permits facilities-based carriers to enter enhanced services markets without cross-subsidization by users of the carrier's basic services.

Comsat Corporation The former exclusive U.S. signatory (investor) in INTELSAT and Inmarsat that operated as a "carrier's carrier" by leasing wholesale satellite capacity to other carriers for subsequent retailing to end users. Lockheed Martin Corporation acquired Comsat in 1999.

conscious parallelism The deliberate matching of prices and services by carriers to avoid more aggressive competition and price wars.

consent decree A remedy in an antitrust case where the proceeding concludes without a verdict in exchange for an agreement by the defendant to refrain from continuing to engage in certain activities or practices. AT&T agreed to consent decrees in 1956 and 1982 to settle antitrust suits.

consortium A still-predominant ownership structure where multiple carriers from several nations jointly participate in the ownership of a submarine cable or satellite venture.

consultation process Scheduled meetings of international carriers and government agencies to evaluate facilities demand and determine where and when to deploy additional capacity.

consultative process The requirement in Article XIV of the INTELSAT Agreement and Article VII of the Inmarsat Convention that parties to these agreements consult with the cooperative to ensure that another satellite system will not cause technical or economic harm.

CONUS Continental United States. The lower 48 states.

convergence The merger of technologies that previously served discrete markets leading to integrated, additional offerings (e.g., computer terminals will serve entertainment applications in addition to information processing, and television sets will serve information processing applications in addition to entertainment).

cooperative The formerly predominant method for spreading financial and operational risk for operating regional and global satellite ventures.

coordination The process by which nations, carriers, and other service providers meet to resolve potential conflicts, including the potential for radio interference. The INTELSAT Agreement requires nations that have agreed to become parties to the organization to demonstrate the absence of technical and economic harm to INTELSAT when authorizing separate satellite systems. The Radio Regulations of the International Telecommunication Union require nations to coordinate the use of frequencies and the satellite orbital arc.

COPUOS Committee on the Peaceful Uses of Outer Space. A Committee of the United Nations that addresses issues pertaining to space exploration, settlement, and orbiting objects in space, including satellites.

corporatization The conversion of a government-owned carrier into a more business-like enterprise, with or without a change in ownership.

CPE Customer Premises Equipment. Telecommunications equipment, including telephones and private branch exchanges, located on user premises.

CPNI Customer Proprietary Network Information. Information about a customer's basic service usage that can provide marketing leads to enhanced service providers. The *Computer Inquiries* established rules requiring the BOCs to withhold such information from enhanced service affiliates or competitors unless authorized to do so by the user.

cross-subsidization Using revenues accrued from one service to underwrite the provision of another service at less than fully compensatory rates.

cyberspace A word coined by William Gibson in his novel *Neuromancer* now referring to the broad, global resources available via computer networks.

DAB Digital Audio Broadcasting. The transmission of audio signals in a higher quality digital format rather than the conventional analog method.

DAMA Demand Assigned Multiple Access. A queuing method by which a larger number of users may share transmission capacity by allocating on an as-needed basis.

DBS Direct Broadcast Satellite. The use of medium to high powered signals for direct broadcasting to satellite terminals located on the premises of the recipient.

dedicated Reserved for the use of one or more specified users (e.g., a private line or satellite transponder may be dedicated for the use by a particular lease). Nondedicated capacity is provided on a virtual or on-demand basis.

de facto standard setting The creation of a dominant or single standard by the interplay of market forces rather than promulgation of a standard through the standard setting process.

de jure standard setting The formation of a standard by law or through the rule-making process of the appropriate regulatory agency.

deregulation A reduction in the breadth and scope of regulation with an eye toward allowing marketplace forces to replace government intervention.

destructive competition Short-term price competition at noncompensatory levels resulting in the market exit by some enterprises and the potential for survivors to raise rates above competitive levels to recoup prior losses.

digital The use of a binary form, comprised of on and off pulses, to represent the continuously varying signals of images and sounds. Information, entertainment, voice traffic, and other forms of communication can be encoded, stored, processed, and transmitted in a numeric, as opposed to analog, form.

digitization The use of computer readable bit streams for transmitting, switching, and routing information.

direct access The opportunity for end users and service providers to acquire satellite capacity directly from the operator (e.g., INTELSAT and EUTELSAT instead of having to deal with an intermediary).

direct broadcast satellite The use of a satellite to transmit programming, including audio and video, directly to end users equipped with receiving dishes and the necessary electronic components (also known as direct to home broadcasting).

disintermediation The ability of new information services, like the Internet, to eliminate channels of distribution for a product or service.

divestiture Severing part of a corporation and creating a separate business entity, on a voluntary basis, or as part of an antitrust settlement or verdict.

DNIC Data Network Identification Codes. ITU-recognized coding system for identifying and routing traffic to specific network operators.

dominant carrier An FCC classification of common carriers who, because of their market power, require closer regulatory scrutiny. Such carriers have less flexibility to change tariff terms and conditions. The classification applies to AT&T, foreign-owned carriers on a route-specific basis, Comsat in its INTELSAT and Inmarsat signatory functions, and carriers operating the only Earth station in an offshore point (Guam, for example).

DOV/DUV Data-Over-Voice/Data-Under-Voice. A technique for transmitting voice and data over a single channel through multiplexing or other signal splitting technology.

downlink The transmission of traffic from a satellite to receiving locations.

duopoly A monopoly shared by two enterprises.

E-1 The European standard for transmission capacity handling 2.048 Mbps.

Earth station Terrestrial equipment used to transmit and receive satellite telecommunications. Some Earth stations provide receive-only functions.

economies of scale A measurement of economic efficiency that identifies who produces a good or service at the lowest per unit cost and the optimal amount of production.

economies of scope A measurement for assessing how efficiently an enterprise will operate when serving adjacent markets (e.g., whether a telephone company can efficiently use its network to provide cable television and information services).

EEC European Economic Community. A trade organization and governance structure for European nations now known as the European Union.

EFTA European Free Trade Association. A trading association of European nations not part of the EEC, including the Nordic Countries, Austria, and Switzerland.

elasticity of demand The intensity of preference for a good or service based on user reaction to an increase or decrease in price.

elasticity of supply How consumers react to changes in the availability of a good or service.

electronic commerce The use of the Internet and other data communications media to promote faster, better, smarter, and more convenient commercial transactions.

electronic data interchange The use of telecommunications and information processing to conduct business transactions, often in an integrated network combining different media (e.g., voice, text, and data processing).

electronic funds transfer The use of telecommunications and information processing to achieve a transfer of money, typically from one bank to another.

e-mail Computer-mediated text communications.

end office The telephone company switching facility closest to a particular user.

end-on-end routing The segmentation of a route into two or more segments typically priced individually.

end-to-end routing A complete traffic routing arrangement typically priced at one composite rate.

enhanced services Enhancements to basic common carrier transmission services involving computer processing that acts on the code, content, protocol, or format of the information in such a way as to change the output and possibly to store it as well for subsequent retrieval and manipulation.

enterprise networks Diversified, customized, and complex international telecommunications and information networks primarily used by multinational enterprises to serve particular requirements. These networks often require the assistance of system integrators and outsourcers who plan, procure, and manage the network.

Erlang A unit for measuring telecommunications traffic and capacity needed to support an acceptable level of service during the busiest usage times.

ETSI The European Telecommunications Standards Institute. A regional standard-setting body organized by the EU.

European Union The Union of European nations formerly affiliated in a less integrated European Community.

EUTELSAT The European Telecommunication Satellite Organization. A regional satellite cooperative providing satellites for delivery of telecommunications and video programming.

facilities-based competition Marketplace entry by enterprises that have installed their own transmission, switching, and related facilities rather than resell the facilities of another carrier.

FCC Federal Communications Commission. The expert regulatory agency created by federal law to allocate, allot, and assign spectrum and to oversee broadcasting and common carrier telecommunications.

FDC Fully Distributed Costs. Costing that includes a contribution to fixed costs by all consumers.

feeder link A radio link to and from satellites for conveying information, including the tracking, telemetry, and network control needed to maintain a satellite in proper orbit.

fixed satellite service The use of satellites to provide service between users at fixed locations.

footprint The range of geographical coverage of a satellite transmission.

forbearance The FCC-articulated concept of refraining from regulating certain common carriers lacking market power thereby affording greater flexibility in pricing and providing service.

foreign correspondent A carrier that has entered into a foreign operating agreement with another carrier entitling it to originate and terminate international traffic.

frame relay A new data transmission technology that quickly switches and routes digital packets with a low bit error rate.

gateway A satellite Earth station or submarine cablehead where domestic facilities access international transmission facilities.

GATS General Agreement on Trade in Services. The process for extending GATT trading principles to services in addition to goods.

GATT General Agreement on Tariffs and Trade. A multilateral agreement by most nations to reduce barriers to trade. The provisions of the GATT were integrated into the governance documents of the World Trade Organization.

geostationary The location above the Earth where launched objects appear stationary relative to Earth. Satellites become geostationary at 22,300 miles in altitude. Placing satellites in geostationary orbit above the equator maximizes the scope of geographical coverage.

geosynchronous The orientation of a satellite's orbit in relation to the Earth's orbit.

gigahertz (GHz) A measure of radio frequency equivalent to 1 billion hertz (cycles per second).

GII Global Information Infrastructure. A concept extending globally national initiatives to promote widespread deployment and access to broadband digital technologies and services requiring high speed digital transmission (e.g., video programming, telemedicine, large data file transport, and Internet access).

globalization Expanding the perspective, strategic business planning marketing, and operational scope of an enterprise to target and serve new international opportunities (e.g., PTTs, facing lost market share domestically as a result of market entry, may seek market share in foreign markets).

GNP Gross National Product. An aggregate measure of income generated from all goods and services.

green paper A vehicle for publishing future policy objectives. The Commission of the European Community prepared a green paper in 1987 articulating telecommunications harmonization goals as part of the Single Europe Act.

GSM Global Systems for Mobile Communications. A standard for digital cellular radio systems that originated in Europe.

half-circuit The smallest unit of international transmission capacity matched by foreign correspondents to route traffic.

harmful interference Spectrum usage that endangers, degrades, or deteriorates the proper functioning of another registered spectrum use.

HDTV High Definition Television. The development of a video transmission and production standard calling for higher resolution and better sound quality.

Heads of Agreement A preliminary agreement by governments and business enterprises articulating general terms followed by a more comprehensive document.

HLC An ad-hoc High Level Committee created by the ITU Plenipotentiary to provide recommendations to make the Union more efficient and effective.

hosting Serving as a repository for content made available via the Internet.

hubbing The aggregation of traffic from throughout a region thereby achieving circuit loading efficiency and economies of scale. Hubbing in telecommunications can be analogized to airport hubbing where long haul routes are loaded from a number of short haul traffic originating in nearby localities.

ICANN Internet Corporation for Assigned Names and Numbers. An organization seeking to establish uniform policies to resolve conflicts relating to the structure and names used to identify Internet sites.

ICAO International Civil Aviation Organization. A specialized agency of the United Nations that establishes rules, standards, and policies affecting international aviation, including navigation and telecommunication issues.

ICO Intermediate Circular Orbit. The installation of satellites into non-geostationary orbits closer to Earth to support applications using small, low powered handheld terminals.

IEEE Institute of Electrical and Electronic Engineers. An international organization that participates in the telecommunications standard-setting process.

IFRB International Frequency Registration Board. The part of the ITU's Radiocommunications Sector, now known as the Radio Regulation Board, that registers spectrum and orbital arc uses and resolves potential interference problems.

IMTS International Message Telephone Service. The conventional dial-up international long distance telephone service.

in-band signaling The customary process of including switching and routing information as headers preceding the call, increasingly replaced by out-of-band signaling.

incentive regulation Alternatives to the conventional rate base regulation designed to reward innovation and efficiency by allowing carriers to capture financial gains rather than automatically flow them to ratepayers through refunds or lower rates.

inclined-orbit Satellite orbits that conserve station-keeping fuel to extend the operational life of a satellite by allowing it to deviate slightly from a geosynchronous orbit, thereby requiring Earth station tracking.

incumbent carriers Carriers that have installed an extensive network and heretofore have faced limited, if any, competition.

infrastructure An essential system to the public health and welfare (e.g., water, sewerage, electricity, currency, telecommunications, and roadways).

Inmarsat International Maritime Satellite Organization. A cooperative of over 70 nations organized to provide ubiquitous maritime services, including safety and distress functions. The parties of Inmarsat expanded Inmarsat's function to include aeronautical and land mobile services on an ancillary basis. Inmarsat expects to privatize, having already spun off part of its assets into a competitive venture.

intelligent network The architecture and plans for a future telecommunications infrastructure with advanced features made possible by expanded network switching, signal processing, and intelligence.

INTELSAT International Telecommunication Satellite Organization. A cooperative of over 120 nations organized to provide ubiquitous international satellite capacity. INTELSAT expects to privatize, having already spun off part of its assets into a competitive venture.

interactive The ability of users to request, manipulate, process, and change data through on-line commands. Interactivity converts one-way information sources into two-way media.

interconnect The physical connection of equipment and lines to secure a complete route and service arrangement. Users physically interconnect CPE to a telecommunications network through a plug physically attached to a jack.

interexchange telecommunications Long haul services that link local exchanges. In the United States, the Modification of Final Judgment precludes Bell Operating Companies from providing those interexchange services that cross local access and transport boundaries.

interface A shared boundary (e.g., between CPE and the telecommunications plant owned and operated by a telephone company).

international comity Recognition that each nation has sovereignty and that mutually beneficial results accrue when nations relinquish a degree of independent decision-making to reach a single, consensus policy.

Internet A multifaceted, international network of databases and users, organized initially for military, scientific, and academic users.

Internet service provider An enterprise that owns or leases facilities to provide end users with access to the Internet. ISPs also may provide or host content.

Intersputnik The former international satellite cooperative organized by the former Soviet Union.

INTUG International Telecommunications Users Group. A trade association of international telecommunications users.

IRAC Interdepartment Radio Advisory Committee. A committee, organized under the auspices of the Commerce Department and comprised of 20–25 representatives from various U.S. federal agencies, with a mission to coordinate on spectrum management issues affecting the U.S. government.

IRC International Record Carrier. International carriers that concentrate on providing textual services like telegraphs and telexes.

IRU Indefeasible Right of User. A method for conveying the rights, but not the title, to use international telecommunications capacity.

ISDN Integrated Services Digital Networks. A telecommunications standard envisioning interconnected digital networks capable of simultaneous delivery of voice and data services.

ISL Inter-Satellite Links. Extremely high frequency communication links between satellites that substitute for a round trip from a satellite down to an Earth station and back up to another satellite.

ISO International Standards Organization. A specialized agency of the United Nations with a broad standard-setting portfolio that includes data processing.

ISP International Settlements Policy. An FCC policy requiring all U.S. carriers to apply identical accounting and settlement rates for IMTS. The FCC considered the ISP essential to prevent foreign carriers from whipsawing U.S. carriers, but now permits deviation from the ISP by WTO member nations lacking government monopoly carriers.

ITU International Telecommunication Union. A specialized agency of the United Nations that formulates policy, regulations, and recommendations in telecommunications, including spectrum allocation, satellite orbital arc registration, telecommunications development, standard setting, and conflict avoidance/resolution.

ITU Council An elected body of the ITU that performs executive board functions including the scheduling of conferences.

IVAN International Value Added Network. An enterprise that enhances leased private lines with customized features (e.g., credit card verification and airline reservation systems).

IXC Interexchange Carrier. A provider of long haul telecommunicatiosn services.

Ka-band That relatively unused portion of the radio spectrum (30 GHz uplink, 20 GHz downlink) allocated for new satellite requirements.

KDD Kokusai Denshin Denwa. The dominant Japanese international carrier.

kilohertz A measurement of radio waves equal to 1,000 cycles per second.

Ku-band An increasingly used portion of the radio spectrum (14 GHz uplink, 11–12 GHz downlink) allocated for satellite requirements.

LATA Local Access and Transport Area. A geographical region sharing common cultural, social, and economic interests within which the divested Bell Operating Companies may provide local and interexchange services.

L-band That portion of the frequency band (1–3 GHz) where a variety of mobile services are provided by geostationary and LEO satellites.

LDC Lesser Developed Country. A classification of nations with economies and telecommunications infrastructure below that of industrialized nations. Also known as Newly Industrialized Nations.

Leaky PBX The ability of users to interconnect on-premises equipment to the public switched telecommunications network, thereby linking private lines, designed to link two points only, with often unmetered switched local services.

LEC Local Exchange Carrier. The provider of local services, often considered a natural monopoly and bottleneck through which most interexchange traffic traverses.

LEOs Low Earth Orbiting satellites. Satellites operating in orbits below the geostationary orbiting arc where service can be provided to low powered handheld terminals. Big LEOs provide a variety of voice and data services while small LEO provide data and emergency position location services.

level competitive playing field The goal of market entrants who claim that regulators and other decision-makers must take affirmative steps to reduce unfair market access opportunities accruing to incumbent carriers as a function of their earlier market entry, bottleneck control, customer base, and regulatory status.

liberalization The relaxation of rules and service obligations. For example, subsidized and under-priced local services, imposed on incumbent carriers, often occurring contemporaneously with privatization, deregulation, and market entry initiatives.

line-side interconnection Interconnection that takes place at a low level in the switching hierarchy typically for end users.

link budget The determination whether an adequate signal can be received by calculating the transmission power and factors that attenuate the signal.

link margin The estimated degree of signal strength above noise at a particular location under various circumstances.

local area network A network of work stations and personal computers linked via wiring installed by the operator or provided by the telephone company, or via wireless applications.

local loop Local exchange facilities comprising the first and last legs of a telecommunications route.

low Earth orbit The use of nongeostationary orbits for satellites primarily providing mobile telecommunications to lightweight transceivers. LEO satellites operate only a few hundred miles above the Earth surface, thereby reducing launch costs and the power needed to transmit to and from the satellites. On the other hand, the closer proximity to Earth requires a larger number of satellites to achieve desired regional or global coverage.

LRIC Long Run Incremental Cost. A measure of the additional cost incurred over the long term to provide a particular amount of a good or service. This measure accounts only for the additional costs incurred without regard to embedded fixed costs that could be shared by additional users.

market power The ability to affect the price or supply of a good or service.

MAOU Minimum Assignable Ownership Unit. The smallest unit of submarine cable capacity, typically a 64-Kbps circuit with associated capacity for multiplexing, allocatable to an investor or available to others through an Indefeasible Right of Use conveyance.

MEO Middle Earth Orbit. Nongeostationary satellite orbits between low Earth orbits below and geostationary orbits above where operators attempt to balance the benefits accruing from operating closer to Earth (e.g., access by lower powered terminals yet operating high enough to provide broad geographical coverage).

MFJ Modification of Final Judgment. Modifications to the 1956 Consent Decree agreed to by AT&T and the Justice Department thereby settling an antitrust suit. The MFJ required AT&T to divest its Bell Operating Companies in exchange for the opportunity to serve data processing markets. The spun-off Bell Operating Companies were limited to local exchange services.

mid-ocean The concept that international half-circuits are matched by foreign correspondents at the centerpoint of a submarine cable.

MITI Ministry of Trade and Industry. The powerful Japanese Ministry with trade and development portfolios.

modem A device that modulates and demodulates data signals enabling computers and other digital devices to operate over an analog network.

most favored nation A basic trade principle requiring nations associating with the WTO to extend to all nations any market access benefit accorded other nations.

MOU Memorandum of Understanding. A preliminary agreement setting forth basic terms and conditions typically followed up with a more comprehensive agreement.

MPT Ministry of Posts and Telecommunications. The government agency in many nations with telecommunications operational and regulatory responsibilities.

MSS Mobile Satellite Service. An ITU-recognized service designation for the provision of voice and data services to mobile users via geostationary or nongeostationary orbiting satellites.

MTS Message Telephone Service. Conventional dial-up voice telephone service also known by the acronym POTS (Plain Old Telephone Service).

multimedia The integration of various media previously considered available via a separate pipeline or marketing channel (e.g., using the television for entertainment and new interactive data processing, consumer order entry, and utility monitoring).

multinational enterprise A venture doing business in a number of nations making it a candidate for turnkey services from systems integrators and outsourcers.

multiplexing Subdividing a circuit into more than one channel to derive more throughput and capacity.

multipoint Using telecommunications services to serve more than one physical location (e.g., originating a video program for delivery to numerous cable systems using the broad geographical coverage of a satellite).

NAFTA North American Free Trade Agreement. A trade agreement between the United States, Canada, and Mexico resolving to reduce or eliminate barriers to trade between nations with an eye toward creating a regional trading bloc.

national hero A domestic carrier or manufacturer benefiting from policies designed to protect it from foreign competition including closed procurements and other policies that handicap, or prohibit, market entry.

national treatment Subjecting foreign manufacturers and service providers to the same regulatory treatment as afforded domestic enterprises.

natural monopoly A single manufacturer or service provider who singularly can operate most efficiently and who can maximize scale economies.

NIC Newly Industrialized Country. A nation with an expanding economy and, typically, with growing telecommunications requirements.

NII National Information Initiative. A plan for using government to stimulate primarily private ventures that will upgrade and expand the telecommunications and information-processing infrastructure in the United States. The government initiative includes a vision for ubiquitous access to a feature-rich information highway.

NIST National Institute of Science and Technology. An agency of the U.S. Department of Commerce that coordinates standard setting for the federal government and whose standards affect the parallel private standard-setting process.

node A point in a telecommunications or information-processing network configuration where two or more lines, routes, or pathways come together (e.g., at a switch, Earth station, or PBX).

noncommon carrier A classification that reduces or eliminates regulation in recognition of the carrier's lack of market dominance and the nonessentialness of the services offered.

NTIA National Telecommunications and Information Administration. The U.S. executive agency, located in the Department of Commerce, that serves as the President's researcher and advisor on telecommunication and information processing policy issues.

NTB Non-Tariff Barrier. A barrier to trade that is not documented in a customs duty or tariff.

NTSC National Television Standards Committee. The developer of a broadcast color television standard in the United States and implemented in North America and Japan.

NTT Nippon Telephone and Telegraph. The dominant local exchange and long distance carrier in Japan.

number portability The ability to designate a single telephone number for accessing individuals with several telecommunications devices operating in different locales.

NWIO New World Information Order. A vision for establishing closer parity of access to information between developed and developing nations.

OECD Organization for Economic Cooperation and Development. An organization that supports shared views on economic development, including compiling statistics and polling member nations on telecommunications and information policies.

offshore branching unit A device, located on the ocean floor, that is used to interconnect submarine cables, thereby expanding coverage and providing for redundant or alternative routing of traffic.

Oftel Office of Telecommunications. The United Kingdom independent regulatory agency.

one+ dialing A calling arrangement for direct connections with a prearranged billing commitment.

one-stop shopping The provision of a number of functions, like telecommunications network, design, procurement, and management by a single enterprise as an alternative to the end user performing these functions or securing the services from a number of enterprises.

OPEC Organization of Petroleum Exporting Countries. An organization seeking to manage the price and supply of petroleum.

open network architecture The FCC-articulated blueprint by which the Bell Operating Companies will revamp their facilities to provide for equal access to network facilities by affiliated and unaffiliated enhanced service providers.

open network provision The EC-articulated blueprint for harmonizing local exchange facilities access within the Community.

orbital arc Locations above Earth where satellites are located after launch. Most communication satellites occupy the geostationary orbital arc to maximize stability and geographical coverage.

OSI Open Systems Interconnection. A seven-level model for organizing data processing and telecommunications functions.

out-of-band signaling The use of a channel, separate from the one carrying traffic, to set up, supervise, and bill the call.

packet switching A transmission technology that reduces messages and data into individually routed packets and reassembles them before reaching the final destination.

PAL Phased Alternation by Line. A broadcast color television standard implemented in Europe and Asia.

PanAmSat Pan American Satellite. The first licensed U.S. facilities-based international satellite carrier that competes with INTELSAT. In 1997, Hughes Communications acquired the company.

party A nation that has acceded to an international agreement, thereby accepting its terms.

PBX Private Branch Exchange. A telephone line switching device located on the premises of users with a number of telephones and lines. The PBX switches and routes all inbound and outbound calls.

PCS Personal Communication Services. A variety of new terrestrial and satellite-delivered services to handheld terminals operating at low power.

peering A commercial arrangement among Internet service providers to accept traffic from onward routing at no cost.

photonics The use of specially treated glass fibers, lasers, and mirrors to provide a medium for extremely high-speed, high-capacity telecommunications and information processing.

point-to-multipoint Traffic routing from a single location to many recipients (e.g., video programming distribution to numerous cable operators).

portal A point of access to the Internet and specific content sources. Portals like Yahoo! and America Online simplify the process of accessing content by operating as a one-stop shop.

POTS Plain Old Telephone Service. Conventional telephone services provided to residential and small business users.

power flux density A measure of power radiated from a transmitter, like a satellite.

preemption Assertion of jurisdiction by one regulatory authority that would replace and dislodge the assertion of jurisdiction by another agency.

price cap A form of incentive regulation where carriers and regulators agree to replace conventional rate base regulation for a ceiling on customer charges and the requirement that these rates decline by a certain percentage each year, to reflect productivity improvements. The required annual reductions are possibly offset by a measure of overall increased costs to producers. Carriers are permitted to capture all or part of increased profits.

primary rate interface The large standard unit of capacity in an Integrated Services Digital Network comprised of twenty-three 64-Kbps bearer channels and one 16-Kbps data channel totaling 1.54 Mbps.

primary status The designation for a particular radio service that qualifies users for maximum permissible protection from interference by users of services holding a lower priority.

private line Dedicated capacity designed to link a single user with requirements in two international locations. Carriers now provide "virtual" private lines by partitioning capacity from public networks through the application of software.

privatization Wholly or partially converting the incumbent PTT to private ownership. The privatized enterprise may maintain a monopoly or face varying degrees of competition from market entrants.

protocol A standard operating procedure or format.

PSTN Public Switched Telecommunications Network. The publicly available local and long haul facilities of the incumbent carrier.

PTO Public Telecommunication Organization. A new designation for Post Telegraph and Telephone administrations.

PTT Post Telegraph and Telephone administration. The incumbent and often exclusive carrier in most nations that provides services on a ubiquitous basis, some of which may be subsidized.

RACE Research and Development in Advanced Communications in Europe. An initiative in the European Community to expedite and emphasize telecommunications research and development.

Ramsey pricing Charging users on the basis of their demand elasticity. For example, users with plenty of options, including dedicated facilities and leased lines from a number of carriers, qualify for rates below those charged to users with fewer options.

RASCOM Regional African Satellite Communication System. An affiliation of African telecommunications administrations with plans to operate a dedicated satellite system after having leased capacity from INTELSAT.

RBOC Regional Bell Operating Company. Upon divestiture from AT&T, the Bell Operating Companies were reformulated into seven new corporations, each representing a particular geographical region of the United States.

redundancy The availability of backup capacity to restore service in the event of a facility outage or peak demand condition.

resale The acquisition of bulk transmission capacity and other services for subsequent resale to individual users who singularly do not generate the demand for large capacity offerings. Resellers perform an arbitrage function and profit by acquiring discounted bulk capacity from underlying facilities-based carriers and repackaging it at rates less than what an individual user could secure directly from the carrier.

RFP Request For Proposals. A tender offer soliciting bids by parties interested in securing a contract to perform some form of work.

roamer An individual desiring to use a mobile telecommunications device, perhaps a cellular radiotelephone, while away from the location where service is usually provided.

rolloff The manner in which signal strength drops off at locations increasingly distant from the targeted service location.

RPOA Recognized Private Operating Agency. An ITU-approved designation of nongovernmental telecommunications service providers that confers official recognition and enhances the ability of such organizations to participate in ITU forums.

rulemaking The process by which a U.S. administrative or regulatory agency establishes a binding rule, regulation, or policy. Procedural due process requirements obligate the agency to notify the public of proposed actions and to provide opportunities for participation through filed comments or testimony.

satellite beam types There are four satellite beam types: Global, Hemispheric, Zone, and Spot. Global: Transmission beams with maximum geographical illumination, approximately one-third of the entire globe. Hemispheric: Transmission beams shaped to cover an entire hemisphere (e.g., North America, Central and South America, or Europe). Zone: Transmission beams shaped to cover one or more portions of a hemisphere with higher signal strength. Spot: Transmission beams shaped to concentrate signal strength over relatively small, but highly populated locations.

S-band The portion of the frequency spectrum around 2 GHz used for terrestrial microwave and some mobile satellite applications.

SDR Special Drawing Rights. An aggregation of national currencies administered by the International Monetary Fund and often used by international carriers in settling toll revenue accounts.

SECAM Sequential Color with Memory. A French-developed broadcast television standard adopted in France, French-speaking African nations, and the former Soviet Union.

sender keep all A toll revenue division arrangement whereby the originating carrier keeps all charges.

sender pay all A charging mechanism whereby the calling party pays all charges.

separate system A satellite system, separate from the INTELSAT cooperative, providing international services.

server A device that houses content accessed via the Internet.

settlement rate The unit of currency and percentage division negotiated by carriers for settling accounts regarding carriage of inbound international traffic. Typically, international carriers on direct routings equally divide the negotiated rate.

signatory A company or government that has executed a binding legal commitment (e.g., Comsat serves as the sole U.S. signatory to INTELSAT and Inmarsat).

Single Europe Act Legislation adopted by the constituent members of the European Community agreeing to closer collaboration and harmonization of laws, policies, and regulations.

smart card A credit-card–sized instrument containing microchips and associated electronics for providing credit and other data for facilitating transactions.

software defined network The use of software to partition transmission capacity from public networks for private use, making it appear as the functional equivalent to capacity dedicated for a single user.

SONET Synchronous Optical Network. Protocols for operating fiber optic transmission facilities at very high, multimegabit speeds.

sovereignty Recognition of independence and legitimacy in a nation's determination of policies, laws, and regulations affecting its citizens and internal affairs.

space station A satellite or other device that operates in space.

spectrum A term depicting the range of frequencies that are used in telecommunications.

spectrum allocation The designation of a frequency band to one or more specified services by the International Telecommunication Union and by individual nations.

spectrum allotment The designation of specific channels of operation to particular localities.

spectrum assignment The grant of operational authority that includes a license to operate on a specific frequency.

spread spectrum A transmission technique whereby signals are transmitted over a range of spectrum at low power with messages handled by assigning a discrete sequence of code or by hopping over different frequencies.

SS7 Signaling System 7. An advanced method for providing software defined signaling necessary to set up telecommunications links.

station keeping The use of jets aboard a satellite to keep it "on station" (in the prescribed orbital location with antennas properly aimed). A station-kept satellite eliminates costly tracking requirements for each Earth station pointing to the satellite. Satellites typically reach end of life when station-keeping fuel has been used up, even though the electronic and power generation components remain operational. Operators may conserve station-keeping fuel to extend a satellite's usable life through the use of maneuvers that allow the satellite to drift slightly.

STM-1 Synchronous Transport Module Ones. A unit of submarine cable transmission capacity equivalent to 155 Mbps of throughput.

store-and-forward The use of computerization to store traffic temporarily for subsequent transmission in batches or efficiently loaded packets. A satellite may store messages until it reaches a point in its orbit where the traffic can be downlinked.

strategic alliances Corporate business relationships designed to tap the comparative advantages and expertise of the participants as well as to achieve a broader geographical marketing presence.

Submarine Cable Landing License Act An act of Congress requiring the FCC and State Department to grant licenses to land a submarine cable on U.S. soil. Foreign carriers will receive such grants only if U.S.-owned and operated submarine cables can make landfall in the foreign country.

substitutable products Products that can replace and compete with each other (e.g., tea instead of coffee and Centrex instead of private branch exchanges).

switch The portion of the telecommunications infrastructure where traffic is received by a device that identifies the intended destination and selects the routing for delivering traffic onward to that point.

T-1 A unit of transmission capacity equal to 1.544 Mbps in throughput.

tail circuits Domestic facilities used to transport traffic from an international caller to a gateway, and from another gateway to the call recipient.

tandem switch A telecommunications switching facility with management control over one or more end offices.

tariff A contract for service setting out the terms and conditions under which the general public will secure service.

TCP/IP Transmission Control Protocol/Internet Protocol. The basic technical standards for Internet traffic routing and addressing.

TDMA Time Division Multiple Access. A technique for deriving more throughput and accommodating increased demand by assigning channels to users in discrete blocks of time.

Telecommunications Act of 1996 An Act of the United States Congress that substantially revised and amended the Communications Act of 1934 to promote competition.

Telecommunications Development Bureau (BDT) A permanent organ of the International Telecommunication Union formed in 1989 to promote telecommunications development, including technical and managerial training.

teledensity The penetration of telecommunications access lines typically measured per 100 inhabitants.

telepoint A new generation of pay telephones and one-way wireless communications based on microcellular radio technology.

teleport A satellite Earth station with extensive access to terrestrial facilities and a broad base of users, situated on real estate developments adjacent to the Earth stations or at other locations.

teletext Slow speed transmission of textual information over the vertical blanking intervals of broadcast television signals.

throughput The amount of data carried over a particular amount of capacity in a specified time span (*see also* bit rate).

traffic The messages, programming, files, and other intelligence requiring transport from one location to another.

transborder data flow The transmission of data across national borders, alternatively viewed as a vehicle to share the wealth of information, or a source for cultural imperialism.

transceiver A device that contains components that transmit and receive radio signals.

transiting The carriage of international traffic via the facilities of an intermediary, third country between the sender and recipient.

transponder The components in a satellite able to transmit and receive traffic. Satellite capacity is typically stated in the number of available 36 MHz equivalent transponders.

Treaty of Rome The treaty enacted in 1957 that created the European Community.

trunking A computer assisted process for queuing channel demands thereby more efficiently using spectrum and accommodating episodic demands for service. For example, a trunked radio system will provide channels of communications for users on an as-demanded basis from the available inventory.

trunk side interconnection The interconnection of facilities and lines between carriers at a level higher up the hierarchy than typically accorded end users.

TVRO Television Receive-Only. Satellite dishes equipped for one-way reception of television signals.

unbundled Separation of services and equipment into discrete elements available on an individual, à la carte basis.

uplink The process of transmitting traffic from Earth to a satellite.

USISC United States International Service Carrier. A U.S. carrier licensed by the FCC to provide international services.

USTR United States Trade Representative. The U.S. Executive Branch agency responsible for engaging foreign nations in bilateral and multilateral trade negotiations.

VAN Value Added Network. An enhanced services provider that typically leases transmission capacity and adds value by performing customized services (e.g., credit card verification and electronic funds transfer).

video dial tone The provision of transparent common carrier transmission capacity for delivery of video programming.

videotext Slow speed transmission of textual information over closed circuits.

virtual networks Software derived networks typically involving lines partitioned from public switched facilities.

virtual private line Using software to partition capacity from switched public facilities to make it function the same as capacity dedicated to a single user.

VSAT Very Small Aperture Terminal. Very small satellite Earth stations, typically situated on user premises, that promote diversification in services, easier access, and lower costs.

WAN Wide Area Network. An interconnected network of personal computers and work stations situated in different locations.

WARC/WRC World Administrative Radio Conference/World Radio Conference. Periodically convened conferences of ITU member nations to address regional or global spectrum planning and policy issues.

WATS Wide Area Telephone Service. Toll-free calling to the call originator provided by most retailing and service organizations.

whipsawing The practice of a monopoly carrier to extract concessions from foreign correspondent carriers operating in a competitive market who must vie for return inbound traffic from the monopoly carrier.

windowing Segmenting the distribution of a product (e.g., a movie) into sequences based on consumer demand elasticities and time from initial availability. Movie producers attempt to extract the greatest amount of rent by calibrating availability and charges: $6–8 at a theater, $2–3 on rented video tape, and no charge to viewers on broadcast television.

wireless local loop The use of unwired transmission facilities to provide connections to the wireline public-switched telecommunications network; many of these technologies also provide mobile services.

wireless networks The use of radio technology to serve increasingly diverse applications, including mobile services and some applications heretofore provided via wireline facilities.

wireline The conventional wire-based telecommunications infrastructure installed and managed by incumbent carriers.

World Wide Web That portion of the Internet that combines text, sound, video, and other graphical material accessible via broswers.

WTO World Trade Organization. The permanent trade policy and dispute resolution forum created in 1994 by nations that had formed the GATT.

WTTC World Telephone and Telegraph Conference. A periodically convened conference of ITU member nations to address issues pertaining to the rules of the road in telephony and other telecommunications services.

X.25 An ITU-recognized protocol for user access to data networks.

X.75 An ITU-recognized protocol for interconnecting data networks.

X.400/X.500 A series of protocols governing electronic mail and directory functions.

X-band A portion of the frequency spectrum in the 7–8 GHz range used primarily for satellites with defense and intelligence gathering applications.

zero+ dialing A calling arrangement involving verification of credit or payment through intervention by an operator or computer.

About the Author

Rob Frieden is a professor of telecommunications at Penn State University, where he teaches courses in management, law, and economics. He also provides legal, management, and market forecasting consultancy services in such diverse fields as personal and mobile communications, satellites, and international telecommunications business development. Professor Frieden has written several books, published dozens of articles in academic journals, and provided commentary in a variety of trade periodicals. In 1999, he and three colleagues completed a 10-year project culminating in the publication of a three-volume comprehensive treatise on communications law.

Professor Frieden has previously served as deputy director of international relations at Motorola Satellite Communications, Inc., where he provided a broad range of business development, strategic planning, policy analysis, and regulatory functions for the IRIDIUM mobile satellite venture. He has also held senior policy-making positions in international telecommunications at the Federal Communications Commission and the National Telecommunications and Information Administration. In the private sector he practiced law in Washington, D.C., and served as assistant general counsel at PTAT System, Inc., where he handled corporate, transactional, and regulatory issues for the first U.S. private undersea fiber optic cable company.

Professor Frieden received a B.A., with distinction, from the University of Pennsylvania in 1977 and a J.D. from the University of Virginia in 1980.

Index

Recent Titles in the Artech House Telecommunications Library

Vinton G. Cerf, Senior Series Editor

Access Networks: Technology and V5 Interfacing, Alex Gillespie

Achieving Global Information Networking, Eve L. Varma, Thierry Stephant, et al.

Advanced High-Frequency Radio Communications, Eric E. Johnson, Robert I. Desourdis, Jr., et al.

ATM Switches, Edwin R. Coover

ATM Switching Systems, Thomas M. Chen and Stephen S. Liu

Broadband Access Technology, Interfaces, and Management, Alex Gillespie

Broadband Networking: ATM, SDH, and SONET, Mike Sexton and Andy Reid

Broadband Telecommunications Technology, Second Edition, Byeong Lee, Minho Kang, and Jonghee Lee

The Business Case for Web-Based Training, Tammy Whalen and David Wright

Communication and Computing for Distributed Multimedia Systems, Guojun Lu

Communications Technology Guide for Business, Richard Downey, Seán Boland, and Phillip Walsh

Community Networks: Lessons from Blacksburg, Virginia, Second Edition, Andrew M. Cohill and Andrea Kavanaugh, editors

Component-Based Network System Engineering, Mark Norris, Rob Davis, and Alan Pengelly

Computer Telephony Integration, Second Edition, Rob Walters

Desktop Encyclopedia of the Internet, Nathan J. Muller

Digital Modulation Techniques, Fuqin Xiong

E-Commerce Systems Architecture and Applications, Wasim E. Rajput

Error-Control Block Codes for Communications Engineers, L. H. Charles Lee

FAX: Facsimile Technology and Systems, Third Edition, Kenneth R. McConnell, Dennis Bodson, and Stephen Urban

Fundamentals of Network Security, John E. Canavan

Guide to ATM Systems and Technology, Mohammad A. Rahman

A Guide to the TCP/IP Protocol Suite, Floyd Wilder

Information Superhighways Revisited: The Economics of Multimedia, Bruce Egan

Internet E-mail: Protocols, Standards, and Implementation, Lawrence Hughes

Introduction to Telecommunications Network Engineering, Tarmo Anttalainen

Introduction to Telephones and Telephone Systems, Third Edition, A. Michael Noll

IP Convergence: The Next Revolution in Telecommunications, Nathan J. Muller

The Law and Regulation of Telecommunications Carriers, Henk Brands and Evan T. Leo

Managing Internet-Driven Change in International Telecommunications, Rob Frieden

Marketing Telecommunications Services: New Approaches for a Changing Environment, Karen G. Strouse

Multimedia Communications Networks: Technologies and Services, Mallikarjun Tatipamula and Bhumip Khashnabish, editors

Performance Evaluation of Communication Networks, Gary N. Higginbottom

Performance of TCP/IP over ATM Networks, Mahbub Hassan and Mohammed Atiquzzaman

Practical Guide for Implementing Secure Intranets and Extranets, Kaustubh M. Phaltankar

Practical Multiservice LANs: ATM and RF Broadband, Ernest O. Tunmann

Protocol Management in Computer Networking, Philippe Byrnes

Pulse Code Modulation Systems Design, William N. Waggener

Service Level Management for Enterprise Networks, Lundy Lewis

SIP: Understanding the Session Initiation Protocol, Alan B. Johnston

SNMP-Based ATM Network Management, Heng Pan

Strategic Management in Telecommunications, James K. Shaw

Strategies for Success in the New Telecommunications Marketplace, Karen G. Strouse

Successful Business Strategies Using Telecommunications Services, Martin F. Bartholomew

Telecommunications Department Management, Robert A. Gable

Telecommunications Deregulation, James Shaw

Telephone Switching Systems, Richard A. Thompson

Understanding Modern Telecommunications and the Information Superhighway, John G. Nellist and Elliott M. Gilbert

Understanding Networking Technology: Concepts, Terms, and Trends, Second Edition, Mark Norris

Videoconferencing and Videotelephony: Technology and Standards, Second Edition, Richard Schaphorst

Visual Telephony, Edward A. Daly and Kathleen J. Hansell

Wide-Area Data Network Performance Engineering, Robert G. Cole and Ravi Ramaswamy

Winning Telco Customers Using Marketing Databases, Rob Mattison

World-Class Telecommunications Service Development, Ellen P. Ward